Desert Ecosystems

M. S. Rathore · Tej Partap · Laxman Singh Rathore
Editors

Desert Ecosystems

Rescripting Development Paradigms

CP

Springer

Editors
M. S. Rathore
Centre for Environment and Development Studies
Jaipur, Rajasthan, India

Laxman Singh Rathore
Director General of India Meteorological Department
Ministry of Earth Sciences
New Delhi, India

Centre for Environment and Development Studies
Jaipur, India

Tej Partap
Kulu, Himachal Pradesh, India

ISBN 978-3-032-03236-2 ISBN 978-3-032-03237-9 (eBook)
https://doi.org/10.1007/978-3-032-03237-9

Jointly published with Capital Publishing Company
ISBN of the Co-Publisher's edition: 978-93-81891-90-2

This Springer imprint is published by the registered company Springer Nature Switzerland AG
The registered company address is: Gewerbestrasse 11, 6330 Cham, Switzerland

Foreword

In recent decades, desert ecosystems worldwide, including those in India, have faced continuous socio-ecological transitions. Existing frameworks with established parameters must be reassessed to comprehend the new horizons of desert ecosystems and their spatial and temporal implications for both human and non-human species. This necessitates a collective approach to analyze the socio-economic, ecological, cultural, and political variables shaping the emerging layers of deserts. Such a complex task can only be fully understood through an interdisciplinary perspective. This edited volume on desert ecosystems aims to achieve this goal.

This volume is unique in many respects. Firstly, all contributors are distinguished scholars in their respective fields. Each original paper was specifically invited to focus on a particular desert ecology, and all chapters underwent rigorous peer review. Additionally, the collective insights of esteemed scholars, representatives of civil society, and farmers have enriched the deliberations held during the conference. As a result, this volume offers numerous actionable recommendations for hot, cold, and coastal desert ecosystems.

My acquaintance with almost all the contributors makes it especially pleasurable to be asked to write this Foreword. This comprehensive and detailed treatment of complex issues provides an indispensable roadmap for researchers, managers, and policymakers working in desert ecosystems.

Prof. Dinesh K. Marothia
President
Indian Society of Agricultural
Economics
Mumbai, India

Preface

All over the world, the 33 major desert lands have always been places of profound mystery, beauty, and stark contrasts. They hold within their expansive arid landscapes, stories of survival, resilience, and adaptation. Among these world's deserts, the deserts of India, both hot and cold, stand out with their unique blend of natural splendour, historical richness, and cultural diversity.

From the golden dunes of the Thar desert, to the coastal desert and the cold desert of Ladakh with its breathtaking high-altitude terrain, India's deserts offer an extraordinary journey through time and space. These deserts are not just barren wastelands but are vibrant ecosystems. Beneath their seemingly harsh and unyielding surfaces lies an astonishing array of life forms, uniquely adapted to thrive in the extreme conditions of desert ecosystems from hardy flora to an incredible fauna and the human communities, all uniquely adapted to the harsh desert conditions. All the desert dwellers have mastered the art of living with water scarcity, extreme temperatures, higher wind velocity, and sparse vegetation. The traditional knowledge of human societies inhabiting these deserts has passed down through generations, invaluable understanding about how to coexist with resilient biota in harsh arid environment of the desert.

Today, deserts are under numerous threats that jeopardize their fragile ecosystems; Climate Change, impacting the flora and fauna of desert regions; Human Activity, Urbanization, overgrazing, industrialization, over exploitation of natural resources and unsustainable agricultural practices lead to habitat destruction and degradation, threatening the delicate balance of desert ecosystems.

The book is aimed to inspire a deeper understanding and respect for the desert land ecosystems, fostering a sense of responsibility towards their biodiversity and people.

The chapters authored by several experts on different aspects of desert land ecosystems, hot and cold, Indian and foreign, seek to explore the hidden intricacies of desert ecosystems, the vital roles the desert land flora, fauna, microbes and people play, and the delicate balance that sustains them. Growing threats posed by climate change, overgrazing, urbanization, and the unsustainable practices disrupt these fragile environments have also been discussed in many chapters. It is an invitation to recognize the deserts not as empty voids, but as vibrant, living landscapes rich with diversity, mystery, and beauty.

Reading the book will help readers to traverse through the deserts' vast stretches, to uncover their secrets, and gain insights into the innovative solutions being developed to address the challenges faced by these unique ecosystems. The scale of the problems and issues are elaborated by authors in respective chapters, underlining the criticality of water scarcity. The impact of climate change on desert ecosystems is a pressing concern. Rising temperatures and changing precipitation patterns threaten the delicate balance of these fragile environments. Research needs about sustainable land management practices, soil conservation, afforestation, and conservation of native flora and fauna which have been stressed by almost all authors. The socio-economic development of desert communities is crucial to sustainable living in this environment. Access to education, healthcare, and employment opportunities remain limited.

The book is an outcome of the papers presented in a conference on Desert Ecosystems: Status, Emerging Challenges and Perspectives in Jaipur on November 15–16, 2022 organised by the Centre for Environment and Development Studies, Jaipur (CEDSJ) and the Indian Society for Agriculture Economics with the financial support of GIZ and NABARD.

This collection of presentations made by the experts from different fields of knowledge, revised by authors into written papers, is a tribute to the captivating landscapes, plant and animal life forms and the human communities that call these deserts their home. Readers will be inspired by the enduring spirit of the desert dwellers and the remarkable beauty of deserts and appreciate the resilience and ingenuity required to thrive in such harsh conditions. Book Editors, appreciate and are thankful to each author for his/her efforts invested in contributing to enhance the understanding of academia and general public about the ecosystem dynamics and the challenges of sustainable development of the desert ecosystems.

Jaipur, India
April 2025

M. S. Rathore
Tej Partap
Laxman Singh Rathore

About This Book

The book deals with ecosystems of hot, cold, and coastal deserts, their present status and emerging challenges in the context of economic development and climate change issues.

The book characterises the Thar desert, which is influenced by Southwest Monsoon as well as mid-latitudinal westerly weather systems and is facing rapid changes in agricultural land use resulting in land degradation. It further describes the mechanisms and pathways for sustainable farming and conserving the ecosystem in the drylands to achieve sustainable farm production in rain-fed ecosystem. It advocates revival of desert bioresources through science-based management of desert biodiversity and its water resources along with traditional knowledge systems. The challenges and opportunities for livelihoods and environmental security in desert ecosystems are also discussed. It envisions and maps the ecological concepts onto the landscape of rural economy keeping in view the resilience of agropastoral system.

The Himalayan Cold Desert agroecosystems and understanding of the fundamentals behind its uncertain future along with details on climate change process in the Himalayan Cold Desert and farmers' adaptation strategies are captured by the professionals who devoted their lifetime in mountain areas. While discussing climate change, the focus is also on degrading land resources and their sustainability through subsistence agriculture in the cold arid region.

The book also deals with the so far neglected area, i.e., management of desert ecosystems in coastal regions. Overall, the book contains useful scientific information for professionals as well as students and also provides good guidance material for policymakers to address the emerging challenges in the hot, cold, and coastal desert ecosystems for planning sustainable development and ensuring food, water, and livelihood security to the population residing in these areas.

Contents

About the Editors

Dr. M. S. Rathore Ph.D. (Agricultural Economics), selected for the Jawaharlal Nehru Award of the ICAR 1981 in recognition of meritorious research work done for Ph.D. degree at ICRISAT, Hyderabad, and Himachal Pradesh University, Shimla Associate Professor, Himachal Pradesh University, Shimla, 1973–1983, Senior Fellow/Professor, Institute of Development Studies, Jaipur, from 1983 to 2008, Professional Staff at International Centre for Integrated Mountain Development (ICIMOD), Kathmandu, Nepal, during 1987–1989, Visiting Professor, Institute of Development Studies, University of Sussex, UK, 1994–1995 and 1997. Presently, Dr. Manohar Singh Rathore is Director, Centre for Environment and Development Studies, Jaipur (CEDSJ), and published more than 100 research papers in various academic journals, seminars, and conferences. Dr. Manohar Singh Rathore completed 81 research projects and written seven books and many book chapters.

Dr. Tej Partap is an ecologist and established mountain agriculture systems scholar and expert on Himalayan agriculture systems and farmers livelihood systems. He served as Professional Staff and Programme Leader at the Mountain Farming Systems Division of the International Centre for Integrated Mountain Development (ICIMOD) from 1987 to 2000. Since 2001, he has been serving as Vice Chancellor of several Universities, including Sher-e-Kashmir University of Agriculture Sciences and Technology, Srinagar; Chaudhary Sarwan Kumar Himachal Pradesh Agriculture University, Palampur; G. B. Pant University of Agriculture and Technology, Pantnagar. Currently, he is serving as Vice Chancellor of Sri Sri University, Cuttack, Odisha.

Dr. Laxman Singh Rathore is Chairman, Centre for Environment and Development Studies, Jaipur, and former Director General of Meteorology, India Meteorological Department (IMD), Ministry of Earth Sciences. He served various international bodies such as: Chair, (Acting 2016–2019) Inter-governmental Board for Climate Services (IBCS), WMO, United Nations (UN). Vice-Chair, (2012–2016) IBCS, WMO. Permanent Representative (Ambassador) of India with the World Meteorological Organization (WMO), United Nations. Member, Executive council WMO, UN. Vice President, Commission for Agriculture Meteorology, WMO. Chairman,

Governing Council, SAARC Meteorological Research Centre, Dhaka. Scientific Member, 3rd Indian Scientific Expedition to Antarctica (1983–1984) and established the first permanent Indian station.

Chapter 1
Overview

M. S. Rathore, Tej Partap, and Laxman Singh Rathore

1.1 Introduction

Desert ecosystems are a part of the larger universal ecosystem. These ecosystems have their own local significance contributing to regional and global atmospheric circulation, besides acting as sources of minerals, natural gas, oil, food, fodder, and also a niche of numerous medicinal and aromatic plants. Deserts also serve as carbon sink, for the bacterial population in the sand helps in storing the carbon dioxide.

Indian desert regions are home to hot, cold, and coastal desert ecosystems and all of these show signs of natural and anthropogenic degradation of different degrees. Occurrence and intensity of heat, drought, dust storms, and episodic heavy rainfall, has been degrading these ecosystems for a long period and changing their character. Loss of a little soil of these ecosystems possess threat to intensified biodiversity loss, which in turn would trigger accelerated soil erosion favouring further desertification. Deserts have an oppressive environment subjecting its fauna and flora to be resilient and adapt to survive under extreme conditions and meagre resources. The natural process of desertification gets further accelerated because of human activities such as: overgrazing, intense/unsustainable farming, industrialization, natural resource exploitation, etc. In recent times irrigated agriculture, urbanization and industrialization have caused much damage to desert ecosystems of India. The increased water

M. S. Rathore (✉)
Centre for Environment and Development Studies, Jaipur, India
e-mail: msrorama@gmail.com

T. Partap
Ecologist, Himalayan Farmers, Farming and Livelihoods Scholar, Srinagar, India
e-mail: tpartap52@gmail.com

L. S. Rathore
Formerly of India Meteorological Department, GOI, New Delhi, India
e-mail: lrathore@gmail.com

M. S. Rathore et al. (eds.), *Desert Ecosystems*,
https://doi.org/10.1007/978-3-032-03237-9_1

requirements and over exploitation of ground water has resulted in drastic depletion in the ground water table.

Arid lands of hot, cold and coastal regions are in general fragile ecosystems that face vagaries of extreme hot and cold climate causing unbearable stress on its biota. Based on centuries of adaptation process, desert inhabitants have evolved unique traditional knowledge and wisdom about sustaining their livelihood systems. The livelihoods of the residents rest on the balanced use of own land as well as societal land which is common for use making conservation and sustainability the way of life. But in the recent past the livelihoods and economy of communities have been undergoing transformation, particularly in agriculture. The green revolution technologies have helped change the focus of farmers from sustenance to cash crops farming. There is the faster growth of human and livestock population in both arid and semi-arid areas, leading to increased pressure on the fragile lands to provide food, fodder, and livelihood security. This has necessitated evolving new approaches and **a desert perspective** for sustainable living on these lands/ecosystems. Native plants and animals' biodiversity of these arid lands have unexplored/under harnessed potentials for bringing economic prosperity to the people.

Western Rajasthan hot desert is primarily a region that has been getting 299 mm average annual rainfall. However, in recent decades, this region has experienced increased rainfall and longer duration of monsoon. Multiple factors have been held responsible for this change in weather pattern during monsoon. Locally, the changes in land use and land cover, deforestation, and urbanization are also considered as contributing factors for change in weather pattern of this desert ecosystem.

Thar desert ecosystem is changing, mainly because of change in the increasing density of human population and land use pattern, caused by the invent of Indira Gandhi Irrigation Canal and Narmada irrigation canal in the Thar desert. Interestingly, there is lack of research on the changing pattern of southwest monsoon and its impact on this desert ecosystem and consequently on the economy of Rajasthan. Most of the studies available on the subject relate to the physical features, geomorphology of landforms, rainfed agriculture, livestock, climatic variability and change and these are mostly independent in nature. Though the Thar desert community's traditional knowledge and practices do support them to adapt to living under the vulnerable environmental conditions but the winds of change towards faster economic development discourages public about making wider use of traditional knowledge and practices. There appears emerging need for research on development strategies for sustainable living in desert ecosystems.

Among the many desert ecosystems of India, focus here is on hot, cold and coastal areas, arid and semi-arid desert ecosystems. Chapters on few other desert ecosystems, particularly Australia and Africa, from around the world are added to help gain more insights into the dynamics of desert ecosystems in general.

The cold desert ecosystem/region of India falls in the Himalayan region. About 80 per cent of the cold desert region is in the union territory of Ladakh, while the remaining part is in Himachal Pradesh and Uttarakhand. The total geographical area of Ladakh (Leh and Kargil District) is 96701 sq. kms. The ongoing changes in the rainfall pattern of the Himalayan cold desert region are now becoming a key influencing factor of ecological change process. As per the India Meteorological Department (IMD) there is a shift in precipitation from winter to summer which sometimes leads to a very high and intense rainfall causing flash floods, cloud burst etc. Such changes are enhancing erosion of rare fertile soil and destruction of agriculture lands. Ladakh, a water scarce region faces serious threats, such as, inadequate precipitation, depletion of water table, and drought like conditions during early cropping season. It is more frequently witnessed in valley-based villages. While the melting of glaciers will lead to excess water in the intermediate time period, but in the long term it will cause shortages of water. Therefore, adaptation to the changing conditions in cold deserts and mountains is an emerging challenge. Local traditional knowledge somehow supports the mountain communities to adapt to the vulnerable situations. Local communities are already taking initiatives to address the emerging challenges, such as building ice reservoirs to provide water for irrigation and households during periods of scarcity.

The Coastal Desert areas, form a unique ecosystem. This ecosystem and its problems are less known. There is absolute lack of research so as to highlight its status, existing conditions and likely future challenges. Even for the present initiative about gathering information on Indian deserts, it was difficult first to identify suitable researchers who could contribute a paper. Finally, a scientist from the Tamil Nadu Agriculture University was able to contribute an introductory paper on this topic.

The other important aspect of research on desert ecosystems is desertification, i.e. persistent degradation of dryland ecosystems by variations in climate and human activities. Desertification affects the flora, fauna, and livelihoods of millions of people who for generations have benefited from the dryland ecosystems. Desertification involves many environmental and socio-economic aspects. Scientists have identified many causes and consequences. Unsustainable agricultural practices are one of the cause factors, having consequences for food security. In order to fully comprehend this two-way relationship, it requires understanding of how agriculture affects land quality, what risks land degradation poses for agricultural production and to what extent a change in agricultural practices can reverse the trend. Cropland expansion and intensification of agriculture are among the basic drivers of land degradation processes and desertification. But it is also true that agriculture itself can provide solutions to land degradation. Therefore, the issue is to know the rate and scale of desertification with its likely future consequences? Desert ecosystem needs to be studied with a wholistic approach. Researchers should use wider definition of desertification, i.e. "Desertification means land degradation in arid, semi-arid and dry sub-humid areas resulting from various factors, including climatic variations and human activities". There is also need to enhance our understanding of the ecosystem services provided by the cold and hot and coastal deserts in India and abroad.

In total 17 papers were contributed covering three themes, i.e. Hot Desert Ecosystem, Cold Desert Ecosystem, and Coastal Desert Ecosystem. Two papers were from outside India covering Hot Desert of Australia and West Africa—Sahel. The conference was organized in Jaipur and attended by many subject matter specialists from different parts of India and outside the country. The papers presented were peer reviewed and organized in three parts covering three different ecosystems and are now being presented in the book form.

1.2 Part I: Hot Desert Ecosystem

Thar Desert covering western part of Rajasthan and Gujarat state was the focus area addressed by the paper contributors. To enrich desert perspective and experiences, beyond Indian deserts, two invited papers have been included in this book - one from the hot desert in Australia and second from West Africa - Sahel. In total, 10 papers were contributed covering diverse themes, such as agro-ecosystem, flora and fauna, biodiversity and traditional knowledge, agriculture farming system, livestock, agro-pastoral system, climate change, etc.

1.3 Part II: Cold Desert Ecosystem

Cold Desert area in India covers the entire Himalayan Mountain Range. The focus of contributors was on the areas, namely union territory of Ladakh, Himachal Pradesh, and Uttarakhand. Cold desert being the far-flung area thinly populated, inaccessible, lack of infrastructure and hard life makes it difficult for researchers to address the problems and it leads to marginalization of area. There were four papers covering the aspects, such as cold desert ecosystems, climate change, land, agriculture-livestock-based livelihoods of cold desert.

1.4 Part III: Coastal Desert Ecosystem

Most people may be unaware of the fact that there is an area called coastal desert in India. Hence lack of research on Coastal Desert areas and their problems. One paper was contributed highlighting the problems and challenges of the area and the second paper covering all the three ecosystems.

1.4.1 Part I: Hot Desert Ecosystem

Hot deserts are characterized by unique climatic, topographic, edaphic, and biotic features, with aridity being a common feature across all deserts for most or all the year with significant fluctuations in day and night temperatures, and between seasons. Low humidity during the day and relatively higher humidity at night are a characteristic feature. Precipitation deficiency is a defining feature, characterized not only by the low amount of rainfall but also by its higher temporal-spatial variability. Rain in the desert is often seasonal, unpredictable, inconsistent, and restricted to specific areas and time periods.

Deserts located around 30°N and 30°S are particularly hot due to the global circulation patterns, that is the descending air from the Hadley circulation. This descending air creates semi-permanent high-pressure areas at these latitudes, leading to abundant sunshine and high surface temperatures. Subtropical deserts form due to these air circulation patterns. Deserts in rain shadow areas develop on the side of mountains that doesn't face the prevailing wind, causing arid conditions due to blocked wet weather systems. Deserts in the interior of continents form because they are not reached by moisture-carrying winds. Coastal deserts, affected by cold ocean currents, may receive very little rain but can be moist with fog as the chilly ocean air warms up over the land. Hyper-arid and arid regions are defined by an aridity index (P/PET) lower than 0.20, indicating less than 20% of water requirement is met by rainfall for optimal plant growth. Key meteorological indicators used for monitoring agricultural drought include the Standardized Precipitation Index, aridity anomaly index, and combined drought indicator.

In India, the rainfed area spans about 78 million hectares and accounts for around 55 percent of the entire arable land in the country. The dry lands, which include the arid and semiarid regions cover more than 50% of the rain fed region. The Indian hot arid zone spans approximately 32 million hectares, predominantly encompassing the northern Western states of Rajasthan (constituting 70% of the total area), Gujarat, Punjab, and Haryana. Additionally, it includes scattered land masses within the states of Maharashtra, Karnataka, and Andhra Pradesh. Due to a changing climate, **India's arid and semiarid area is predicted to expand in size in the coming decades.**

Western Rajasthan is one of the great desert regions of the world with a long history of human occupation, complex and sophisticated culture and institutions. The Thar Desert in Rajasthan has low and unpredictable rainfall, along with high rates of evaporation, resulting in frequent occurrences of drought. The desert experiences very high daytime temperatures, ranging from 45 °C to 49 °C. Despite increasing trends in annual and monsoon rainfall, variability persists across districts. Although it is probably the most densely populated desert region in the world, but as compared to other parts of India it is much less densely populated. As in most other desert environments the landscape is largely dominated by agro-pastoral communities and resource use. Agriculture in recent years has become dependent on deep ground water. Historically however the region was mostly rain-fed-based cropping. A far more important component of this subsistence economy was the livestock based open grazing of desert range lands. Pastoralism of varied degrees formed part of almost all the strategies of adaptation. There existed a range of mobility among herders extending from short migration to long distance migration, over many months. Camels, sheep, goat, and cattle were the primary species herded and all of them invariably incorporated some degree of movement and in any case were primarily maintained on open range grazing.

Most households depending on agriculture in western Rajasthan faced an unpredictable and risky future. Today, the area is confronted with recurrent droughts, soil erosion, desertification, excessive groundwater depletion, declining soil and water quality, diminished productivity, institutional weaknesses, malnutrition, and higher transactional costs, largely attributed to an extensive production system.

The ecological landscape of western Rajasthan has over the long term evolved or has been shaped by the way in which it has been used and manipulated by people keeping livestock. While agriculture forms historically a very important part of the rural land use system, but it played less influential and subordinate role. Although the period after independence saw a reversal of this relationship, it is important to understand at why pastoralism and in general livestock had such a dominant role in shaping the landscape and the institutions of the region. Only after we understand this, especially the ecological foundations of pastoralism that we will be able to visualize a framework that goes beyond the standard model of growth and development. As we have sketched above, the form in which life and adaptation in this landscape had evolved prioritized the dimensions of livelihood security and capacities for coping with extremes of variability. These capacities, evolving over extended periods of time, were an expression of the character of the people, the values of their society expressed in the systems of resource utilization. The historical engagement with 'development' was illustrative of how this 'sustainable' socio-ecological regime was replaced by a regime far more vulnerable to perturbation.

The first paper in this section is by Dr. Amal Kar titled "Agricultural Land Use and Land Degradation in Thar Desert: Observations, Monitoring Methods and Emerging Trends for Future Strategies". Author has highlighted the specific features of Rajasthan's desert area with his long experience of living, travelling and conducting studies on different aspects of Thar Desert. Thar Desert, situated between the Aravalli Hill Ranges in the east and the Indus River in the west, is an integral part of the arid western part of Indian Subcontinent. The northern boundary runs almost along the dry and under-fit valley of the Ghaggar (Hakra-Nara) River. It once flowed as Saraswati River (Kar, 2014a), its southern boundary is along the Great Rann of Kachchh. About 72% of the Desert's 29.07 million ha (mha) area lies in India, and the rest 28% is in Pakistan. Thar Desert has a long history of sustainable arid land agricultural practices, especially recorded from the Harappan Civilization period (~5800–3500 years before present, or 5.8–3.5 ka, [kilo annum]). The land surface in Thar Desert being mostly covered with aeolian sand sheets and sand dunes, the rainfall being meagre and the wind during summer months being strong, the landscape here is highly vulnerable to wind erosion. Rain-fed agriculture, livestock and agro-pastoralism systems remain the main sources of food, incomes and employment for the poor populations.

The author highlighted three major environmental issues of the Thar Desert that need serious attention of the stakeholders; and those are water management, land quality and dust emission. Due to Global Warming the South-West monsoon has started behaving more erratically than before, and is getting modified due to the overbearing presence of the Westerly winds. Despite the setbacks there is now a growing awareness that keeping the land in good shape is the key to sustained production from it. As new uncertainties in the land use decisions are being introduced by the

Global Warming, author suggests it is time that a robust system of land degradation assessment and monitoring is developed.

The second paper listed in this section is titled "Transforming Drylands: Systems-Based Sustainable Agriculture in Arid Western Rajasthan" contributed by Shalander Kumar, Abhishek Das, Anthony Whitbread and M. L. Jat. The paper highlights the major determinants to shaping the configuration of agricultural production systems across western Rajasthan. Along the West–East rainfall gradient there is availability of water and moisture. With the advent of canal water and as groundwater extraction intensified, there has been a noticeable shift from the system of cultivating traditional millets and legume crops along with livestock (cattle and sheep) towards a system of cultivating commercial crops such as cumin (*Cuminum cyminum* L.) and Isabgol (*Plantago ovata*), horticultural crops like pomegranate and fig, or raising livestock such as buffalo and goats.

The paper also reveals that historically, the off-farm and non-farm income and migration have been some of the major coping strategies of the farm and other rural households. The arid farming and livelihood systems, which are relatively most fragile due to very high climate risk, market risk, and other natural threats like locust attacks, etc. need to use resources synergistically by adopting an integrated farming and food systems approach that aligns well with the market demand to manage the system's risk and optimize the resource use under such unfavourable conditions. Despite high fragility, the region also presents potential opportunities for sustainable rural livelihoods. Crop production in arid regions faces very high climate risk but has unique opportunities to integrate resilient crops and cropping systems. The agro-ecology and resource situation are highly diverse and therefore need context-specific crop planning and related innovations. An integrated farming systems approach may also hold relevance in promoting circular economy. But it entails efficient extension services.

The paper highlights that the state's agriculture extension network is poor due to inadequate staff, lack of funds, improper capacity building, and motivation. Although Farm Science Centres (KVKs) are located in each district of Rajasthan, considering the size of districts in the desert region, KVKs probably find it hard to provide services to all farmers. Similarly, the farmers' groups organized under different public programs are either non-existent or are not properly functioning. Author advocates developing a framework and assessment of ground realities capturing location-specific needs and prospects, and then develop action plans, strategies, and policies for each of these arid zone sub areas and people.

The third paper in this section is contributed by Robert N. Truelove, Stephen C. Lellyett and Abul K. S. Huda titled "Australian Desert Ecosystems: Perspectives and Challenges for Agriculture". The authors have attempted review of literature about Australian desert ecosystems and the social and economic systems there in, with a focus on the perspectives and challenges for agriculture based sustainable livelihoods. The Australian deserts are continental in scale. They stretch 2000 km from the arid coast of Western Australia to western Queensland and NSW. In the case of most of Australia's arid and semi-arid regions, another defining feature is highly variable timing of rainfall events and great variation in the amount of rainfall per event. As well low atmospheric humidity, high daytime temperatures, and winds contribute to the general 'dryness' of the Australian deserts. The Australian desert regions are home to over half a million people, or about 2.0% of the current national population. The per capita gross regional product from the Australian desert regions was $AUD80, 360 compared to around $AUD52, 700 in the broader economy in 2008. The major contributors to total output in the desert regions of Australia, were the mining sector (39.5%), followed by the manufacturing sector (12.6%) and agriculture, forestry and fishing (9.3%). Most soils of the arid and semi-arid regions are leached, and their infertility is a dominant feature of the ecology and potential land use. The sources of water in these arid and semi-arid zones are: (i) rainfall, the northern monsoonal rains in particular, which generate high seasonal precipitation in the north of the continent and are the source of the large water flows that occur from time to time along the central Australian rivers and creeks; (ii) ephemeral and permanent surface waters including rock holes and rainfall fed non-porous clay pans; (iii) sub-surface or sub-artesian soaks; and (iv) wells, (including native wells) and bores which access deep artesian waters. Climate projections for late to end-century indicate an increase in frequency and severity of hot weather events for this region. Some of the challenges and opportunities for sustainable agriculture and improved livelihoods are explored in this paper.

Biodiversity is an important feature of desert ecosystem. The fourth paper titled "Phytodiversity of Thar Desert: Traditional Knowledge, Sustainable Utilization and Conservation" by Narpat Singh Shekhawat, et al. highlights the status and threats to plant resources in the Thar Desert. Scientific literature on plant resources of Indian Thar Desert is limited. Anthropogenic pressures on valuable plants like *Calligonum polygonoides* and *Ziziphus* species are threatening. The vegetation of Thar has so far attracted least attention of the botanists. It was neglected as compared to plants of other deserts. Biodiversity of medicinal plants that are numerous and endemic and edible plants are dwindling. Cycles of ignorance, and heavy machines/earth-movers/bulldozers-enabled land clearing for developmental activities, together are harming ecological landscape. Impact assessments/evaluations of development project(s) in fragile ecosystems/zones are non-existing. On the other hand, policy decisions are taken and implemented by plant-blind and eco-insensitive agencies for land use and introduction/promotion of cultivation of exotic/invasive plants species thus, endangering reserve forests of Aravalli/Desert. Invasive non-native plant species are causing havoc to the native flora.

The fifth paper in this section is by S. S. Burark, M. S. Rathore and G. L. Meena titled "Status of Temporal Changes in Agriculture in Western Rajasthan". The western region of Rajasthan characterized by changing prospects of agriculture due to high variability in rainfall, shrinking agricultural land and diversification towards plantation and other commercial crops. Agriculture including livestock is very important for livelihood and economy of the region. The factors that directly and indirectly affect the growth of agriculture such as: declining size of land holding, change in cropping pattern especially towards horticulture and cash crops, depletion of ground water for irrigation, changing composition and population of livestock, etc. The national movement to address climate change and sustainability, the intervention in the area of energy sector by installing solar and wind energy structures in common and grazing lands affecting water and composition/productivity of livestock. The population of sheep and camel is declining, gap in demand and supply of fodder is increasing and consequently affecting the ecosystem of the region. Livestock sector is being neglected or is at low priority despite that it plays an important role in supporting livelihoods of large number of households.

The other important factors that significantly affect the hot desert ecosystem, is the present paradigm of development based on exploitation of natural resources. For example, rampant mining activity (sand stone, granite and china clay), encroachment of common lands and water bodies, green energy projects, etc. There is an urgent need to re-examine the policies and programs from desert perspective. The problems of desert regions are likely to be further aggravated by human activities such as: unsustainable farm practices (overgrazing, excessive use of chemicals in agriculture and mechanization), heavy industry and mining, deforestation, over exploitation and contamination of water and other natural resources, urbanization and increase in artificial surfaces, etc. All these activities are leading to desertification. More specifically, the change in land use because of urbanization, industrialization and use of fertile land for development needs are the reasons responsible for the imbalance created in the ecosystem. Further, with fast increasing human and livestock population resulting in higher water demand for human use, for livestock, in urban areas and for the development activities has led to drastic depletion of ground water level.

This section also has a paper from Africa by Boubié Vincent Bado and, André Bationo titled "Sustainable Management of Dry Lands of Africa: A Framework of the Holistic Participatory Approach". The dry lands of the semi-arid and dry subhumid areas, covering 41 percent of the land surface, produce 44 percent of the crops, and contain over 2 billion people and half of the world's livestock. Two-thirds of the African continent is desert or dry lands. About 73% of the African dry lands are degraded and 51% are severely degraded. The population of Africa will double by 2050. The fragile land resources of Africa will face these high pressures from the growing population and livestock. However, rain-fed agriculture, livestock and agro pastoral systems remain the main sources of food, income and employment to the poor inhabiting these dry lands. Livestock plays a crucial role in generating income to invest in agriculture, purchase more diverse diets, paying for education, health and other household needs. In farming systems where the availability of fertilizers is limited and not affordable, manure from livestock is at the same time the main provider of nutrients to soils for staple crops. Food insecurity and poverty continue to be significant challenges in meeting the needs of a growing population. The primary drivers of food insecurity, malnutrition, and poverty in sub-Saharan Africa are the depletion of soil fertility, land degradation, desertification, and deforestation.

Over the 1960s and 1970s, there was focus on increasing crop production through "external inputs". The concepts of Low External Input for Sustainable Agriculture (LEISA), Integrated Nutrient Management (INM) and the concept of Integrated Soil Fertility Management (ISFM) were developed only later as strategic adaptations to the new challenges for sustainable management of soils fertility and land resources.

This includes the important role of nutrients as key factor for a better use of the limited rainwater in dry lands. This paper analyses the main achievements of integrated approaches, including the role and participation of rural communities for sustainable management of land resources in dry lands. The aim is to suggest new participatory approaches of sustainable management of dry lands for the resilience of farming systems. A critical review of the main successful achievements and failures could help in understanding the constraints and opportunities for sustainable management of dry lands. This includes crop-oriented or livestock-oriented approaches combining water harvesting techniques, N_2-fixing legume crops and agroforestry systems. The aim is to investigate new strategies for sustainable land use in dry land agroecosystems and resilience of smallholder farmers living therein. It was observed that local communities could play an important role for sustainable management of dry lands in a participatory and holistic approach.

The seventh paper in this section is by Rajendra Prasad Pandey titled "Challenges of Drought and Water Resources Management for Resilient Ecosystem". The author points out that the sustenance of ecosystem and life on the earth is governed by presence of water. The water bodies, landscape and soil provide habitat a number of aquatic species. Changes in various components of hydrologic cycle due to temperature rise and increasing Green House Gases (GHG) concentration in the atmosphere are causing threat to the ecosystem. It is widely believed that the climate change is primarily responsible for intensification of the components of hydrologic cycle and consequently, the hydrologic extremes (i.e. droughts and flood) and heat waves are expected to be more frequent and severe. This paper presents current level of understanding of occurrence of drought in different climatic regions of India and the challenges in drought mitigation. Various drought indicators and indices suitable for drought monitoring and assessment have been conversed including region specific suitability of mitigation options. The relationship of regional climatic parameters with frequency, severity, and persistence of drought have been discussed in detail. The prevailing approach of drought management in India needs a shift from a reactive crisis management to proactive mitigation planning. It has emerged as an important need to create region specific regular drought monitoring system for assessment and dissemination of information in a useable manner. A planning of areas specific mitigation measures and risk reduction strategy may pave a way for proactive actions to cope with drought.

The eighth paper titled "Challenges and Opportunities for Livelihoods and Environmental Security in Desert Ecosystems of Rajasthan, India" is by V.P. Singh; highlighting the complex problems of Thar or western region of Rajasthan. Though thinly populated, it is characterized by recurrent droughts, low and erratic rainfall, high air and soil temperature, intense solar radiation, and high wind velocity, and wide spread poverty. The author suggests designing the required vegetation combinations for Livelihoods (food, nutrition, fuel, fodder, timber, and other purposes), and Environmental Security, and developing technologies for growing them on the farm and landscape level, on lands owned by the small holders. The sequential restoration of vegetation and ecological services would be a multiplier benefit of this suggestion. Appropriate measures for adaptation to climate change can substantially reduce many of its adverse impacts, and thus contribute to food and livelihood security of the vulnerable rural population. The author suggests more focus on arresting natural resources degradation and their over exploitation, ecological restoration at scale and conflict resolution. Reduce soot, smoke, and GHG emissions and build larger constituency through federating processes, bringing various stakeholders together.

The ninth paper is by Dr. Purnendu Kavoori titled "Mapping Ecological Concepts on to the Landscape of Rural Economy: What Makes for Resilience in Agropastoral Western Rajasthan?" The paper highlights the issues pertaining to uncertainty and dry land adaptation in the context of western Rajasthan. There is, an increasing apprehension that we are probably entering a long-term phase of uncertainty/unpredictability in our ecological and social environment. At the global level, evidence indicates that as a whole earth's ecosystem is likely to become increasingly volatile with consequences that are difficult to anticipate. This uncertainty is likely to be a part of the scenario for the Century that lies ahead. It is important to note however that this insecurity is not equally distributed geographically, as also are those who would be affected by it. We may expect a considerable unevenness in the way 'uncertainty' is likely to be socially and ecologically distributed, which aspect often gets masked when we look at the problem at global scale. Desert ecosystems have a wide range of landscapes, temperatures, altitudes and aridity. These landscapes have however one feature in common, namely the presence of a high degree of variability. Arid and semi-arid landscapes invariably have considerable fluctuation in temperature, rainfall, and local bio-physical gradients. There are two consequences of this from the point of climate change. One that these habitats are likely to experience the impact of uncertainty in ways that will probably be different from other kinds of landscapes. Secondly, the nature of adaption in these habitats has historically evolved within to deal with uncertainty, and consequently, there is much that we can learn from the ways people have adapted to risk and the institutions and practices that enabled people living in these habitats to cope with.

The last paper of this section, by P. M. Chauhan and V. P. Joshi, titled "Towards Achieving Sustainable Development in Agriculture with Rainfed Ecosystem" of semi-arid area of Gujarat. Kachchh in Gujarat is the second-largest district of India, accounting for 25% of the state's area. Positioned in a semi-arid zone, it experiences severe weather extremes, with summer temperatures soaring to 50° Celsius and winter temperatures dropping to 3° or below. Water resources are scant, constituting only 3% of Gujarat's total, which significantly impacts the agricultural practices and lifestyle of the people of this region. Here agriculture coupled with animal husbandry, forms the backbone of local livelihoods. Despite the erratic monsoon and limited rainfall, averaging 300–350 mm annually. Rivers in Kachchh are non-perennial, leading to frequent droughts and water scarcity. The proximity of the district to the Arabian Sea exacerbates groundwater salinity, further complicating farming due to high levels of fluoride, chloride, iron, and nitrate, as reported by the Central Ground Water Board (CGWB).

The harsh climatic conditions of the district surprisingly benefit the farmers by reducing the prevalence of pests and diseases, making it conducive for growing a variety of horticultural crops such as mango, sapota, citrus fruit, date palm, banana, dragon fruit, guava, pomegranate, papaya, and custard apple. Areas, namely Anjar, Abdasa, Nakhatrana, Bhuj, and Mandvi, have relatively better aquifers, support cultivation of variety of these crops. Introduced in 2014, Dragon fruit (kamalam) has quickly become a favoured crop among Kachchh farmers due to its resilience to extreme temperatures. Today, over 200 farmers across the Kutch area of Gujarat have embraced dragon fruit farming, covering an area of approximately 700–800 acres.

The other important intervention leading to transforming fortunes of Kutch farmers is Date Palm Cultivation. In the Kutch-Bhuj region of Gujarat, local date palm varieties have traditionally been grown from seeds. However, this method has not been widely successful due to the dioecious nature of the plants, which complicates natural seed-based propagation, and the unavailability of high-quality planting materials. Presently in Kutch alone about 1.89 lakh hectares is under Date palm cultivation producing 1.79 lakh metric tonnes of fresh dates in the year 2020–21. Remarkably, Kutch is home to around 20 lakh date palm trees, with 17 lakhs being indigenous varieties, predominantly grown from seedlings.

1.4.2 Part II: Cold Desert Ecosystem

The Himalayan Cold Desert, as an ecological zone, represents an extremely fragile ecosystem that shows a complex relationship of the climatic and geomorphological processes, and exhibits very less but highly endemic diversity. Many rare species of flora and fauna are found in the cold desert. In total, four papers are there in this section. The first paper is by Tej Partap titled "Himalayan Cold Desert Environment, People and Change Process". The Himalayan cold desert ecozone is a vast highland mountain area having cold, dry and sunny climate and few people. The region has an area of 79,388 sq kms and includes whole of the Union Territory of Ladakh and two districts in Himachal Pradesh, namely Lahul & Spiti and Kinnour. Small area with cold desert conditions also falls in Uttarakhand and Sikkim too. The mountain communities that live here have been making a living on nomadic pastoral and agro-pastoral farming systems, supplemented by tourism, services and other nonfarming livelihood systems. Observations and analysis presented in this paper, are based on the published accounts and more often on the first-hand knowledge and information gathered by the author, over a period of three decades (1990–2022) presenting a situation analysis and a perspective of the dynamics of change process of the land, environment and people of this cold desert zone.

Signs of climate change impact in cold arid zone are visible and these can be both positive and negative. Looking at the negative signs, glaciers are receding faster than expected, snow fall is declining in terms of annual snow fall quantity as well as widening cycle of years when heavy snowfall is experienced. These events may become severe. Summers are warmer now and if the trend continues, that will mean availability of longer period for plant growth—good for agriculture. In future Ladakh may see upward movement of agriculture and opportunities for growing new crops.

In Himachal Pradesh, warming of local weather has been not only observed but strategies to benefit from it in farming have already been put in place. The success story of vegetable and fruit (Apple) farming in cold and dry zone of Himachal highlights the fact that with sincere institutional efforts, farm economy of cold arid zone can be turned into a profitable proposition. The cold and arid area was a remote area and the highland community lived in isolation evolving their own social institutions and mechanisms that were necessary to live in this ecozone. The national attention is drawn to this zone only when there has been Chinese aggression, whether in 1962 or now in May 2020. With these incidents, one thing has happened and that was felt need for road infrastructure.

Since past few decades one finds increasing national attention to these areas for various reasons. Consequently, there is enhanced institutional support in all spheres be it physical infrastructure, social infrastructure, human welfare schemes and flow of outside population into the cold arid hinterlands. Consequently, a change process in economy and ecology, both, is visible. The changed ecological and economic conditions may offer new opportunities for development, and it is for the national, provincial and local institutions to have better understanding of the new normal of the ecological and economic environment for designing better development policies for this zone.

The second paper in this section is by Anurag Saxena, et al. titled "Climate Change, Farmers Adaptation Strategies and Technological Interventions Led Agricultural Development Process in the Himalayan Cold Desert". A number of invasive pest species have arrived in the High Himalayas in recent years. Climate change may be one reason for it. These include various species of aphids on cereals, vegetables, and horticultural crops; yellow, brown, and black rust in wheat and barley; locust outbreaks in the Zanskar Valley and Nyoma block; outbreaks of brown tail moth in Kargil; and outbreaks in the lower belt of district Leh. The author argues that further research is needed, primarily on agro-climatic components. This includes developing more straightforward methods for simulating adaptation responses, such as improvements in agro-technology that allows farmers to reduce risks and/or maximize profits in the face of changing climates. Researchers are working to create new seeds, crops, and germ plasm that can withstand novel environments. Scientists in research centres are striving hard to create the kind of varieties that can withstand climate change.

The third paper in this section is by M.S. Raghuvanshi titled "Indicators of Changing Climate, Natural Landscape and Agriculture Dependent Livelihoods in the Desert Region of Ladakh". Ladakh, the cold desert area of Indian Himalayan region has primarily agropastoral economy. It is topographically mountainous with Martian landscapes, which has significant bearing on the climate processes. Soil here is shallow to very shallow. It makes the region vulnerable to natural degradation processes, caused by varied climatic, edaphic and anthropogenic activities. The most significant processes of land degradation in the region are: water erosion and vegetation degradation followed by wind erosion. The region is under active weathering state which would foresee deviations under the impact of climatic change. Landscape is not in cosmogenic steady state due to topographic effect. Vegetation has been observed as the most important factor in barren ecosystem. People, here are exposed to multiple drivers of change, such as climate, economic changes led pressure on land and resources and the transforming livelihood systems. During, 1960s, changes in the economy had significant impact on the dominant economic activities of Ladakh, i.e. agriculture and pastoralism. The introduction of various cash crops such as apples, apricots, potatoes, peas, mustard and other vegetables occurred in the late 1980s has seen significant impact on improving the economic wellbeing of the people of this region.

Ladakh is undergoing various biotic (BS) and abiotic stresses (AS), which are associated to each other. Abiotic stresses such as drought, high and low temperature and salinity play an important role in influencing the occurrence and spread of pathogens, insects, and weeds. Consequent upon climate changes Ladakh is experiencing infiltration of exotic floral and faunal species and significantly influencing the native vegetation.

In Eastern Ladakh, there is another emerging conflict. Here the pastoral communities of Changthang Plateau face the increasing population of wild ass “Kiang”. They are overstocking in the less productive Trans-Himalayan pastures (moist sedge meadows) and resulting in forage competition with livestock. Phenomenon is causing degradation of the pasture lands.

The fourth paper in this section is by Mahesh K. Gaur titled “Land Resources and Agricultural Economy of Leh District of Ladakh: A Case Study”. The author argues that there is a major challenge to understanding the relationship between various land attributes, land degradation, agricultural productivity, and food security in the cold arid region of Ladakh. This research paper highlights the underlying issues. Author argued that the agricultural system of Ladakh will have to respond to the land use drivers such as population growth, changing food habits, tourist in-flow, and future opportunities. Pathways to a better future for agriculture, land use, and land cover changes and the people who depend for their livelihoods on agriculture in this cold arid region have been studied. The results show that to meet future food production demands, the area under agricultural land is likely to remain almost static, however existing farmlands see crop intensification and diversification to generate additional income. It is argued that agricultural intensification would be crucial to ensure food security and employment generation. It is equally important to slow down the degradation of cropland and pasture area. The analysis of land resources and agricultural economy in Ladakh showed a shift from production-oriented to a supply-oriented food system, driven by socioeconomic change. The current transition affects the three interrelated food system components: availability, access to, and utilization of food. Changes in agricultural land use have led to a remarkable decline in regional food production. The trends indicate that agriculture economy of Leh in particular and Ladakh in general will continue to face challenges of climate change, shrinking land resources and increasing future demand. Future efforts should focus on strengthening infrastructure, improving water management, and fostering market access to fully realize the potential of its land resources. With continued support from governmental and non-governmental organizations, Leh can pave the way for a robust, resilient agricultural economy that benefits both its people and the environment.

1.4.3 Part III: Coastal Desert Ecosystem

As noted earlier, coastal desert ecosystem of India is less known and under researched. There are two contributions in this section. Globally, one-fifth of the landmass is occupied by desert regions. Each continent on the planet contains a desert, and each desert ecosystem is characterized by its own features and India broadly has three ecosystems, namely, hot desert, cold desert, coastal desert ecosystem. The coastal desert is restricted to the state of Tamil Nadu and Gujarat, each having their own peculiarities. The first paper addressing coastal desert in India is by Dr. V. Geethalakshmi titled "Management of Desert Ecosystems in Coastal Tamil Nadu". Although, the state of Tamil Nadu gets precipitation from south-west as well as north-east monsoons, some places in Tuticorin district and adjoining areas receive less precipitation, hence being dry and arid, resulting in desert. ***Locally called Therilands, coastal desert area spreads over*** south coastal districts of Tamil Nadu, Thoothukudi and Tirunelveli. It has active and stabilized sand dunes. Coastal desert area of Tamil Nadu is characterized by low and inconsistent precipitation, high evaporation, and invariable occurrence of drought stress. It is characteristic of deserts owing to moisture scarcity. Coastal desert patches are sharply demarcated by vegetation comprising of plants which are either short-lived annuals or xerophytes. The sparse vegetation is a consequence of limited water availability and unfavourable climatic conditions. Plants develop adaptation mechanisms in order to grow in this particular zone. The soils are deficient in nutrients and poor in water holding capacity. The exploitation of groundwater has led to the seepage of brackish sea water into the coastal area, deteriorating the quality of the groundwater aquifer along the coast of Tirunelveli district. It indicates the necessity of formulating groundwater policies for the coastal desert ecosystem.

The soils of coastal desert can be transformed into productive soils through applying various soil management techniques and selection of suitable plant or silvimix. These include, improving soil organic carbon, managing irrigation to enhance water use efficiency, use of microbial cultures and inputs to create favourable biological conditions in soil and improving the soil moisture holding capacity. The selection of crops and tree species can also help in improving ***theri soils***.

The paper by Nabansu Chattopadhyay, K. Malathi and Laxman Singh Rathore, titled "Climate Dynamics and Agriculture System in Deserts" summarizes the characteristics of all the three types of desert ecosystems, i.e. Hot, Cold and Coastal deserts. Global climate change is having a pronounced impact on desert regions, affecting them more significantly than many other areas. Deserts are experiencing higher temperature increases compared to other regions. This intensifies evapotranspiration rates, further drying out already arid areas. While deserts typically receive minimal rainfall, climate change is causing more erratic and extreme precipitation events leading to flash flooding and increased erosion, disrupting the fragile desert ecosystems. The frequency and severity of droughts in desert regions are increasing. Increased temperatures and prolonged droughts put additional stress on desert vegetation, which is already adapted to harsh conditions. Higher temperatures and reduced vegetation cover can lead to soil degradation. Communities living in or near desert regions are facing greater challenges due to water scarcity, reduced agricultural productivity, and increased heat stress. This can lead to migration and socioeconomic instability. Efforts to monitor and mitigate these impacts include advanced remote sensing technologies, incorporating new data, methodologies, and metrics improves the capability to monitor regional-scale agricultural drought conditions accurately. The use of geospatial technology allows for precise mapping and tracking of drought conditions across different regions, facilitating targeted interventions.

The hot Thar Desert has seen significant changes in recent years, with increase in both human and animal populations. Animal husbandry is crucial due to challenging farming conditions, leading to issues like overgrazing and environmental degradation. The region has a high livestock-to-person ratio, with cows, buffaloes, sheep, goats, camels, and oxen being the primary livestock. Sustainable practices are essential to mitigate environmental degradation and support the local economy. Dryland regions need an integrated approach for enabling current and future resilience in view of their fragile agro-ecosystem under climate change. An integrated farming systems approach oriented to harness the market-led opportunities may be the best way to support the development of sustainable farming and livelihood systems in arid western Rajasthan as well as other desert agroecosystems. Policy planning and actions should frame a desert perspective for long-term planning, taking into account trade-offs and synergies. It should also ensure considering fragility of desert ecosystems along with the possibility of increased shocks.

Ladakh's cold desert environment presents unique agricultural challenges. Agricultural areas are typically confined to river valleys, especially in central and western Ladakh. However, innovative techniques can significantly improve agricultural productivity and sustainability in the region. These include: Water Management and Solar Energy Utilization, Protected Cultivation and Precision Farming and promoting natural farming practices and educating farmers about soil health and nutrient management.

In the coastal desert of Tamil Nadu's Thoothukudi district, the soil has low water and nutrient retention capacities, making it inhospitable for most vegetation. To protect land from wind erosion, maintaining vegetative cover and constructing wind breaks or shelter belts is crucial. Tree species for wind breaks should be selected based on soil type, climate, rainfall, and moisture availability.

1.5 Conclusion

The concluding points emerging from the papers on these three types of desert ecosystems are listed below;

1. There is clearly a lack of DESERT PERSPECTIVE in our understanding of these desert ecosystems and consequent formulation of development strategies / policies / planning.

2. Government desert development policies have so far failed to address the emerging challenges in the cold, hot and coastal desert ecosystems. May be the present sectoral development policies are largely based on natural resources exploitation rather than their sustainable use.
3. There is over emphasis on livelihood issues rather than conservation and protection of desert ecosystems. Even the planned infrastructure development is not ecofriendly affecting the biodiversity of the desert regions.
4. Increasing human interventions in desert regions are leading to changes in desert ecosystems.
5. Increasing agriculture and non-agricultural activities in the desert regions are leading to degradation of soils triggering the process of desertification.
6. It is increasingly becoming difficult to adapt or mitigate the impact of climate change.
7. The changing climate is impacting flora and fauna, livestock and livelihood of human population.

To sum up, all papers support the statement that the present pattern of economic development in the desert regions is seriously affecting the socio-cultural fabrics of people residing in the three different desert regions, i.e. cold, hot, and coastal desert regions. Market forces are playing dominant role in triggering these cultural changes. State has failed to understand the ecological role of desert ecosystems of the country and does not have a *desert perspective framework for guiding sustainable development areas falling under the three kinds of desert ecosystems.*

Part I
Hot Desert Ecosystem

Chapter 2
Agricultural Land Use and Land Degradation in Thar Desert: Observations, Monitoring Methods and Emerging Trends for Future Strategies

Amal Kar

2.1 Introduction

The Thar Desert, situated between the Aravalli Hills Ranges in the east and the Indus River in the west, is an integral part of the arid western part of Indian Subcontinent, where the moisture availability index of −66.67 defines the eastern limit of the arid land (Krishnan, 1968; Anon, 2001). The northern boundary runs almost along the dry and under-fit valley of the Ghaggar (Hakra-Nara) River, through which once flowed the Sutlej, or the Saraswati River (Kar, 2014a), while the southern boundary is along the Great Rann of Kachchh (Fig. 2.1). About 72% of the Desert's 29.07 million ha (m ha) area lies in India, and the rest 28% is in Pakistan. Sand dunes, sandy plains of variable depth and underlain by layers of carbonate nodules in alluvium or colluvium, barren hills and uplands, small ephemeral saline lakes in the wake of many hills and gravelly pavements are the major landforms. The Luni and its tributaries in the south-east, and the wide valleys of the erstwhile Saraswati and the Drishadvati rivers that are now canalised, constitute the only major drainage systems worth mentioning, but the wanderings of these streams in the geological past and under fluctuating climates built up the plinth and the soil cover of the landscape (Kar, 2020a).

For the sake of a synoptic long-term regional assessment of land use and its effects, we divided the region into six major traditional agro-ecological sub-zones, which summarise over space the broad patterns of natural landscape variability and the dominant socio-economic conditions (Fig. 2.1). While the patterns of landscape variability were summed up from field observations, satellite images and thematic maps, assimilation of the cultural aspects needed some tweaking of the boundaries

A. Kar (✉)
Formerly of Central Arid Zone Research Institute, Jodhpur, India
e-mail: akarcaz50@gmail.com

M. S. Rathore et al. (eds.), *Desert Ecosystems*,
https://doi.org/10.1007/978-3-032-03237-9_2

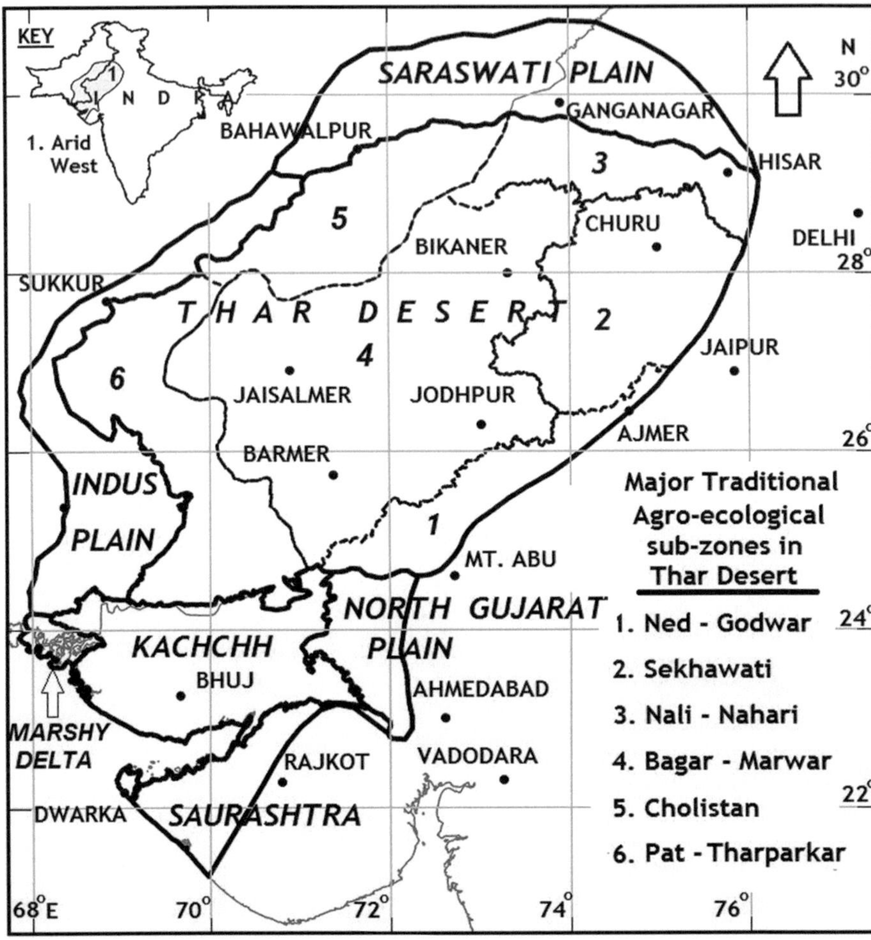

Fig. 2.1 Arid western part of Indian subcontinent, showing its seven major physical sub-divisions and the six major traditional agro-ecological sub-zones in Thar Desert

to accommodate the dominant administrative units within which the socio-economic data are collected consistently for the last three centuries, and which form a basis for assessing the major variables. The mapped sub-zones within India are Ned-Godwar (AEZ-1), Sekhawati (AEZ-2), Nali-Nahari (AEZ-3) and Bagar-Marwar (AEZ-4), while in Pakistan these are: Cholistan (AEZ-5) and Pat-Tharparkar (AEZ-6).

2.2 Scoping Natural Variability for Land Degradation

Being located in a transitional zone between the South-west Monsoon winds and the Westerly and North-westerly winds, the weather here experiences wide variability, often leading to extremes of temperature, precipitation and wind regimes that episodically accelerate the normal rates of water erosion/deposition (i.e., fluvial processes) and wind-blown sand activities (aeolian processes). Aeolian sand cover is ubiquitous, but Aeolian processes are more dominant in the west due to stronger wind in a drier environment. Natural plant cover is dominated by small shrubs and grasses, interspersed with occasional trees. About 85 percent of the annual rainfall (about 500 mm in the east and 100 mm in the west) is received during the monsoon months of July to September but with wide spatial variability of the amounts (long-term coefficient of variation, cv, ranging from 35% in the east to 80% in the west). Recurrent droughts (once in 3 years to every alternate year; Narain & Kar, 2005; Rathore, 2005) and moderate to high sand/dust storm activities (Goudie & Middleton, 2000; Middleton, 1986) are normal across the desert, while localised floods (2–3 in a decade) take place due to very high rainfall in short spells of 3–4 days (Rao, 2009) when stagnant water in the interdune plains partially recharge the groundwater aquifers that are usually deep and more often brackish to saline.

Superimposed on the broad natural framework and its resource endowment are the millennia of human occupations and resource utilisation, which tend to modify the landscape through land cover changes and land uses. Often such activities, when carried out at a pace that exceeds the capacity of Nature to keep balance through its normal rates of processes, lead to accelerated erosional activities, which in a desert typically get manifested through faster appearance of wind erosional features (e.g. formation of deflation hollows, sand sheets and sand dunes), water erosional features (e.g. sheet-wash, gullies), waterlogging and salinity build-up, degradation of natural vegetation, and other forms of chemical and physical degradation (e.g. polluted surface and groundwater, soil fertility decline, hard-setting of soil, mine spoil, etc.). Since such manifestations can also take place due to faster changes in the atmospheric variables (e.g. climate change), it often becomes difficult to identify a singular driver for the observed land degradation. Commonly, land degradation in the arid, semi-arid and dry sub-humid areas is called 'desertification' (UNCCD, 1995), which is expected to lead to a slow but perceptible loss of land's biological productivity, affecting the socio-economic conditions of the people whose livelihoods depend on those lands (Kar, 2018). Agriculture has been the dominant land use in Thar Desert since at least the dawn of Civilisation. The stakeholders in this land use, especially practicing crop cultivation and livestock rearing, had for millennia depended on the quality of the land's conditions for their production systems, and by implication had a system of caring for the land's well-being. Before discussing the land degradation scenario in the desert we propose to provide a brief history of its agricultural land use.

2.3 A Brief History of Land Use in the Context of Land Degradation

Thar Desert has a long history of sustainable agricultural practices, especially recorded from the Harappan Civilisation period (~5800–3500 years before present, or 5.8–3.5 ka, [kilo annum]), which mostly came up on the older alluvium of the Bangar landscape in the north under a climate that was transiting from a wet phase (10–6 ka) to a dry aeolian phase (~5.0–3.5 ka). Mean annual rainfall during the Early Harappan period (~6 ka) was roughly twice the present mean annual, received mostly during the SW monsoon period (June–September, or JJAS; Fig. 2.2). Contribution of winter and spring rains (DJF and MA) began to increase substantially since 4.0 ka, as has been elaborated from studies in the Kalibangan-Bhirrana-Rakhigarhi area (Bisht & Prabhakar, 2020; Lightfoot et al., 2020; Neogi et al., 2019; Petrie & Bates, 2017; Sarkar et al., 2016) and validated through the Arabian Sea sediment core data (Burdanowitz et al., 2021; Giosan et al., 2018). These studies suggested that both summer and winter cropping were prevalent, especially during the Mature Harappan period (~4.5–3.9 ka), the dominant crops being the millets, rice, wheat, barley, pulses and notably cotton. The higher winter rains, however, proved to be a tipping point for increased aridity since ~3.9 ka (Late Harappan period) and led to mass migration of people eastward from the Ghaggar-Hakra Valley, especially from the cities like Ganweriwala and Kalibangan, and finally the collapse of the Civilisation by ~3.5 ka.

As aridity began to put stress on the water and vegetation resources, inhabitants started developing innovative technologies to conserve soil and water and to

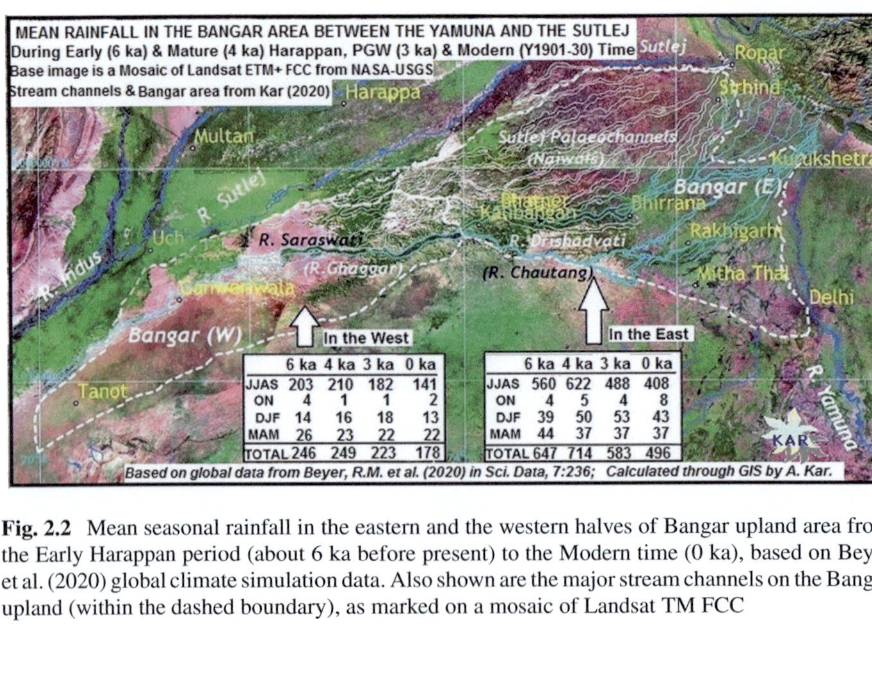

Fig. 2.2 Mean seasonal rainfall in the eastern and the western halves of Bangar upland area from the Early Harappan period (about 6 ka before present) to the Modern time (0 ka), based on Beyer et al. (2020) global climate simulation data. Also shown are the major stream channels on the Bangar upland (within the dashed boundary), as marked on a mosaic of Landsat TM FCC

adapt to the evolving resource availability pattern. One of these was in the use of fuel resources. As fuelwood became gradually scarce, archaeo-botanical remains revealed the increasing use of dung and crop residues (Lancelotti, 2018). Our analysis of recent climate simulation data from Beyer et al. (2020) indicates that with the increased aridity by 3 ka, the wind speed also increased sharply in the region, both during winter-spring (mainly Western Disturbance) and pre-monsoon months (SW monsoon wind). It implies very high wind erosion from the region for about 6 months annually, leading to high atmospheric dust load and adverse feedback for the summer monsoon. We may add here that contrary to some popular belief that the intensive agriculture (including animal grazing) during the Civilisation period might have led to very high land degradation that led to aridification and consequent collapse of Civilisation, archaeological findings do not endorse that the scales of those operations were anywhere comparable to those of the modern times, and so whatever enhancement in sediment accretion rates in ocean cores are noticed (Burdanowitz et al., 2021; Giosan et al., 2018) were mainly due to the secular climatic variation of the time. Summer monsoon began to peak again at ~3.2 ka, which led to the re-establishment of agriculture and trade from the Painted Grey Ware and Rangmahal periods along the Valley (Giosan et al., 2018; Kathayat et al., 2017). These findings have important lessons for us to understand the impacts of future climate on land use and land degradation.

Utilising the memories of the received knowledge since the re-establishment of SW monsoon during Early Historical period, and learning through a long process of judicious use of the region's sand-dominated land resources, as well as from an understanding of the adaptation regimes of local plants and animals to the extremes of temperature and meagre rainfall, the desert inhabitants gradually developed a robust and integrated farming system of rain-dependent crop diversification and risk aversion measures. It involved appropriate crop-livestock mix at household level, maintaining trees and shrubs in crop fields, conservation practices like land fallowing and mulching to maintain soil fertility level, erection of fences around crop fields to trap the fertile silt from blowing away during sandstorms, lopping of trees (rather than felling) for fuel and fodder, efficient rainwater harvesting in ponds and covered tanks, etc. (Bhati & Joshi, 2007; Dhir, 2003). Mixed cropping used to help in negotiating the risks from small monsoon aberrations. Animal husbandry, including a system of animal migration, not only provided a safety net to cope with the droughts when crops failed, but also had commercial utilities (Bharara, 1993; Kavoori, 1999; Robbins, 2004). During the pre-independence era, the major agricultural export items in the princely states of Marwar, Bikaner and Jaisalmer were either live domestic animals or animal products like high-quality wool and ghee (Erskine, 1909). Marwari bullocks, because of their stamina and strength, used to fetch a very high price for long-distance transportation between Delhi and Gujarat. Camels of Jaisalmer were in high demand for ploughing. To maintain such valuable herds, it was necessary to develop sacred forests and pastures through proper management systems. Lakhi Jungle to the NW of Hanumangarh and Pali Jungle in Sultana-Bahla area near Mohangarh were famous for their managed utilisation (Kar, 2014b). For small village pastures and the partially fallow lands a payment-based system of rotational/deferred grazing and fuelwood collection was developed, and most village pasture lands were given the status of permanency in revenue records. Unfortunately, the system gradually became a victim of high exploitation and huge corruption (Jodha, 1985). Despite such setbacks, the 'Traditional Wisdom' helped the farmers to withstand the worse of climates, so much so that there is no evidence for climate-driven mass exodus since the Harappan collapse (Kar, 2022).

Examples of deviations from the rain-dependent system were very few and negligible. For centuries till the independence in 1947, only a few small areas under irrigated rabi crops existed in the desert tract of India. Notably, in the traditional Khadins within the rocky uplands of Jaisalmer, the conserved moisture from meagre monsoon rainfall in the obstructed ephemeral channel bed of the Kankni River used to irrigate wheat and mustard crops. In the Rajgarh-Taranagar area in the NE, a few ha areas around dug-wells in the downstream end of the Kantli River used to be under irrigation; so were some small plots in the delta of the Luni River near Sanchor. Canals were first reported in 1355 AD, which benefited the Bhadra-Bhatner (Hanumangarh) area (Anon, 2002; Kar, 2014b). Several other canals were constructed during the British era. Tharparkar in Sindh and possibly the Cholistan desert area (both in Pakistan) had a longer history of canal irrigation, going back to ~300 BC at the least, as understood from the descriptions in Chanakya's Arthasashtra and Megasthenes' Indika (Panhwar, 2011).

The situation began to change faster after the independence in 1947 when large-scale transhumance between India and Pakistan led to extremely high pressure on the natural resources of the region. Land fallowing became a difficult proposition, while permanent pastures and open rangelands became strong targets for uncontrolled grazing and fuelwood collection. A series of recurrent droughts of moderate to large intensities in the first two decades after independence, and three severe droughts in the third decade put so much added pressure on the sparse plant population that the shrubs and the ground flora became extremely scarce. During 1961–1970 the average food grain production declined by about 26%. Human suffering forced the state to distribute a significant area under the marginal 'wastelands' to farmers for cropping (Kar, 2018).

The period of acute agrarian crisis was followed by the Green Revolution, an agricultural technology that was developed in the USA to enhance crop production, especially using high-yielding varieties of food crops, irrigation water, chemical fertilisers and pesticides. Farmers in the neighbouring state of Punjab were using the technology since the 1970s with huge success, and so it was introduced in the deserts of northern plains, first using the diesel pump sets to exploit the groundwater. Soon rural electrification began to encourage farmers to energise their wells for irrigation, while the construction of the Indira Gandhi Canal system and R&D on advanced agro-technologies provided impetus to canal irrigation. Difficult dune-infested terrain that formerly constituted vast open grazing lands, or low-yielding kharif croplands with a long fallow system, began to be converted into irrigated rabi croplands. The net sown area within the arid western Rajasthan increased from 41% of the total land in 1965–70 to 55% in 2010–15. In 1965–66 about 9% of the net sown area used to be irrigated, of which 38% area was under well irrigation and the rest 62% under canal irrigation (in the north). By 2015–16, the wells irrigated 62% area and the canals 38%. Cultivation spread far and wide into the erstwhile sandy wastes, and left fewer fallow and grazing lands. The old fallow lands have declined from 1960–61 to 2014–15 by 39%. Culturable wastes have declined by 29% during the same period, while the open rangelands, consisting mostly of barren and unculturable lands, have declined by 60%. Production of food crops reached unprecedented levels (Fig. 2.3). By 2010–11, the rabi-season food crop production in western Rajasthan registered an increase of about 78% over the1980-81 level, while the spices and medicinal crops increased by 68% and industrial crops by 97%. Land productivity in the irrigated croplands also increased manifold. In the non-command western part of the desert, however, dry farming still continues to be dominant, where average land productivity is 150–250 kg grain per ha, plus 500–1000 kg fodder per ha (as compared to about 5000 kg grain per ha in irrigated rabi croplands, but with little fodder production).

During the time of independence an average household with 8–10 ha land in the eastern part of the desert used to get about 70–80% of his total agricultural income from crop cultivation, while the animal products used to provide him the rest of the income. In the central part crops used to fetch about 60–70% of an average farmer's total income, while in the west crops could hardly fetch more than 30–40% of the total. This reflected a strong environmental control over the components of agricultural land use. Green Revolution, and especially irrigated farming, has so far been able to change the paradigm. It has helped to not only feed better the growing human and livestock populations in a region with limited natural resources, but also to produce enough marketable surpluses to make agriculture more viable than before. One of the complementary factors that may have assisted in this progress

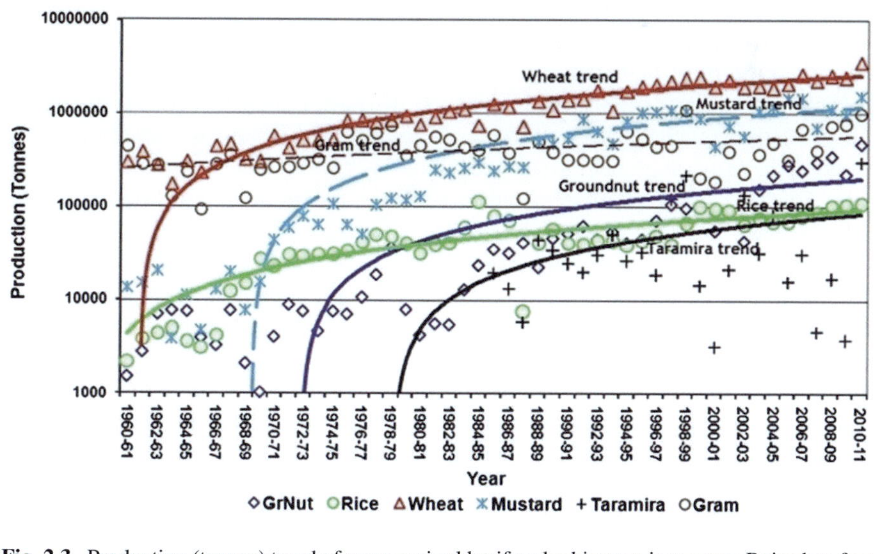

Fig. 2.3 Production (tonnes) trend of some major kharif and rabi crops in western Rajasthan from 1960–61 to 2010–11. Trend lines are the 4th degree polynomial curves

is lesser number of moderate to severe droughts per decade, which have declined sharply after the long drought of 1984–87, although occurrence of droughts (av. 4 per decade) remains a constant worry. Rainfall is still a major determinant of land productivity here, so much so that a 1% increase in rainfall roughly contributes to 0.43% increase in agricultural productivity (Chand et al., 2009). Irrigation through groundwater has taken the most crucial pivotal position in this development story, which needs to be taken a very serious note of, as the progressive emptying of the aquifers might lead to a sudden collapse of the whole system, especially through sharp changes in the land degradation scenario, drastic fall in the crop production and disastrous consequences for the human life and livelihoods. We shall elaborate this factor in the next section on land degradation scenario.

2.4 Present-Day Land Degradation: The Proximal Causes

The land surface in Thar Desert being mostly covered with aeolian sand sheets and sand dunes, the rainfall being meagre and the wind during summer months being strong, the landscape here is highly vulnerable to wind erosion. When kept in an undisturbed natural condition, the sandy landscape is always in a semi-stable condition, the natural stability being provided largely by the apparently sparse cover of grasses, shrubs and trees. A few layers of nodular calcium carbonate within the sand profile and a thin mat of cyano-bacterial colony just below the surface of many dunes also provide some protection to the mass of fine sand, but their roles are often understated. Several trade-off narratives provide evidence that the Green-Revolution-mediated agricultural success story is also a root cause for some emerging negative trends in the environment and the society. In the drier and windier west that is now under the command of the Indira Gandhi Canal network, a slow but perceptible marginalisation of its traditional land users, the pastoralists, has pushed them out of the interdune plains, and on to the flanking high sand dunes, whose scarce plant resources they have to plunder to survive (Ghai, 2021). A thick colony of the nutritious grass, *Lasiurus sindicus*, which used to be harvested after the monsoon from the Sultana-Bahla area in north Jaisalmer to feed the cattle across Rajasthan, was put under the canal network. This, as well as the large-scale destruction of other naturally-occurring and deep-rooted plants like *Calligonum polygonoides* to meet the higher demands of fuelwood in the expanding settlements, and the gradual cutting back of the dune slopes to expand the area under irrigated cropping, has accelerated sand reactivation on the dune slopes, followed by episodic forward movement of the dunes. At Sam, near Jaisalmer, linearly arranged trees and shrubs for forest plantation and camel safaris for the tourists on the dunes, have encouraged high mobility of a linear dune complex, which progressed by more than 2 kms between 1987 and 2020, partly overwhelming the rocky upland downwind (Fig. 2.4).

Irrigated cropping in the interdune plains across the desert is accelerating sand reactivation on the dune slopes and crests. As groundwater exploration progressed, the animal-driven wooden ploughs got gradually replaced almost everywhere by

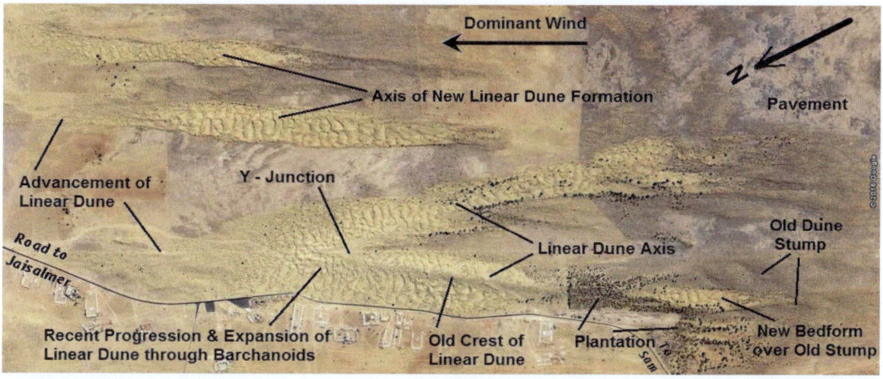

Fig. 2.4 Google Earth image of several 10–16 m high, Y-junctioned linear dunes at Sam, advancing north-eastward over a thinly veneered rocky limestone cuesta

the deep-ploughing tractors. At Dechu between Jodhpur and Jaisalmer the above technological changes led to high mobilisation of a parabolic dune complex that overwhelmed a highway several times, forcing the state to relay the road further downwind, while the blown sand from the dune encouraged the fast growth of another dune complex downwind (Fig. 5a,b). Farmers now prefer tractor for land levelling to usher in irrigation and for fast tillage-cum-sowing operation after the first sowing rains before it evaporates (about 30 mm in a day). However, tractor operation is antagonistic to the random distribution of plants in a field, which need to be uprooted. The naturally-grown plants that formerly were part of a traditional agro-forestry system, thus, have become a non-entity in the irrigated farming system, save and except along the field bunds.

The combined effect of widespread tractor-ploughing, shrub removal, disintegration and dying of the cyano-bacterial colonies, and shift back from irrigated cropping to rainfed cropping is noticed in the dune landscape as a gradual rise in the thickness and spread of loose sand mass that facilitates higher wind erosion and sand/dust storms during the dry summer months (Fig. 2.6).

Broadly, the western-most part with finer soils, higher wind strength, more dryness and poorer plant cover is a major source of atmospheric dust. Once air-borne the dust gets carried eastwards and gradually settles in the eastern part of Rajasthan and adjoining parts of Uttar Pradesh and Madhya Pradesh. We gathered the wind-field data for a major dust event on 27 May, 2010, whose movement as a 7-km-thick dust cloud was recorded by the Calipso satellite. Using the Hysplit model of NOAA we mapped the dust's likely flow path through different locations in the desert, and with increasing height from near Bikaner, which helped in better understanding of the spatial context of such events (Fig. 7a,b; Kar, 2020b).

With each episode of dust/sand storm, the soil gets gradually depleted of the already poor reserve of silt and clay as well as of the plant-available nutrients. A recent study has shown that the soils in western Rajasthan are predominantly deficient in OC, P, Zn and Fe, especially in the rainfed croplands (Kumar et al., 2021), where

(a)

(b)

Fig. 2.5 **a** Cultivation pressure in the interdune areas at Dechu, leading to high sand reactivation and its accumulation along the arms and nasal parts of a parabolic dune complex, as seen on Google Earth image of 2015. **b** Field photograph of a part of the above dune, showing forward movement of the dune nose from 1984 (approximately along the tall bushes on the crest) to 2007, burying an old road

continuous cropping and emission through dust could be some critical factors. The other major impacts of severe wind erosion are the damages to crop plants, burial of good agricultural lands and infrastructures and disruption of transportation network. The east-moving dust cloud also impacts the progress of the south-west monsoon rains.

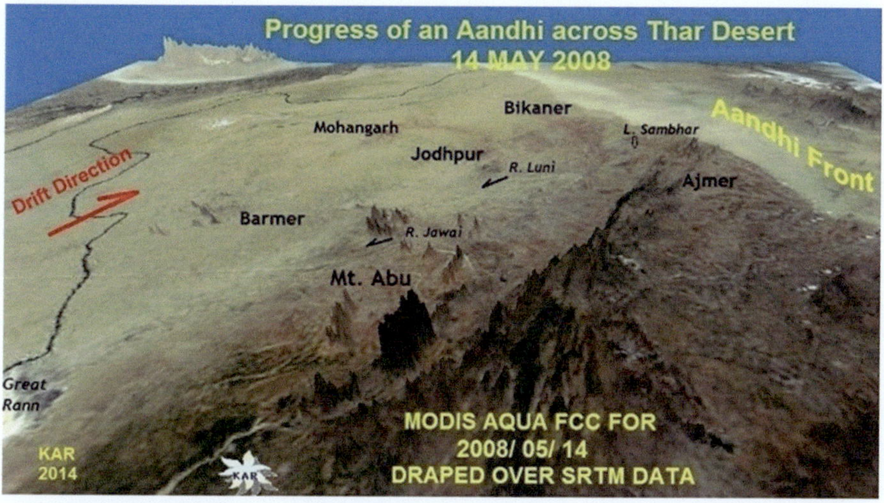

Fig. 2.6 MODIS Aqua image of 14 May, 2008, draped over SRTM 1 km DEM, showing the advancing front of a sandstorm in the Thar Desert, which moving north-eastward without any wet was scavenging. This sandstorm event persisted for about a week in the western part of the desert, which is common during the summer

Too much exploitation of groundwater for irrigation has led to near-exhaustion of many aquifers across the desert. The overall groundwater exploitation in the region during 1991 was 48% of the rechargeable reserve, which rose to 120% by 2001 and 149% by 2011. About 87% of the exploitation is used for irrigation and only 13% is available for drinking and other purposes. By 2020 a total of 83 groundwater development blocks out of 115 non-saline have been designated as over-exploited, leaving 6 under critical and 5 under semi-critical categories. The saving grace of 21 blocks under safe category includes all the blocks of Ganganagar (9) and Hanumangarh (6) districts due to canal irrigation (CGWB, 2021). As farmers dig deeper at a higher energy cost, they often draw poor quality water from a greater depth, which renders the soil saline-alkaline and reduces the crop yields. Such land quality degradation is taking place in parts of Jhunjhunun, Nagaur, Jodhpur and Barmer districts. Conversion of the irrigated rabi cropland to un-irrigated kharif cropland then becomes the only viable option, as has also been found in parts of Jhunjhunun district (Fig. 2.8).

In the canal-irrigated areas the partial leaching of the residues from applied chemical fertilisers and waterlogging are affecting the soils in Hisar-Sirsa-Suratgarh area. Excessive canal irrigation in the northern alluvial plain since the mid-1960s created an enormous problem of waterlogging. In the Sirsa-Dabli-Hanumangarh area, which was traditionally the granary of the region, widespread waterlogging compelled the construction of a Diversion Channel to drain the excess water to a number of interdune plains to the west of Rawatsar (Fig. 2.9), in the hope that the sand thickness would absorb the impounded water. This proved to be a grave mistake in policy

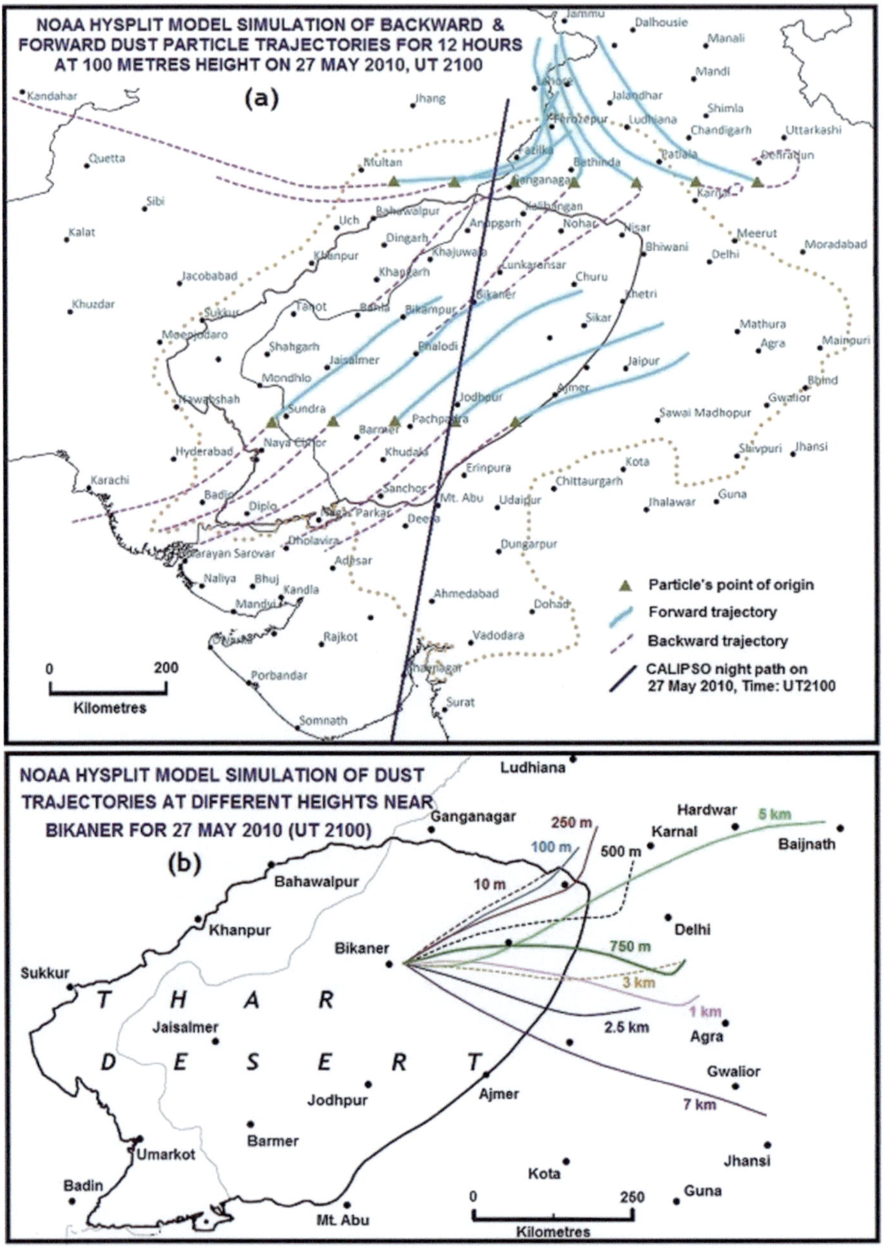

Fig. 2.7 **a** NOAA HYSPLIT model simulation of backward and forward dust particle trajectories from selected locations in Thar Desert for 12 h at 100 m height for 27 May 2010 (time UT 2100). Also shown is the CALIPSO satellite's night-time footprint through Mt. Abu, Jodhpur and Bikaner. **b** NOAA HYSPLIT model simulation of forward dust particle trajectories at different heights near Bikaner for 27 May 2010 (time UT 2100)

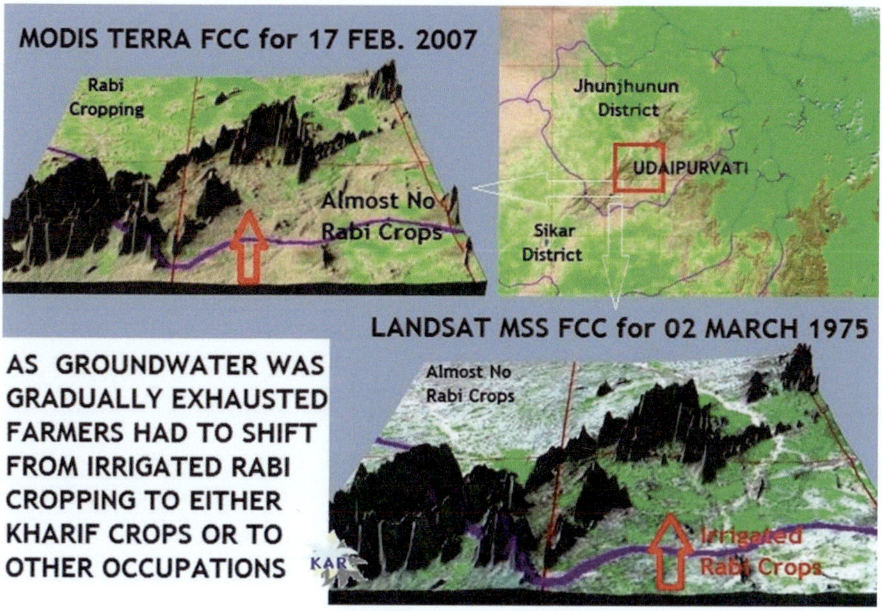

Fig. 2.8 Temporal changes in cropland use in the Udaipurvati area of Jhunjhunun district over time, as revealed from the Landsat MSS FCC (1975) and the MODIS Terra FCC (2007), both draped over SRTM DEM. Over-exploitation of groundwater within the Aravallis led to a shift from rabi to kharif crop lands, while new exploitation areas in the western sandy plain led to an increase in rabi-cropped lands

planning and implementation, with enormous socio-economic consequences for the inhabitants.

Although the landscape in Hanumangarh—Rawatsar area looks like a vast stretch of thick sand, the top sand layer is in fact underlain by a massive and impervious gypsum layer (Fig. 10a), and the land slopes imperceptibly northward to a fertile valley of the Chautang River (erstwhile Drishadvati) that holds a palaeochannel. Shallow groundwater along the valley used to support both kharif and rabi crops for generations. As the impounded water began to seep northward to gradually fill up the valley, farmers first noticed a rise in the groundwater table, followed by water in the root zone of their crops. They initially responded by shifting to crops that can withstand such a situation (e.g. rice), but soon this was not enough as water began to come up to the surface, inundating the prime lands. Satellite monitoring has shown that Baropal area was the first to be under deep water, followed gradually up the valley to Jakhranwali in the east (Fig. 10b). By the year 2000 Rawatsar and Gandheli got drowned under deep water (Fig. 10c). Villagers pleaded for discontinuation of the water inflow, but by then another influential group of stakeholders appeared from elsewhere, who began to strongly resist the demand, as they had started to invest heavily in the artificial lakes for pisciculture, with the aim to meet the ever-growing demands of fish in the markets of Delhi and NCR. Finally, after nearly three decades

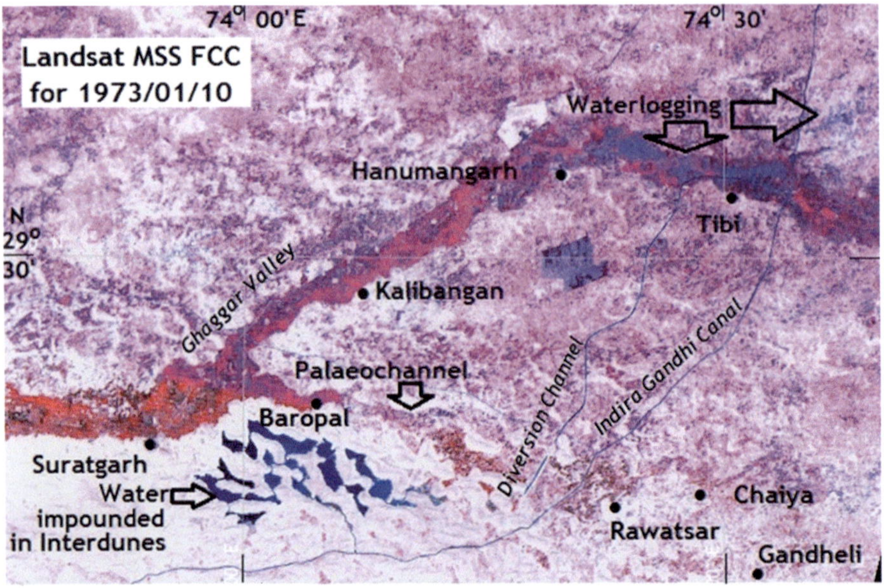

Fig. 2.9 Landsat MSS FCC for 10 January 1973, showing waterlogged areas along the Ghaggar Valley between Tibi and Hanumangarh, and along the Indira Gandhi Canal near Dabli, as well as along the Diversion Channel that carried the excess water from Tibi area into some interdune plains to the south of Baropal. Healthy plants (crops) are in deep red, which appeared abundant along the palaeochannel through Gandheli, Rawatsar and Baropal

the villagers got justice, and as the water level dropped the farmers began to restore their land for double cropping. Satellite images since 2021 do not show any sign of waterlogging in large parts of the valley.

Water erosion is not so prominent in our desert, and is episodic, especially in the Pali-Jalor area in the south-east, where scalding and gullies provide evidence for erosion in the upland areas. Downslope of the uplands a number of ephemeral streams were embanked since the early-20th Century for localised tank irrigation of rabi crops. This severely restricted stream flow in downstream areas and stopped recharging the shallow wells along stream banks. Consequently, the croplands along the banks that used to be irrigated earlier from those wells, shifted back to rainfed farming. As overall land quality degraded due to the changes in hydrology and as the beneficial plant species became scarce, an invasive woody weed, *Prosopis juliflora*, began to aggressively colonise the landscape, forcing the land users to accept reluctantly their dominance as a new normal. Robbins (2001) provides a succinct account of the ecological impact of land use change in the area. The traditional pomegranate orchards of Ahor suffered a collateral damage. The decline in the area of irrigated cropland in Pali district between 1988 and 2004 was by a staggering 39%. Rainfed cropland area increased during the same period by 13% (Ram, 2009).

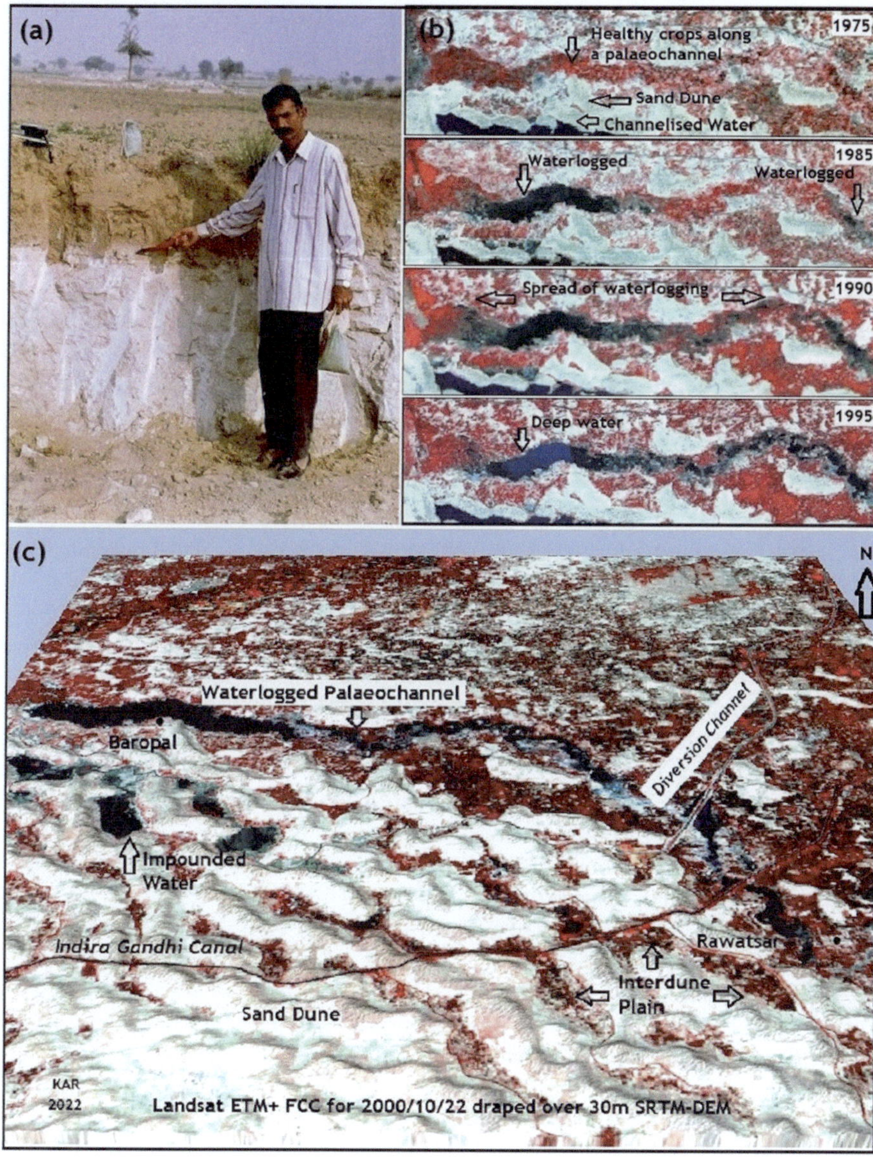

Fig. 2.10 Waterlogging in Suratgarh-Baropal-Rawatsar-Gandheli area, highlighting, **a** a massive gypsum layer below the surface sand which hinders water percolation, **b** sequential satellite images (Landsat) of the progress of waterlogging from 1975 to 1995, which gradually wiped out the healthy crops and, **c** a 3-D view (Landsat ETM + FCC draped over SRTM 30 m DEM) of the waterlogged areas between Suratgarh and Rawatsar in the year 2000

2.5 Mapping Land Degradation and Determination of Its Trend

Conventionally land degradation mapping is carried out on the basis of visual observation of the physical manifestations of the degraded landscape elements (e.g. rills and gullies, mobile sand dunes, salt-affected land, etc.) on cloud-free satellite image FCCs of summer and winter/spring, which is followed by validation of the marked features from random field checks. Singh et al. (1994) developed for western Rajasthan a satellite- and field-based mapping method through visual interpretation and used it on the satellite data of 1988–91 for mapping the degradation categories. Space Applications Centre (SAC), Ahmedabad, improved upon it for mapping the degradation pattern over India during 2003–05, 2011–13 and 2018–19 (SAC, 2013, 2021). According to these maps, western Rajasthan had about 89% degraded land during 1988–91 (75% under wind erosion), which changed to ~83% (73% wind erosion) during 2003–05, ~80% (72% wind erosion) during 2011–13 and ~79% (71% wind erosion) during 2018–19. It is apparent from the figures that wind erosion remains the most dominant process, mostly in the dune landscape and in the rainfed croplands. Water erosion is dominant in a small area along the Aravalli foothills, and salinisation mostly in the irrigated tract. Although the mapping suggested a declining trend in degradation, caution needs to be taken in interpreting the results because interpreter's eye estimation of the colour and tone of the images was, and continues to be a major deciding factor in assigning a degradation class, rather than any surface-reflectance-based independent digital method. Vegetation degradation, although known to be widespread, could not be properly mapped from visual interpretations. A re-look at the mapping methods was, thus, warranted.

Progress in satellite remote sensing and Geographic Information System (GIS) has now opened the vista for quantitative assessment and mapping of all the biophysical and socio-economic variables through some unbiased digital means, and for a holistic understanding of the land degradation problems, especially in the framework of Driving force (D), Pressure (P), State (S), Impact (I) and Response (R). The quantification of the map variables allows their meaningful integration for monitoring at precise pixel levels and for further identification of hotspots and effective remedial measures through land users' active participation. The new goal for the remedial measures is to achieve 'land degradation neutrality' in foreseeable future. Although several methods have been proposed for the purpose, derivation of a common set of D-P-S-I-R indices on land degradation parameters has met with difficulties, first because of the too many natural variables existing at the local level and having somewhat complex relationship among themselves and also because a proper understanding of the symbiotic to predatory relationship of the humans with the natural system at local scale under compulsion of population pressure and climate change will be necessary for factoring them in GIS and for assigning logical arguments. Collection of reliable and unbiased time-series data on all the relevant physical and socio-economic variables at regular grids is also problematic at local level. Fortunately, conditions of some of the physical variables can now be assessed quantitatively and repetitively from the many different proxy data that are gathered by the different satellite monitoring systems at high-to-moderate spatial resolutions and through daily to weekly re-visits. Kar (2018) provides a summary of the recent modelling efforts.

For the Thar Desert, we used the D-P-S-I-R scheme to digitally map land degradation in a sand-dominated southern tehsil of Churu district (Kar, 2011). Sand colour brightness index (CBI), derived from surface reflectance values in the red, green and blue bands of Landsat TM sensor, was calculated to represent the level of sand reactivation. Cropland use intensity index (TCI) was calculated from village revenue data and grazing pressure index (GPI) from the livestock data. Co-evaluating the digital maps on the three variables in GIS it was revealed that the cultivation pressure was the major determinant of sand reactivation in the tehsils and not the grazing pressure (Table 2.1).

Up-scaling of the CBI scheme for a regional-scale mapping of sand reactivation hit a roadblock due to almost overlapping signatures of surface reflectance from aeolian sand and gypsiferous material. We, therefore, developed a different system for mapping sand reactivation across Thar Desert, which is summarised below.

Using the MODIS surface reflectance data at 1 km pixel (from NASA-LPDAAC) for the years 2000–2015, we first constructed colour brightness index maps from the end of March 1st week (Julian day = 65) to the middle of the 3rd week of June (Julian day = 168), summarised into thirteen 8-day periods. Thus, a total of 208 maps were generated for the 16-year period. We then digitally segregated the signatures of aeolian sand brightness from the brightness of other surface sediments, notably the river-borne sand, gypsiferous and carbonate-rich sediments and other surface materials. Since some mixing was noticed with the gypsiferous sediments, we calculated a grain-size index and integrated it with the brightness index to construct a sediment type and brightness index (STBI) for all the periods. Since land surface emissivity varies inversely with the reactivation of sand, and since surface materials other than aeolian sand register higher emissivity values, we fortified the derived STBI maps with the land surface emissivity values, as calculated from the MODIS temperature data, and reported as GLASS-MODIS_BBE (www.glass.umd.edu; University of Maryland, USA). Our end product, called the 'Aeolian Sand Reactivation Index, moderated by BBE' (ASRI_bbe), provided a more convincing estimation of sand reactivation across the desert from early spring to the end of pre-monsoon season. Calculating the total area under sand reactivation pixels in the desert during each 8-day period as percentage of the total desert area, we constructed a graph of stacked

Table 2.1 Index of agreement between sand reactivation classes and pressure variables

CBI class	TCI class	Index of agreement	CBI class	GPI class	Index of agreement
1	1	−0.0008	1	1	0.0000
2	2	0.0026	2	2	0.3474
3	3	0.0060	3	3	0.2353
4	4	0.3088	4	4	0.1579
5	5	0.5395	5	5	0.2832

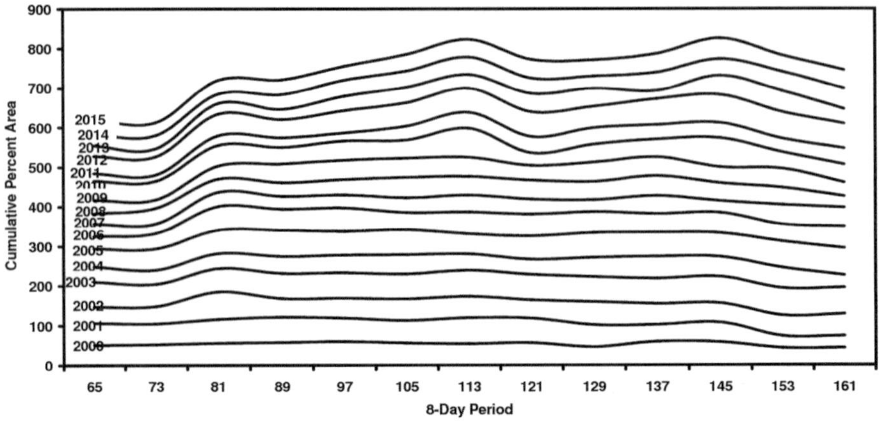

Fig. 2.11 Mean areas under sand reactivation in Thar Desert from ASRI-BBE mapping for 8-day periods from day (D) 65–72 to D 161–168 (X axis) during the years 2000–2015 (Y axis), which are depicted as percentages of the total desert area. The stacking of the plots reveals some gradually emerging inflection periods, especially during D 81–88, D 113–120 and D 145–152

yearly patterns of the 8-day period events over the 16 years that depicted the progress of aeolian sand deposition during summer in each year (Fig. 2.11). Three peaks of activity could be recognised from the graph: first around the 4th week of March (Day 81–88), followed by another about the 4th week of April (Day 113–120) and a third one by the 4th week of May (Day 145–152), after which the reactivation gradually declines as monsoon clouds begin to arrive. As the years progressed, a sharp decline in reactivation began to be noticed around the 1st week of May, which matches with a Western-Disturbance-related rainfall event that is becoming more frequent.

When we stacked the individual 8-day sand-reactivated area graphs over the different years (i.e., years on the X axis), we found a rhythmic behaviour of all the 8-day periods over the years, such that the ASRI_bbe values peaked in the years following a below-normal rainfall year. For example, a major drought year, 2002, did not record high ASRI_bbe, but 2003 did, as the summer wind speed during the drought year was not very high. An opposite situation was noticed after the high monsoon rainfall event of 2010 (Fig. 2.12).

Despite the relationship of mean ASRI_bbe values with the rainfall deviation from normal in a year, however, a consistent pattern of gradual decline in sand reactivation with time has taken place, as the mean summer wind speed has gradually declined (Fig. 2.13). Moving forward and backward in time with this analytical method hit a hurdle due to the non-availability of daily or weekly MODIS-equivalent surface reflectance data for pre-2000 period and also because NASA changed the method for 8-day summary pixels, which generated unrealistic numbers. We are currently trying to negotiate those hurdles.

Using the GLASS-AVHRR_BBE data (~5.6 km pixel resolution), we also constructed a long series of emissivity maps of the desert from 1982 to 2020 for the following 8-day periods: D 65–72, D 81–88, D 113–120 and D 145–152. The

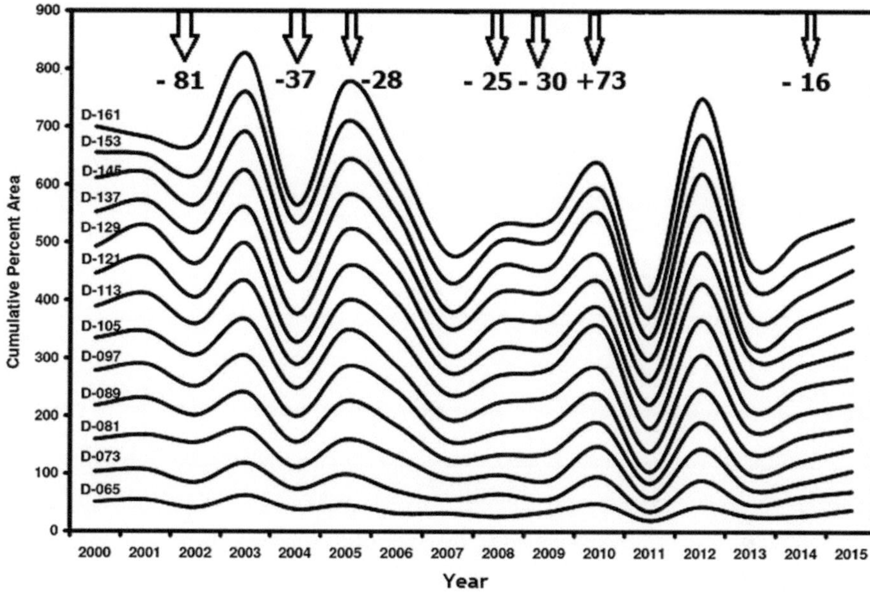

Fig. 2.12 Mean areas under sand reactivation in Thar Desert from ASRI_bbe mapping for the years 2000–2015 (X axis), covering the periods D 65–72–D 161–168 (Y axis), depicted here as percentages of the total desert area. Arrows with numbers show the percent deviation of mean July–September rainfall from the normal. Broadly, higher sand reactivation is noticed in the year following a drought, or low monsoon rainfall

periods represented the early spring, late spring, early to mid-summer and peak summer, respectively. Linear trend analysis (ordinary least squares, OLS) on the series suggested a declining trend in emissivity during the summer months, roughly between Bikaner, Barmer and Jaisalmer, which translates into higher sand reactivation in the sandy areas. Figure 2.14 shows the pattern of change during the peak summer. The R^2 values in sand dune areas tended to cluster around 30–45.

In the OLS method the outlier values in a noisy dataset generally affect the trend-fit adversely. So we used a Median Trend analysis (Sen, 1968; Theil, 1950) to understand the spatial pattern of changes in plant cover condition during the last four decades, especially in the croplands. We used the AVHRR's normalised difference vegetation index (NDVI) and vegetation health index (VHI) data for the 36th to the 40th week (representing the maturity stage of kharif crops) and for the 5th to the 8th week (representing the maturity stage of rabi crops), both series at 16 km pixel from NOAA-STAR website (www.star.nesdis.noaa.gov), assuming that homogenisation of reflectance values at that scale would highlight only the characteristics of the most dominant green crops, except in the rocky uplands and shrub-covered dune country in the west. NDVI has been used by several researchers to map the vegetation condition in our desert and to relate the phenomena with rainfall pattern (e.g. Kundu et al., 2017; Sur et al., 2018). We noticed that the kharif crops here still depend mostly

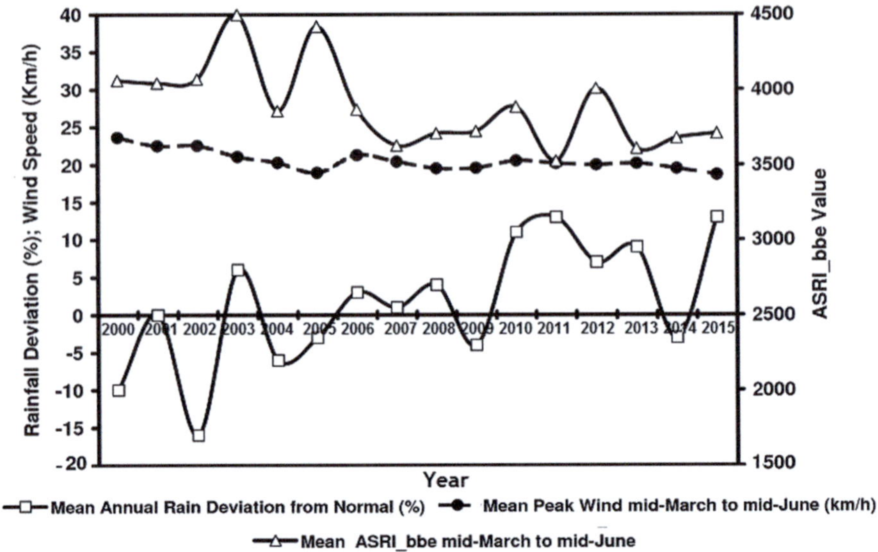

Fig. 2.13 Relationship between mean annual rainfall deviation from normal (%), peak wind speed for mid-March to mid-June (km/h) and mean ASRI_bbe values for mid-March to mid-June in Thar Desert

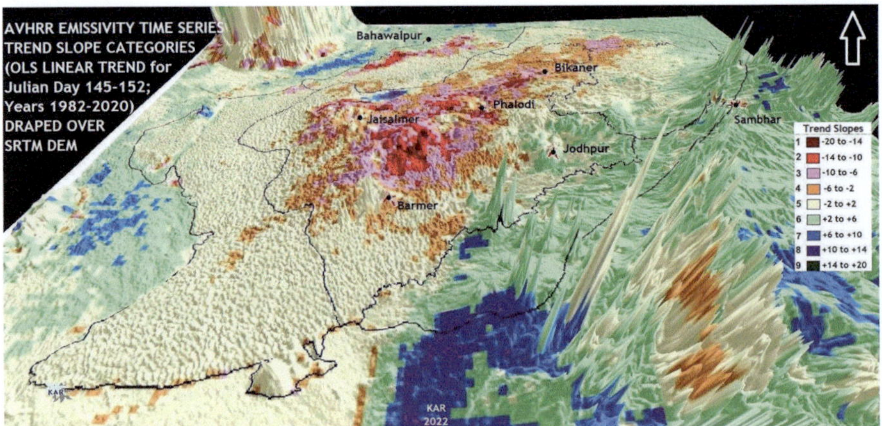

Fig. 2.14 OLS linear trend slope categories for land surface emissivity from the AVHRR TIR band data for D-145–152 (peak summer) during 1982–2020, which are draped over the 1-km SRTM DEM data. Strong negative slope values in the sandy areas suggest a trend of persistently high aeolian sand reactivation

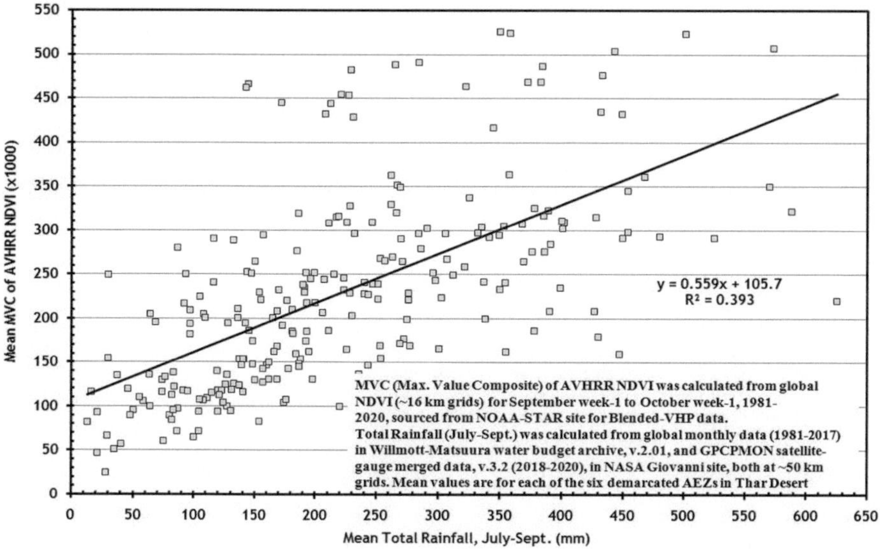

Fig. 2.15 Mean Maximum Value Composite of AVHRR NDVI at ~16 km pixel in Thar Desert and its surrounding areas during September 1st week to October 1st week (1981–2020), plotted against the mean rainfall (mm) for July–September (1981–2020). Data sources are given in the plot

on the rainfall, save and except in few areas of supplementary irrigation (Fig. 2.15; please note the aberrant NDVIx1000 values of >400 across the rainfall gradient along the topmost part of the graph).

We subjected the time-series images of kharif-NDVI to the robust Theil-Sen Median Trend analysis in which a pixel will show a trend only if it shows a rise or fall for at least 29% of the reported time (Eastman et al., 2009). The positive slope values in the south-eastern part of the desert are encouraging, but some pixels in Sekhawati and Ned-Godwar tracts are also registering considerable negative slope values (Fig. 2.16). We also notice that the positivity of slopes is smaller in large parts of Sekhawati, possibly due to falling soil nutrient levels, as was apprehended from a recent soil mapping (Kumar et al., 2019).

The above assumption is strengthened by the Theil-Sen trend analysis of the vegetation health index (VHI) for the period, which also shows a decline in the Sekhawati area (Fig. 2.17). The open rangelands to the west of Jaisalmer appear to be in a meta-stable condition, but beyond it the conditions in parts of Cholistan and Tharparkar are not very encouraging, as revealed from the trends in NDVI and VHI. We apprehend that this western part is set to become a major source of sand mobilisation and dust emission, gradually affecting the areas to the east. The high positive values in both NDVI and VHI in the north (Nali-Nahari and beyond) is especially due to the stability provided by the canal irrigation. When we look at the summer monsoon rainfall trend (Fig. 2.18), this northern area shows a consistent decline; yet the NDVI and VHI trends are kept strongly positive by the canals!

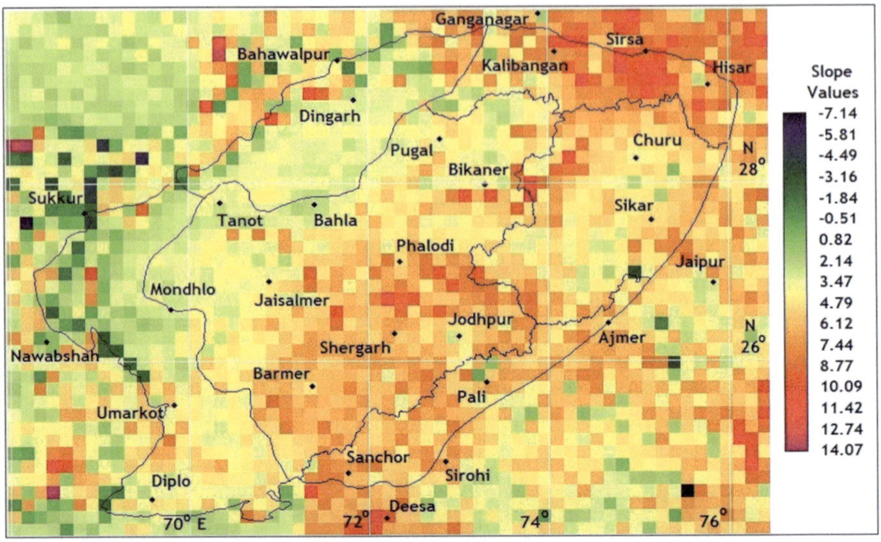

Fig. 2.16 Theil-Sen Median Trend slopes for the time-series maps on maximum value composites of AVHRR NDVI in Thar desert and its surrounding areas during September 1st week to October 1st week (1981–2021) at ~16 km pixel, which were calculated from the basic NDVI maps sourced from NOAA-STAR (www.star.nesdis.noaa.gov). Also shown are the six traditional agro-ecological sub-divisions in Thar Desert

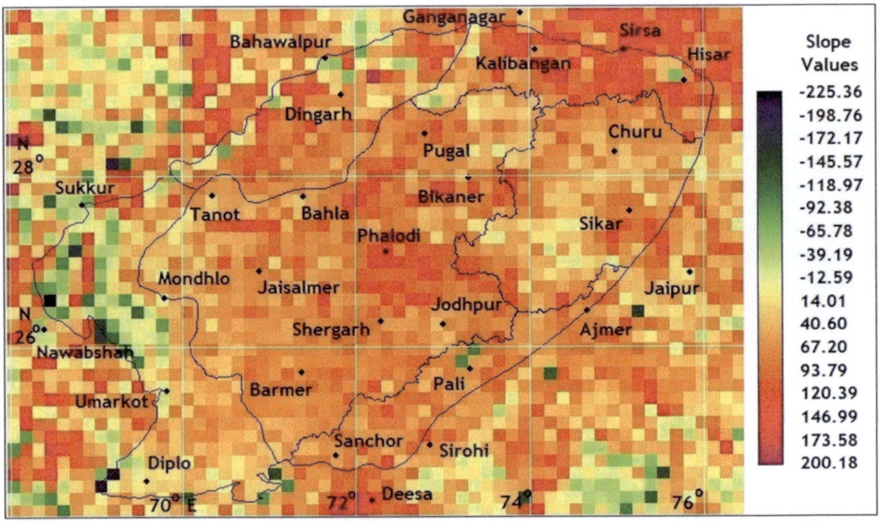

Fig. 2.17 Theil-Sen Median Trend slopes for the time-series maps on maximum value composites of AVHRR vegetation health Index (VHI) in Thar desert and its surrounding areas during September 1st week to October 1st week (1981–2021) at ~16 km pixel, which were calculated from the global VHI maps sourced from NOAA-STAR (www.star.nesdis.noaa.gov). Also shown are the six traditional agro-ecological sub-divisions in Thar Desert

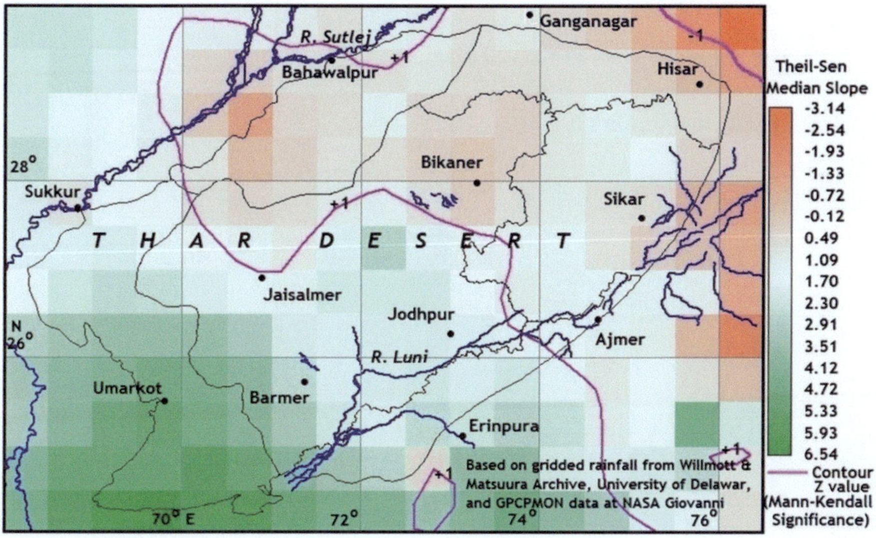

Fig. 2.18 Theil-Sen Median Trend slopes for the time-series maps on July–September rainfall (1981–2020) at 0.5° grids for Thar Desert and its surrounding areas, which were calculated from the monthly global precipitation data (1981–2017) in Willmott & Matsuura Archive in University of Delawar, USA, and from the global GPCPMON data (2018–2020) at NASA Giovanni site. Also shown are the Z-value contours for the Mann–Kendall significance of the calculated slopes, and the six traditional agro-ecological sub-divisions in Thar Desert

The VHI trends for the rabi season further strengthens our apprehension about the degrading situation in the Sekhawati tract (Fig. 2.19). It also validates our observation on the emerging poorer conditions in parts of the Ned-Godwar tract. The Indira Gandhi Canal command area has till now a high positivity rate, even though the sand dunes in the area show a slight decline in emissivity. The groundwater-irrigated central part between Jodhpur and Phalodi, has also witnessed an increasing trend, but this should be a matter for concern, as it may lead to a sudden collapse after the groundwater is exhausted. The broad trends from the above data analyses are now being pursued at higher resolutions. We also calculated the Theil-Sen trend slopes of the February VHI (1982–2018) from the 4-km-pixel AVHRR data and mapped the areas under negative trends in 14 major natural sub-regions within the desert (Fig. 2.20).

2.6 Outlook for the Future

Presently, the three major environmental issues in our desert that need more attention of the stakeholders are water management, land quality and dust emission, all of which are getting inter-twined. The stark realities of fast-declining groundwater

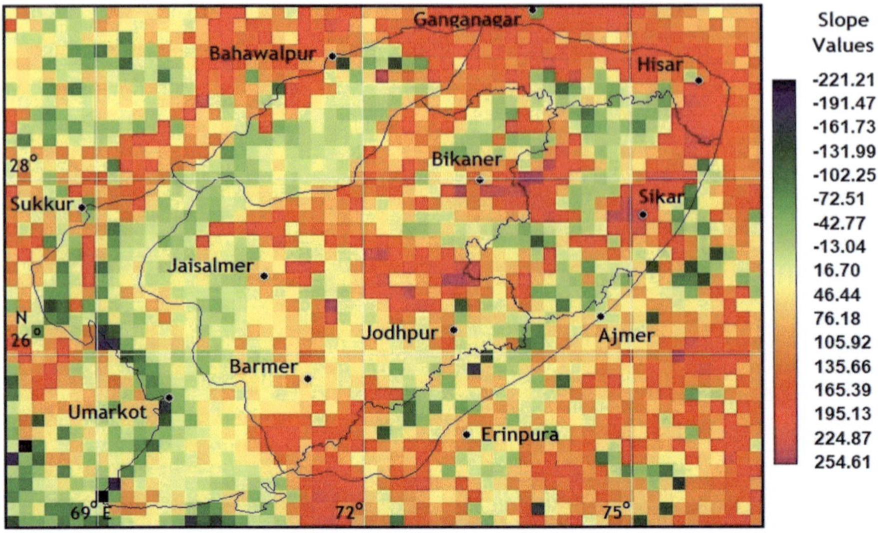

Fig. 2.19 Theil-Sen Median Trend slopes for the time-series maps on Maximum Value Composites of AVHRR VHI in Thar Desert and its surrounding areas during February (1982–2022) at ~16 km pixel, which were calculated from the global VHI maps sourced from NOAA-STAR (www.star.nesdis.noaa.gov). Also shown are the six traditional agro-ecological sub-divisions in Thar Desert

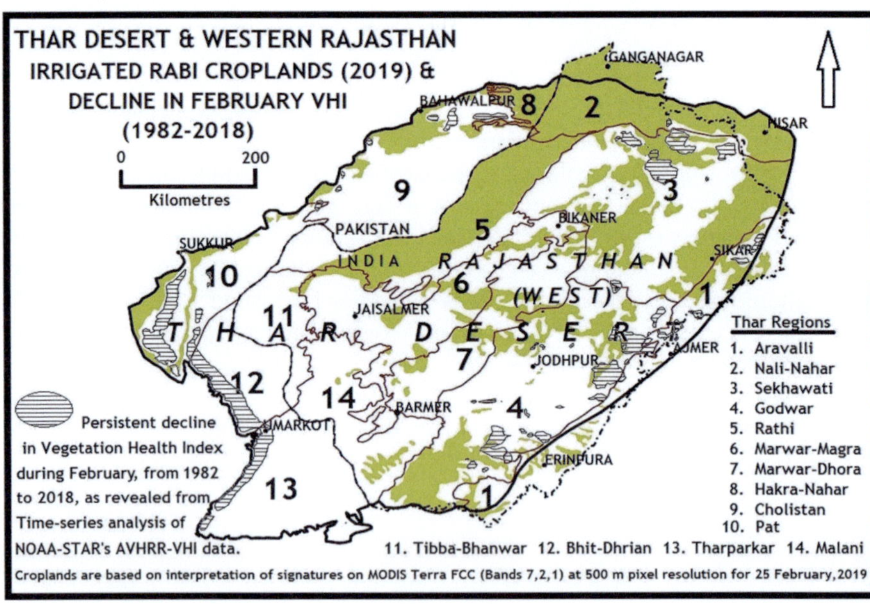

Fig. 2.20 Irrigated rabi croplands (2019) and areas of persistent decline in February VHI (1982–2018) in Thar Desert and the remaining areas of western Rajasthan, as interpreted from the calculated Theil-Sen Median Trend slope values of VHI at ~4 km pixel. VHI data was sourced from NOAA-STAR (www.star.nesdis.noaa.gov)

reserve have forced the state to stop digging for irrigation wells in most of the affected blocks within the Indian part, but more efforts may be necessary to better manage the water and to promote crops and their varieties that demand less water, and have adequate heat and drought tolerance.

Observations across the desert show that due to Global Warming, the SW monsoon has started behaving more erratically than before and is getting delayed due to the over-bearing presence of the Westerly winds. The wind strength has started building up after a fall from the mid-1980s to the 1990s, and surface temperature during both kharif and rabi seasons is increasing at an unprecedented rate (Joshi & Kar, 2009). Calculation of wind erosion index (WEI) has revealed that after the lowest mean values recorded during 2001–05, WEI may gradually rise again and attain a peak in the mid-century (Kar, 2012). Studies elsewhere have shown that with per degree rise in night temperature in northern India the wheat yield is likely to decline by 400–500 kg per ha (Sinha & Swaminathan, 1991; Swaminathan & Rengalakshmi, 2016). The higher evaporative demand of crop plants is often compelling the farmers to arrange for life-saving irrigation. Yet, unable to cope with the rising winter temperature the high-value rabi crops are wilting more often, while the yields from the kharif croplands are declining despite the fertiliser inputs, possibly because a stage of land fatigue has set in. As the cost of cropping is, therefore, becoming more unviable, and success more uncertain, many small farmers, who had invested through loans for well irrigation of their fields and for other inputs to grow high-value crops, are forced either to sell their land or to become share-croppers for some city traders and money lenders. For those traders a parcel of land is a capital investment only, and so the land's resources need to be fully exploited fast to get good returns from the investment in the shortest possible time, be that through over-exploitation of groundwater or through continuous cropping assisted by high doses of chemical fertilisers. Once the land's inherent resources are exhausted and a phase of diminishing returns begins, the investor abandons the plot for newer destinations. This vicious cycle of land development to land degradation is emerging in the region, with enormous environmental and societal costs (Kar, 2014b).

Despite the setbacks, there is a growing awareness that keeping the land in good shape is the key to sustained production. Learning lessons from past mistakes farmers in the groundwater-irrigated areas have started to install sprinkler and drip systems in most of their fields. This should substantially reduce groundwater use over large parts. In the canal-command areas prone to water logging, farmers do not over-irrigate anymore and ask for a choice of crops and soil drainage. Awareness to keep a parcel of the land fallow for a season and to grow beneficial tree/shrubs along the field bunds as wind brake is getting popular in the windy west, especially with the large farmers who can afford them. Along the wetter eastern margin preference is given to winter crops that can continue till March or early-April. Despite the strong temptation of the market prices on the choice of crops, a growing tendency is noticed to factor the soil nutrient availability for sustainable yield and to balance between farm income, groundwater use, land quality degradation and next year's yield potentials, factoring the current year's crop cultivation. Researchers have developed alternate cropping systems that can optimise the groundwater use through improved water and other resource use efficiency of the suggested crops, withstand the heat and the agricultural droughts and improve the net economic returns from the land even if the total crop yields might decline in some cases (Rathore et al., 2022). To reclaim the land degraded by high sodicity in irrigation water, CAZRI has standardised the doses of gypsum application to soils of arid region. A variety of crops are now grown in the region with sodic water after the soil is periodically treated with gypsum. For water erosion control in the eastern part, and for better utilisation of the rainfall a number of soil and water conservation and management practices have been evolved and the traditional systems have been improved upon. These are now packaged into an integrated watershed management programme, involving rainwater harvesting for drinking, farm pond ('diggi') construction for limited irrigation, agro-forestry, agronomic treatments, groundwater recharge and management, contour bunding in upper catchments, etc. Several other low-cost land management technologies are also being tested and provided, but the technology adoption rates are slow in the desert due to inadequate finances, slow land reforms, lack of trained personnel and lack of market facilities, while the growth rates of human and livestock population are high, mismanagement of resources are still widespread and climate variability are yet to be understood properly for quick responses, despite the simulation results. Hopefully, the emerging crises situations will provide the space for better initiatives and coordination to surmount the problems, as has always been noticed here historically.

A glimmer of hope is also provided by the emerging rainfall uncertainties. As climate is changing, the region has started getting slightly higher yet infrequent rain from the Western Disturbances during the spring to the mid-summer. This rain is beneficial to trees and shrubs, and by implication for the rangeland development and livestock rearing. Livestock-based farming, which was a traditional strength of the region, especially in the western half, has now become more dependent on crop wastes than on natural forage, especially because of widespread rangeland degradation. Livestock sector did not receive any encouragement during the Green Revolution, although farmers know that during the periods of moderate to severe droughts, when crops fail totally, the livestock bails out their economy to a large extent. Now, as Nature has started providing exclusive rain advantage for the sector, investing in its core competence as an industry may provide a much higher dividend under the evolving climate. It is to be recognised that unlike in the case of crop cultivation where growth is partly related to grain storage facilities and market access, in the livestock sector growth depends more on the faster transportation of the products from the hinterlands to the processing plants, as well as on the development of energy-secured milk processing and cold storage facilities, the establishment of industries to process dairy, meat, bone, hide, etc., and efficient marketing strategies (Kar, 2022). Once the infrastructure is developed and the livestock herders start receiving better returns with little risk, the rangelands and permanent pastures will be automatically taken care of. Encouraging the sector will also automatically lessen the pressure on groundwater and croplands. Knowing that this early rain is also beneficial for tree-based farming, especially horticulture, many farmers in parts of Jodhpur, Jalor, Pali and Barmer districts have started to develop orchards.

As new uncertainties in the land use decisions are being introduced by Global Warming, one feels it is time that we develop a robust system of land degradation assessment and monitoring that quantifies the parameters through a D-P-S-I-R methodology and integrates the physical and socio-economic variables reasonably for explaining the changes and then showing the future trajectories of changes under different scenarios, rather than to find 'vulnerabilities'.

References

Anonymous (2001). *India: National action programme to combat desertification in the context of united nations convention to combat desertification (UNCCD)* (Vol. I, p. 294) New Delhi: Status of Desertification. Ministry of Environment & Forests, Government of India.

Anonymous (2002). *History of Indira Gandhi Nahar*. (p. 405) Indira Gandhi Nahar Board, Govt. of Rajasthan.

Beyer, R. M., Krapp, M., & Manica, A. (2020). High-resolution terrestrial climate, bioclimate and vegetation for the last 120,000 years. *Scientific Data, 7*, 236. https://doi.org/10.1038/s41597-020-0552-1

Bharara, L. P. (1993). Livestock migration and pastoralism in Rajasthan Desert. In A. K. Sen & A. Kar (Eds.), *Desertification and its control in the Thar, Sahara and Sahel Regions* (pp. 219–247). Scientific Publishers.

Bhati, T. K. & Joshi, N. L. (2007). Farming systems for the sustainable agriculture in Indian arid zone. In K. P. R. Vittal, R. L. Srivastava, N. L. Joshi, A. Kar, V. P. Tewari & S. Kathju (Eds.), *Dryland ecosystem: Indian perspective* (pp. 35–52). Jodhpur: Central Arid Zone Research Institute and Arid Forest Research Institute.

Bisht, R. S., & Prabhakar, V. N. (2020). Evolution of early human settlements in the Sarasvati River basin: Archaeological evidence and site distribution analysis. In S. K. Acharyya, K. Ghosh, & A. Kar (Eds.), *Saraswati: The River par Excellence* (pp. 159–196). The Asiatic Society.

Burdanowitz, N., Rixen, T., Gaye, B., & Emeis, K.-C. (2021). Signals of Holocene climate transition amplified by anthropogenic land-use changes in the westerly-Indian monsoon realm. *Climate of the past, 17*, 1735–1749.

CGWB (2021). *Dynamic ground water resources, Rajasthan, as on 31.03.2020* (p. 57). Jaipur: Central Ground Water Board, Western Region.

Chand, R., Garg, S. & Pandey, L. (2009). *Regional variations in agricultural productivity: A district level study. Discussion Paper NPP 01/2009*. New Delhi: National Professor Project, NCAEPR.

Dhir, R. P. (2003). Thar Desert in retrospect and prospect. *Proceedings, Indian National Science Academy, 69A*, 167–184.

Eastman, J. R., Sangermano, F., Ghimire, B., Zhu, H., Chen, H., Neetin, N., Cao, Y., & Crema, S. (2009). Seasonal trend analysis of image time series. *International Journal of Remote Sensing, 30*, 2721–2726.

Erskine, K. D. (1909). *Rajputana Gazetteers, Volume III-A: The western Rajputana states residency and Bikaner agency* (p. 399). Allahabad: The Pioneer Press.

Ghai, R. (2021). Understanding 'culture' of pastoralism and 'modern development' in Thar: Muslim pastoralists on north-west Rajasthan, India. *Pastoralism: Research, Policy and Practice, 11*(3). https://doi.org/10.1186/s13570-020-0090-1

Giosan, L., Orsi, W. D., Coolen, M., Wuchter, C., Dunleaa, A. G., Thirumala, K., Munoz, S. E., Clift, P. D., Donnelly, J. P., Galy, V., & Fuller, D. Q. (2018). Neoglacial climate anomalies and the Harappan metamorphosis. *Climate of the past, 14*, 1669–1686.

Goudie, A. S. & Middleton, N. J. (2000). Dust storms in South West Asia. *Acta Universitatis Carolinae, Suppl. 35*, 73–83.

Jodha, N. S. (1985). Population growth and the decline of common property resources in Rajasthan, India. *Population and Development Review, 11*, 247–264.

Joshi, N. L., & Kar, A. (2009). Contingency crop planning for dryland areas in relation to climate change. *Indian Journal of Agronomy, 54*, 237–243.

Kar, A. (2011). Geoinformatics in spatial and temporal analysis of wind erosion in Thar Desert. In S. Anbazhagan, S. K. Subramanian, & X. Yang (Eds.), *Geoinformatics in Applied Geomorphology* (pp. 39–62). CRC Press.

Kar, A. (2012). GCM-derived future climate of arid western India and implications for land degradation. *Annals of Arid Zone, 51*, 147–169.

Kar, A. (2014a). The Thar or the Great Indian Sand Desert. In V. S. Kale (Ed.), *Landscapes and Landforms of India* (pp. 79–90). Springer.

Kar, A. (2014b). Agricultural land use in arid western Rajasthan: Resource exploitation and emerging issues. *Agropedology, 24*, 179–196.

Kar, A. (2020a). Identification and mapping of the Saraswati River system: A review and new findings. In S. K. Acharyya, K. Ghosh, & A. Kar (Eds.), *Saraswati: The River par Excellence* (pp. 29–98). The Asiatic Society.

Kar, A. (2020b). Quaternary aeolian landscape development in Thar Desert and its drivers. *Proceedings, Indian National Science Academy, 86*, 405–417.

Kar, A. (2018). Desertification: Causes and effects. In D. Bartlett & R. P. Singh (Eds.), *Exploring natural hazards: A case study approach* (pp. 159–205). Boca Raton, USA: CRC Press, Taylor & Francis Group.

Kar, A. (2022). Adaptation of agri-based society to environmental changes in the Thar Desert. In A. R. Siddiqui & A. Sahay (Eds.), *Climate change, disaster and adaptations* (pp. 151–171). Switzerland: Springer Nature.

Kathayat, G., Cheng, H., Sinha, A., Yi, L., Li, X., Zhang, H., Li, H., Ning, Y., & Edwards, R. L. (2017). The Indian monsoon variability and civilization changes in the Indian Subcontinent. *Science Advances, 3*, Article e170129.

Kavoori, P. S. (1999). *Pastoralism in expansion: The Transhuming herders of western Rajasthan.* New Delhi.

Kumar, M., Kar, A., Raina, P., Singh, S. K., Moharana, P. C., & Chauhan, J. C. (2019). Spatial variability of available soil nutrients in the Shekhawati region of Thar Desert, India. *Journal of the Indian Society of Soil Science, 67*, 21–33.

Kumar, M., Kar, A., Raina, P., Singh, S. K., & Moharana, P. C. (2021). Assessment and mapping of available soil nutrients using GIS for nutrient management in hot arid regions of north-western India. *Journal of the Indian Society of Soil Science, 69*, 119–132.

Kundu, A., Patel, N. R., Saha, S. K., & Dutta, D. (2017). Desertification in western Rajasthan (India): An assessment using remote sensing derived rain-use efficiency and residual trend methods. *Natural Hazards, 86*, 297–317.

Lancelotti, C. (2018). 'Not all that burns is wood'. A social perspective on fuel exploitation and use during the Indus urban period (2600–1900 BC). *PLoS ONE, 13*(3): e0192364. https://doi.org/10.1371/journal.pone.0192364

Lightfoot, E., Jones, P. J., Joglekar, P. P., Tames-Demauras, M., Smith, E., Muschinski, J., Shinde, V., Singh, R. N., Jones, M. K., O'Connell, T. C., & Petrie, C. A. (2020). Feeding the herds: Stable isotope analysis of animal diet and its implication for understanding social organisation in the Indus Civilisation, northwest India. *Archaeological Research in Asia, 24*, Article 100212. https://doi.org/10.1016/j.ara.2020.100212

Middleton, N. J. (1986). A geography of dust storms in South-west Asia. *Journal of Climatology, 6*, 183–196.

Narain, P., & Kar, A. (Eds.). (2005). *Drought in Western Rajasthan: Impact* (p. 104). Coping Mechanism and Management Strategies. Central Arid Zone Research Institute.

Neogi, S., French, C. A. I., Durcan, J. A., Singh, R. N., & Petrie, C. A. (2019). Geoarchaeological insights into the location of Indus settlements on the plains of northwest India. *Quaternary Research, 94*, 137–155.

Panhwar, M. H. (2011). *Six thousand years of history of irrigation in Sindh* (p. 313). Hyderabad: M.H. Panhwar Trust.

Petrie, C. A., & Bates, J. (2017). 'Multi-cropping', intercropping and adaptation to variable environments in the Indus civilisation. *Journal of World Prehistory, 30*, 81–130.

Ram, B. (2009). Land use in arid lands in India. In A. Kar, B. K. Garg, M. P. Singh & S. Kathju (Eds.), *Trends in arid zone research in India* (pp. 62–77). Jodhpur: Central Arid Zone Research Institute.

Rao, A. S. (2009). Climate variability and crop production in arid western Rajasthan. In A. Kar, B. K. Garg, M. P. Singh & S. Kathju (Eds.), *Trends in arid zone research in India* (pp. 48–61). Jodhpur: Central Arid Zone Research Institute.

Rathore, V. S., Nathawat, N. S., Bhardwaj, S., Yadav, B. M., Santra, P., Kumar, M., Shekhawat, R. S., Reager, M. L., Yadav, S. R., Lal, B., Yadava, N. D., Singh, J. P., Kumar, P., Bhaskar, S. & Yadav, O. P. (2022). Alternative cropping systems and optimized management practices for saving groundwater and enhancing economic and environmental sustainability. *Agricultural Water Management, 272*. https://doi.org/10.1016/j.agwat.2022.107840.

Rathore, M. S. (2005). State level analysis of drought policies and impacts in Rajasthan, India. In *Working paper 93, drought series paper 6* (p. 29). Colombo: International Water Management Institute.

Robbins, P. (2001). Tracking invasive land covers in India, or why our landscapes have never been modern. *Annals of American Geographers, 91*, 637–659.

Robbins, P. (2004). Nomadism now: Cultural survival in a changing desert environment. *Annals of Arid Zone, 43*, 1–19.

SAC (2013). *Desertification status mapping of India—2nd cycle* (p. 49). Ahmedabad: Technical Guidelines. Space Applications Centre.

SAC (2021). *Desertification and land degradation atlas of India* (p. 282). Space Applications Centre, Ahmedabad.

Sarkar, A., Deshpande-Mukherjee, A., Bera, M. K., Das, B., Juyal, N., Morthekai, P., Deshpande, R. D., Shinde, V. S., & Rao, L. S. (2016). Oxygen isotope in archaeological bioapatites from India: Implications to climate change and decline of Bronze Age Harappan civilization. *Scientific Reports, 6*, 26555. https://doi.org/10.1038/srep26555

Sen, P. K. (1968). Estimates of the regression coefficient based on Kendall's tau. *Journal of the American Statistical Association, 63*, 1379–1389.

Singh, S., Kar, A., Joshi, D. C., Ram, B., Kumar, S., Vats, P. C., Singh, N., Raina, P., Kolarkar, A. S., & Dhir, R. P. (1992). Desertification mapping in western Rajasthan. *Annals or Arid Zone, 31*, 237–246.

Sinha, S. K., & Swaminathan, M. S. (1991). Deforestation, climate change and sustainable nutrition security. *Climate Change, 16*, 33–45.

Sur, K., Dave, R., & Chauhan, P. (2018). Spatio-temporal changes in NDVI and rainfall over western Rajasthan and Gujarat region of India. *Journal of Agrometeorology, 20*, 189–195.

Swaminathan, M. S., & Rengalakshmi, R. (2016). Impact of extreme weather events in Indian agriculture: Enhancing the coping capacity of farm families. *Mausam, 67*, 1–4.

Theil, H. (1950). A rank-invariant method of linear and polynomial regression analysis. *Nederlandse Akademie Wetenchappen, Series A, 53*, 386–392.

UNCCD. (1995). *United Nations convention to combat desertification* (p. 71). Interim Secretariat for the Convention to Combat Desertification.

Chapter 3
Transforming Drylands: Systems-Based Sustainable Agriculture in Arid Western Rajasthan

Shalander Kumar, Abhishek Das, Anthony Whitbread, and M. L. Jat

3.1 Introduction

Transitioning to the agri-food systems that satisfy the varied food and livelihood requirements of an expanding population while preserving the environment and creating inclusive income-generating opportunities for farm households is the most significant challenge that humanity is currently facing (FAO, 2014; Garnett et al., 2013; Godfray et al., 2010). The severity of this problem is amplified in the arid and semiarid regions, which cover around 30 percent of the planet and are home to about 20 percent of the human population (Sivakumar et al., 2005).

In India, the rainfed area spans about 78 million hectares and accounts for around 55 percent of the entire arable land in the country. The drylands, which include the arid and semiarid regions, cover more than 50% of the rainfed region. The Indian hot arid zone spans approximately 32 million hectares, predominantly encompassing the northern Western states of Rajasthan (constituting 70% of the total area), Gujarat, Punjab, and Haryana. Additionally, it includes scattered land masses within the states of Maharashtra, Karnataka, and Andhra Pradesh. Due to a changing climate, India's arid and semiarid area has been predicted to expand in size in the coming decades (Rao et al., 2013). Characterised by meagre and unpredictable precipitation ranging from 100 to 450 mm annually, along with extreme temperatures, prolonged sunshine duration (6.6–10 h), diminished relative humidity, elevated wind speeds (9–13 km/h), and substantial evaporative demand (2500–4500 mm/year), the region exhibits distinct climatic attributes. Moreover, the soils are poor in fertility and prone to degradation. Due to high biotic and abiotic pressures, the ecosystem balance of

S. Kumar (✉) · A. Das · M. L. Jat
International Crops Research Institute for the Semi-Arid Tropics Patancheru, Hyderabad, Telangana, India
e-mail: Shalander.Kumar@icrisat.org

A. Whitbread
International Livestock Research Institute, Dar es Salaam, Tanzania

M. S. Rathore et al. (eds.), *Desert Ecosystems*,
https://doi.org/10.1007/978-3-032-03237-9_3

this fragile tract is disturbed and, if not attended to, may cause irreparable loss to the natural resource base leading to desertification. Evapotranspiration is four to five times higher than rainfall, while it is only two times higher in semiarid regions. Hence, the arid region experiences heightened levels of aridity, water deficit, scarcity, and drinking water challenges, exacerbating the severity of the situation.

Most households depending on agriculture in western Rajasthan face an unpredictable and risky future. The area grapples with recurrent droughts, soil erosion, desertification, excessive groundwater depletion, declining soil and water quality, diminished productivity, institutional weaknesses, malnutrition, and higher transaction costs largely attributed to an extensive production system. The extent of moisture deficit, as indicated through the aridity index (Fig. 3.1), land degradation (Fig. 3.2), and vulnerability to climate risk (Fig. 3.3), are extremely high in arid western Rajasthan. Additionally, the population is rapidly expanding and now stands at 28 million. Farmers in Rajasthan are at the forefront of climate change since their region has had a negative water balance for much of the year.

Therefore, in this region, the severe weather conditions have a negative impact on human and animal life, making it more challenging and precarious to live there. The unstable state of agriculture and low productivity and profitability in the region is a source of concern not only for the Government but for the sustainability of ecosystem services and the community impacting human well-being. In the past, the rural communities of western Rajasthan demonstrated a high degree of resilience by adopting various coping and survival strategies in these highly adverse conditions. However, the current climate change trends indicate that this region will be one of the most vulnerable hotspots and thus need special attention.

This paper therefore aims to explore potential development pathways for sustainable integrated farming and agroecosystems that can enhance the system productivity and resilience and livelihood opportunities in arid western Rajasthan. The sections following, describe the agro-climatic status and trends in western Rajasthan.

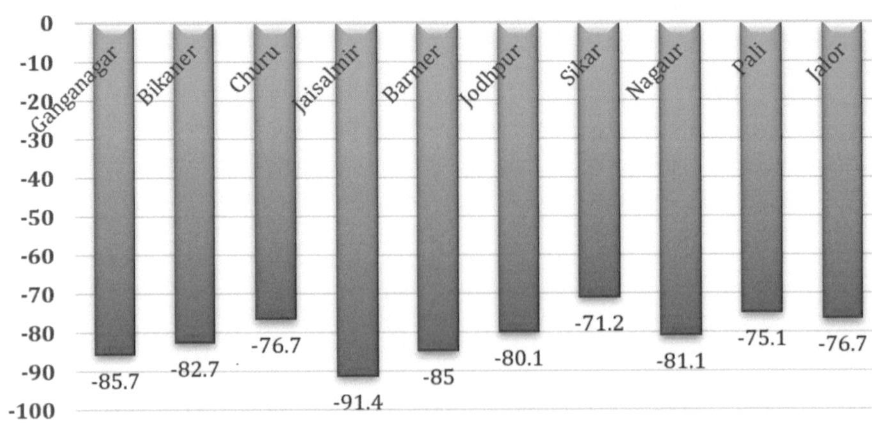

Fig. 3.1 Extent of aridity in western Rajasthan. *Source* Authors calculations

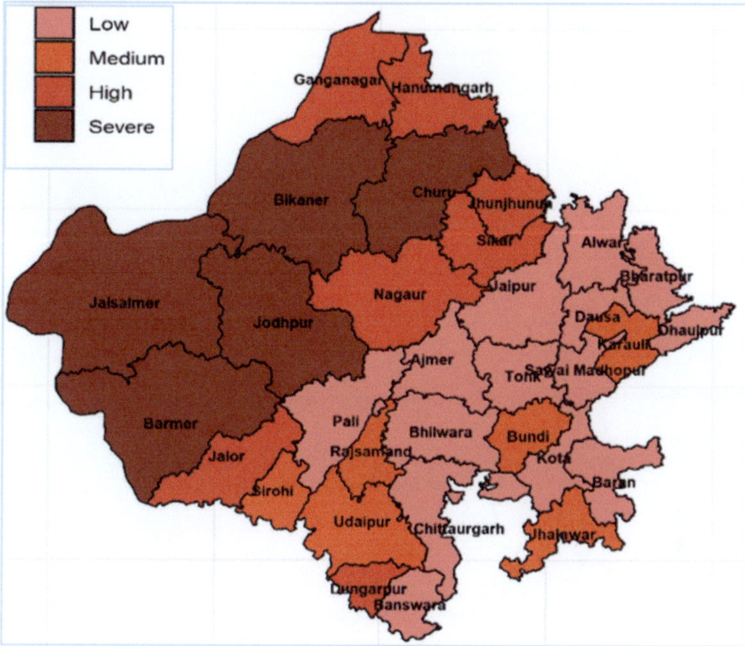

Fig. 3.2 Severity of land degradation in arid western Rajasthan. *Source* Authors calculations

Potential context-specific strategies and policy pathways some conclusions and recommendations for action.

3.2 Farming and Agro Ecosystem in Arid Western Rajasthan

The major determinant shaping the configuration of agricultural production systems across western Rajasthan along the West–East rainfall gradient is the availability of water and moisture. As groundwater extraction intensifies, there is a noticeable shift away from cultivating traditional millet and legume crops alongside livestock (cattle and sheep) towards cultivating commercial crops such as cumin (*Cuminum cyminum L.*) and Isabgol (*Plantago ovata*), horticultural crops like pomegranate and fig, or raising livestock such as buffalo and goats. Considering livestock, buffalo keeping is more common in the eastern part (Jodhpur), having relatively higher rainfall, while a tendency to shift in composition towards small ruminants is observed for the drier, more Western part. In the western region, particularly in Jaisalmer, pastoral and small ruminant systems prevail, often involving seasonal transhumance to Gujarat and central India. These areas are characterised by extensive and degraded natural pastures, along with some wadis or khadins (depressions) where trees and certain

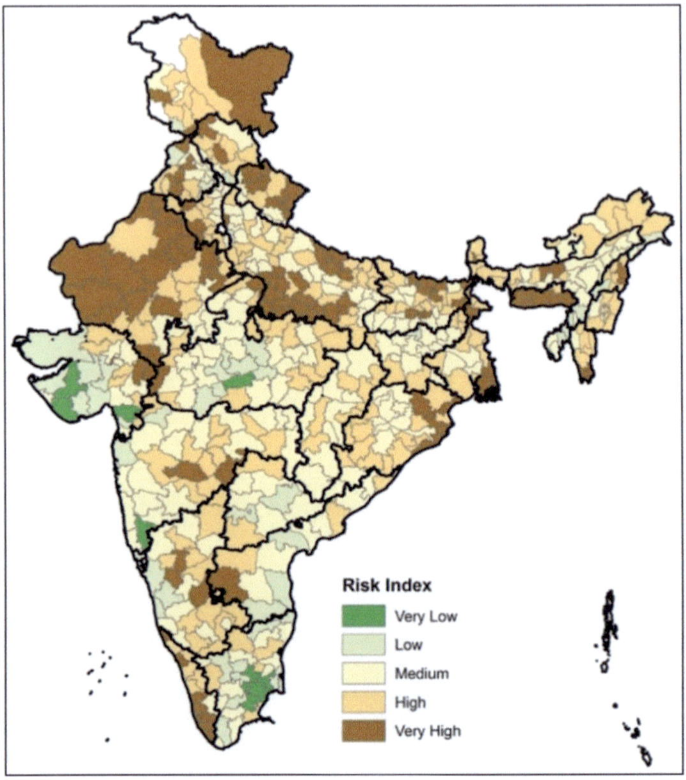

Fig. 3.3 Extent of district-level climate risk, India. *Source* Rao et al. (2013)

crops can be cultivated. Progressing towards the east, where rainfall levels rise, the agricultural systems transition from pastoral to agro-pastoral and eventually to mixed crop-livestock-tree systems. In districts like Jodhpur, parkland systems predominate, featuring Prosopis cineraria (Khejri) as the primary tree species. Livestock serves as a crucial source of rural livelihood, with small ruminants and cattle being prevalent in rainfed areas, and buffaloes in irrigated regions. Horticulture, though limited in scope, is typically practiced on sub-marginal rainfed lands. These farming systems, along with ovine and household industries, contribute significantly to the resilience of low-income rainfed croplands, which predominantly employ low inorganic inputs. Adopting a holistic farming system approach is essential for the development of these three sub-systems, incorporating measures such as improved rainwater utilisation efficiency, stress-tolerant crop varieties, and sustainable land management practices, all interconnected with market dynamics. A visual representation of the prevalent farming systems in the arid lands of western Rajasthan is provided in Fig. 3.4.

The region's agricultural output has increased due to the introduction of high-yielding cultivars, the increasing application of chemical fertiliser, and the expansion of irrigation systems. Though crop yields have increased, agricultural output is highly

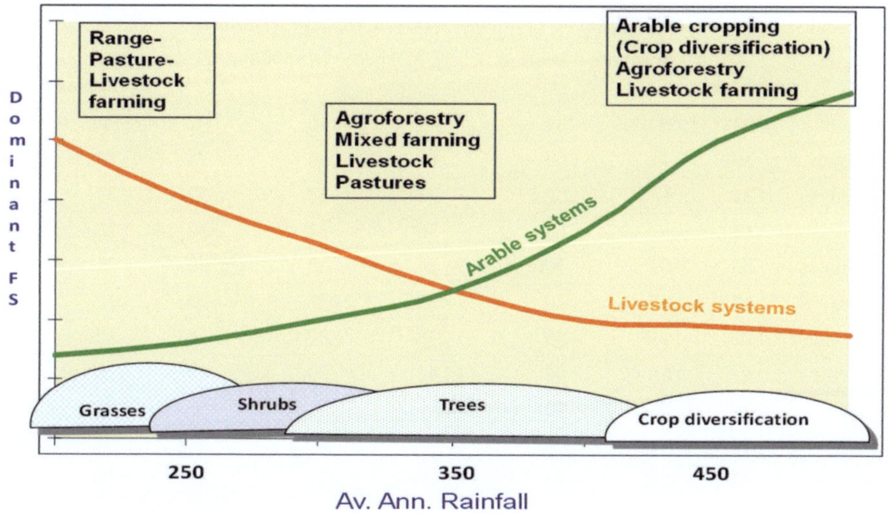

Fig. 3.4 Traditional farming systems in the arid region of Rajasthan. *Source* Bhati et al. (2017)

variable (Table 3.1) leading to negative impacts on investments, production, employment, and income distribution, which in turn hinders economic growth. The irrigated croplands, which play a substantial role in agricultural output and revenue generation, primarily rely on groundwater extraction, a practice that is approaching exhaustion of groundwater in numerous districts. Diminishing groundwater reserves, unreliable canal irrigation in lower areas, escalating input expenses, and salinity issues pose imminent threats to the economic viability of irrigated regions. Priority measures in this context include enhancing irrigation water use efficiency, integrating livestock economy with cropping systems, and implementing a comprehensive soil monitoring and amendment initiative to address the declining soil nutrient balance.

The off-farm and non-farm income and migration have been some of the major coping strategies of the farm and other rural households. The arid farming and livelihood systems, which are relatively most fragile due to very high climate risk, market risk, and other natural threats like locust attacks, etc. need to use resources synergistically by adopting an integrated farming and food systems approach aligning with the market demand to manage the system's risk and optimise the resource use under such unfavourable conditions. The primary socioeconomic challenges in the region encompass illiteracy, land degradation, diminishing forage resources, low agricultural yields, high-risk farming practices, limited economic opportunities, and nomadic lifestyles. Sectors such as mining, tourism, handicrafts, and non-farm employment are emerging as key sources of livelihood. However, the adverse effects of some of these activities on delicate arid ecosystems are exacerbated by substantial population pressures. Consequently, the natural resource base is rapidly deteriorating, and unchecked trends may exacerbate desertification. Addressing the major challenges faced in the chronically drought-affected arid western Rajasthan

Table 3.1 Coefficient of variation (CV) of crop yields in Western Rajasthan

District	Wheat	Sorghum	Pearl millet	Chickpea	Groundnut	Sesame	Rape seeda Mustard
Barmer	33	60	83	40	90	92	24
Bikaner	**30**	41	69	29	36	75	33
Churu	**21**	180	62	72	37	59	37
Ganganagar	**18**	67	68	36	32	49	26
Jaisalmer	**35**	62	106	24	27	49	23
Jalore	**17**	50	79	35	35	73	18
Jhunjhunu	**22**	139	55	36	32	56	32
Jodhpur	**17**	68	72	43	34	67	20
Nagaur	**24**	60	46	33	46	56	20
Pali	**24**	74	64	35	62	70	34
Sikar	**23**	90	46	25	32	45	26

Source Authors' calculation from ICRISAT district-level data set.

necessitates implementing sustainable land and rainwater management practices to ensure adequate provisions of food, fodder, energy, drinking water, and livelihood opportunities, both within agricultural activities and through off-farm employment avenues.

Several studies conducted in arid Rajasthan and other arid regions have shown that the income of farmers can potentially be multiplied and stabilised by diversifying their agricultural practices in an integrated way, ensuring the long-term sustainability and economic viability of the marginal and small farms (Bhati et al., 2008; Kumar et al., 2016a, 2016b; Kumar et al., 2017, 2019; Jayanthi et al., 2001; Manjunath & Itnal, 2003; Singh, 2004; Gill, 2004; Singh et al., 2007). Based on many practical studies, the advantages of this 'integrated farming systems'(IFS) approach for western Rajasthan, are clear and compelling, and the system has the potential to be applied across a range of integrated options based on crop-livestock mixed farming systems with appropriate integration on multi-purpose trees, silvi-pastures, medicinal and spices crops and a combination of legumes and cereals together with suitable soil and water conservation practices. The IFS approach in such fragile regions is important not only for improving the system's productivity but also for managing risk and adaptation (Paramesh et al., 2022; Kumar et al., 2016c). Considering these findings, there is an immediate need to promote sustainable IFS principles for the hot and varied agro-climatic conditions in western Rajasthan. Rajasthan State Agriculture Policy has also been framed in line with the national agricultural policy. It also emphasises the sustainable management of drylands of arid regions through IFS development approaches and the effective integration of livestock components to boost economic gains and resilience and reduce disparity amongst farmers of various categories.

3.2.1 Key Systemic Issues and Potential Opportunities

The arid western Rajasthan which is the most populated desert region in the world characterised by extreme weather conditions and severe natural resource degradation faces several systemic challenges. Despite high fragility, the region also presents potential opportunities for sustainable rural livelihoods provided systemic solutions are identified and promoted by adopting a holistic approach. Major systemic issues and opportunities are listed below:

Land, water, and climate

- Natural resources, including land, water, khadins, and common pastures, are facing significant degradation due to overuse, neglect, and mismanagement.
- After access to canal water in most of the regions' cities for drinking, a significant amount of wastewater is generated in the cities. This wastewater could be recycled after treatment to cultivate suitable plant species near the cities.
- Increasing risk of climate and other shocks.
- Context-specific climate information services and forecasts need to be developed. Besides public agencies, the Government may support agri-tech start-ups to develop and provide robust climate information services in the region.

Valuable biodiversity

- The traditional park-land system which brings resilience and diversification is under attack and needs to be prioritised.
- Underdevelopment of unique medicinal and aromatic herbs potential
- Optimising integration of traditional (ber, gunda, aonla, etc.) and newer high-value horticulture (pomegranate, fig, etc.)

Livestock

- Optimising livestock integration, production, and common pasture resource development
- Controlling and managing stray animals that destroy crops.

Nutrition

- Addressing the malnutrition challenge, especially in children and women.
- Increased integration of traditional nutritious foods like millet, khejri pods, etc., into diets. Moringa can also be cultivated in the home gardens of at least half of the districts of the arid regions.
- There is a good potential to establish 'pearl millet' as a superfood. It needs to be promoted through integrated efforts from research, the processing and equipment industry, entrepreneurs, farmers' producers' organisations (FPOs), school feeding programs, etc.

Mechanisation, solar energy, and carbon markets

- Harnessing the potential of agri-voltaic farming with expanding solar infrastructure- Solar-water harvesting- Need arrangements with solar companies to work with relevant experts and departments to ensure rainwater harvesting is linked to appropriate cropping systems. Initially, more applied research may be needed.
- Mechanisation is specific to the region's agriculture and also for managing agroforestry systems- Context-specific need-based mechanisation will be critical in the women workforce-dominated farming systems through policy support for enhancing the capacity of local manufacturers, custom hiring centres, and individual entrepreneurs.
- Capacity and mechanisms to enable farmers' participation in the emerging voluntary carbon marketplace, especially through agroforestry and horticulture integration.

3.3 Context-Specific Strategies and Development Pathways Toward Sustainable Farming Systems and Value Chains

3.3.1 Optimising Cropping and Farming Systems

Crop production in arid regions faces very high climate risk but has unique opportunities to integrate resilient crops and cropping systems as part of integrated farming systems. However, the agroecology and resource situation are highly diverse and needs context-specific crop planning and related innovations. Optimising cropping patterns entails consideration of various dimensions including water resources, societal needs, economic factors, ecological sustainability, and environmental impacts. Foremost is to undertake crops/cropping and farming system suitability analysis. Also, the different farm household types in the same village need different interventions (Kumar et al., 2019). Such information can help inform decision-making on integrating various potential cropping systems, horticulture, agroforestry, and livestock options in different contexts/niches. The cropping/farming systems suitability analysis may help the farmers and development and extension agencies in promoting optimal farming systems components as a few examples below:

- Rainfed dominated resilient cropping/farming systems aligning with animal husbandry (millets, legumes, etc.), parkland system, and fodder options, considering local nutritious consumption based on soil, water, climate, and market opportunities.
- Niche high-value horticulture systems- pomegranate, anjeer (fig), etc.
- Traditional horticulture-agro-forestry based systems- Ber, gunda (*Cordia mixa*), khejri (*Prosopis cenararia*), and agroforestry systems.
- Cumin, fenugreek, isabgol (P*syllium*), climate information, markets, and good management practices are spice cropping systems.

- Medicinal and aromatic high-value crops- good management practices, quality seed, and linking with the market are critical. Contract farming could be one of the options. Encouraging industry to create value-addition opportunities where a sizable market surplus is available.
- Regular monitoring and maintaining soil health is fundamental to designing sustainable farming and food systems. However, despite many relevant programs, many farmers need to follow soil test-based inputs and cropping systems. There is a need to create a functional infrastructure through public–private partnerships and awareness programs for farmers. Digital mapping of soil health will also help to plan for environmentally responsible intensification of agriculture and management of natural resources.
- Investments and enabling policy for conserving landraces of millets and other local crops and diverse xerophytic flora, including medicinal: Nutritional studies and policy support needed to conserve landraces of the high potential of climate resilience and unique taste and high nutrients.

Case Study 1: Integrated Farming Systems Approach Implemented in Western Rajasthan

Any sustainable intensification strategy for enhancing farmers' income in the risky arid region must also consider ways to minimise the risk. Therefore, any comparative assessment of economic returns of various options needs to undertake a long-term analysis, say about five years' time horizon.

A systems approach considering landscape to farm and household to the whole value chain has the potential to harness the potential of ecosystem services in the region. Since the production is scattered, the action plan for intensification needs to consider the economically viable surplus of production to enable its linkage to the markets and value chains. Through the CGIAR Collaborative Research Program on Dryland Systems (2013–2016), ICRISAT with its partners, implemented an integrated systems approach for building resilience and income in arid Rajasthan (Figs. 3.5, 3.6 and 3.7) that resulted in improved management of natural resources and significantly enhanced farmers' income (Kumar et al., 2016c). However, these pilots need to be scaled widely through regular agricultural development planning and action.

Case Study 2: Medicinal Herb Value Chain

Implementing a systems approach, an innovation platform was established to promote the cultivation and marketing of a locally abundant medicinal herb, Shankhpushi (*Convolvulus pluricaulis*), in Rajasthan's Barmer district. The success stories of farmers who cultivated Shankhpushpi attracted others to join. Starting with 20 farmers in 2014, the cultivation expanded to 250 farmers across five to six villages in 2015 and eventually reached about 1000 farmers in Barmer and Jodhpur districts. Currently, Shankhpushpi sells at INR 24 (US$ 0.36) per kg, while its seeds fetch INR 1,500–1,600 (US$ 22.7–24.2) per kg. A partnership with Dabur India Ltd was established for product buyback, with technical support provided by KVK Barmer.

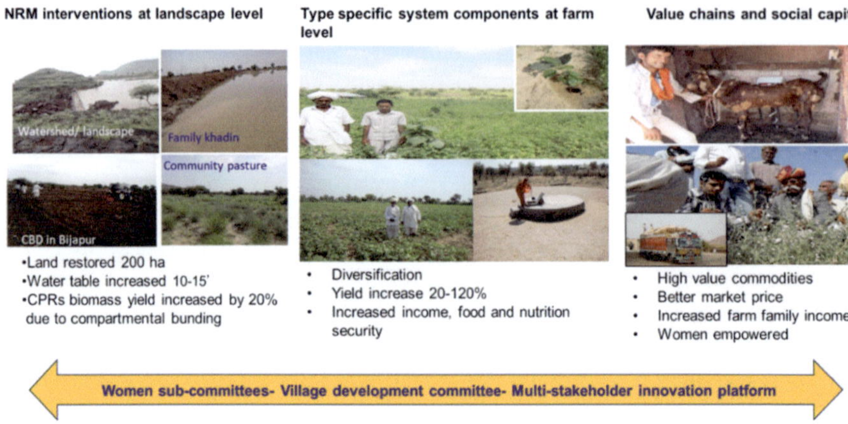

Fig. 3.5 A systems approach from landscape to farm to value chain implemented in western Rajasthan

Fig. 3.6 Medicinal herb shankhpushpi value chain development: on-farm trials, on-field learning, capacity development, primary processing, and market linkage

Encouraged by the success, another medicinal plant, Jeevanti (*Leptadenia reticulata*), was introduced in 2016, and cultivated by about 25 farmers, with potential earnings of INR 300–400 (US$ 4.5–6) per plant. Additionally, Arna (*Clerodendrum phlomidis*), a drought-resistant plant used for fencing and roofing, now commands a buyback rate of INR 15 (US$ 0.23) per kg. This initiative has not only enhanced farmer incomes but has also raised awareness among policymakers. Over 1000 farmers have been integrated into the Shankhpushpi value chain, experiencing additional annual benefits ranging from INR 5000 to 45,000 each. The Principal Secretary of the government

Fig. 3.7 Participatory approach for common pasture development in Dhok village, Barmer district

of Rajasthan for the herbal medicines department visited the field sites and pledged support for upscaling the initiative.

(http://timesofindia.indiatimes.com/city/jaipur/Herb-cultivation-benefits-western-Rajasthan-farmers/articleshow/48417637.cms).

Incorporating high-value medicinal crops into integrated farming systems, along with appropriate market connections, led to a substantial rise (more than double) in the income of dryland farmers (Kumar et al., 2016c). The state government has displayed a strong interest in expanding this initiative. There is a need to identify potential areas, allocate budget and building capacity, and fix targets for the relevant district-level department to upscale such pilots.

Case Study 3: Harnessing Common Pastures Resources- Critical for Pastoral-Dominated Far-Western Part

In the arid zone of Rajasthan, rural households, particularly those with limited resources, heavily rely on Common Pool Resources (CPRs), notably community silvi-pastures (Oran and gochar), given the prominence of livestock rearing within their rainfed agricultural systems. However, these common grazing areas have suffered severe degradation, rendering the rural poor susceptible to climate-related and market fluctuations. Despite various efforts to enhance the management and rehabilitation of community pastures (Conroy & Lobo, 2002), the long-term sustainability of these new management practices remains uncertain. This uncertainty stems from complex social dynamics, insufficient attention to community preferences, and inadequate institutional arrangements (Jodha, 2001; Mishra & Kumar, 2007).

Our research conducted in Jaisalmer, Jodhpur, and Barmer (Kumar et al., 2016a, 2016b) revealed varying perceptions, engagements, and responses to Common Property Resources (CPRs) management interventions among different groups within each village. Small ruminant keepers, cattle keepers, landless individuals, smallholders, and large farmers exhibited distinct approaches. Often, influential individuals emphasising productivity enhancement played a dominant role in management decisions. Groups of higher socioeconomic status tended to advocate for the complete protection of common pastures, whereas resource-poor small ruminant keepers advocated for open grazing. Women engaged in livestock keeping, particularly those practicing the cut-and-carry system, significantly contributed to the improved management of common pastures. Another notable factor is the limited economic viability of livestock keeping. Farmers are hesitant to invest time in pasture rehabilitation due to the perceived low individual benefits and uncertain cooperation from fellow farmers. There exists a perception among the community of a vicious cycle, wherein poor governance leads to CPR degradation, resulting in low productivity and community disinterest, which in turn perpetuates poor governance.

Following our study and the present situation, the following strategies are needed to improve the CPRs:

- Sharing information with the community on the benefits of CPRs and managing it collectively to address the social dilemma.
- Enhance biomass productivity to make CPRs economically viable.
- Integrating more profitable species into the CPRs
- Capacity strengthening of stakeholders on improved practices.
- Engaging all sections of the village through representative VDC, etc.
- Platform to raise concerns for any conflict management.
- Fairness in the distribution of benefits
- All members should get their share even if they are not livestock keepers.
- Panchayats need to be encouraged to take an interest in CPRs management; it will help if are CPRs managed at the hamlet level by the local committee.
- For convergence, the agricultural department needs to be involved in CPRs management & improvement with the revenue department.
- Continuous feedback to enable a favourable policy environment.

3.3.2 Optimising Water Use for System Sustainability

In the arid zone, a perpetual negative water balance prevails, with groundwater typically deep and often brackish. Evaporation losses contribute to approximately 20–25% of total water loss. Unregulated water usage on undulating, highly permeable sandy soils via conventional irrigation methods has led to a significant decline of 3–7 m in the groundwater table (Bhati et al., 2017). Only a mere 10.8% of land in the dry western region of Rajasthan receives irrigation. Historically, runoff water collected through surface structures such as khadins, ponds, and tanks has been integral to farming practices in arid regions like Rajasthan (Kumar et al., 2016a, 2016b; Singh et al., 2022).

Case Study 4: Khadin System- a Unique Method of Water Conservation and Utilisation Under the Arid Region

Traditional Khadin water harvesting and farming systems in West Rajasthan (~170 mm mean annual rain) are commended for their high productivity of crops when compared with typical rainfed in similar ecoregions. The 'Khadin' represents a distinctive land-use system that originated centuries ago in the Jaisalmer district, characterised by the presence of rocky catchments and adjacent valley plains. Runoff water from these catchments is captured and stored on the lower valley floor, contained by an earthen 'bund', facilitating crop cultivation (Kolarkar et al., 1983). However, the policy and institutional weaknesses threaten the sustainability of these time-tested resilience-enhancing systems. Uncontrolled mining and lack of holistic approach are concerns:

- Uncontrolled mining activity: loss of rainwater trapped in mining pits, accumulation of salty silt in beds, soil quality being affected. There is a need to regulate mining activity to minimise externalities.
- Lack of community efforts- organising khadin farmers into FPOs may help improve systems productivity and reduce high transaction costs. Digital tools can help in such remote areas.
- Traditional cultivars and packages of practices
- Wastage of water through evaporation and percolation- needed to be stored in the tanks in the upper part, suitable for arid horticulture and agroforestry.

The overexploitation and wasteful water use have led to a significant drop in the groundwater table, waterlogging, and soil salinisation, which urges immediate action. The use of extensive irrigation should be emphasised over the use of intensive irrigation. It is claimed that as much as 70 percent of water is lost while transporting, applying, and distributing it. In this light, efforts are needed to cut down on water waste and integrate water-efficient crops and cultivars, as a result, improve the effectiveness of irrigation and water use efficiency and productivity. The following approach may help to optimise water usage;

- *The first step is to design water-efficient farming systems, including suitable crops*-rotations, agro-slivo-horti systems, and livestock.
- *A village-level participatory water budgeting* and identifying suitable crops can help design water-efficient systems.
- *Micro-irrigation (drip system):* Though its utility is proven, adoption is relatively low. An excellent economic analysis is needed to identify suitable crops, types of agroecosystems and soils, and the availability of relevant infrastructure where drip systems will be economically viable to harness their maximum potential. Such mapping and information may help spread awareness among farmers and other stakeholders and, consequently, its adoption. Other important requirements are building the capacity of farmers and village youth to handle micro irrigation systems and encouraging public–private partnerships to promote drip irrigation for water-efficient agriculture.
- *Mulching*: It is beneficial for high-value horticultural crops and for growing crops in dryland locations during dry-season cropping. It can potentially increase yield by up to 50–60 percent. The selection of mulch materials largely depends on the

accessibility of the material, the local climate, its durability, and its relative affordability. Plastic mulch, which is often used, negatively impacts the environment and soil ecosystem; thus, research on new mulching materials is necessary.

Harnessing the Potential Regenerative Principles as Part of Integrated Farming Systems

An integrated farming systems approach can also play an important role in promoting circular economy principles and regenerative agriculture contributing to improved economic, environmental, and social sustainability of the production and livelihood systems. Rajasthan, known for its extensive areas of almost naturally organic agricultural production (approximately 69,750 hectares of registered organic farm space), provides a diverse range of possibilities for organic farming. Additionally, the Government offers subsidies to encourage expansion in organic/low chemical input farming. The promotion of organic farming in its true sense in the arid zone may not get the farmers' acceptance because the soils are poor in organic matter and need fertiliser supplements. However, there could be a good scope for low chemical input regenerative agriculture. However, the following policy support will be needed to scale regenerative agriculture (RA):

- RA is highly context-specific- there is a need for district/block level analysis to develop RA suitability maps for regions and typology of cropping/farming systems
- Price incentive- green certification- Branding of RA products will help create markets, maybe produced by FPOs/ RA clusters, and strengthen the value chains for RA/NF products
- Level playing field: A similar policy support is needed for organic manure/inputs as it is given for the chemicals and fertilisers
- RA support fund: farmers face a three- to five-year window to switch to regenerative farming practices, which presents a financial burden- in the form of alternative infrastructure and inputs equitable access to RA/NF solutions, including science evidence-based capacity building on natural (plant and animal-based) products for soil health- such as Jeevamritha, Beejamritha, Neemastra, etc.
- Local-led approaches are paramount. Each watershed, village, and community possess unique characteristics and must be addressed within its specific social and environmental framework. Hence, providing tailored, localised support is vital for the success of regenerative farmers.

Payment for Ecosystem Services and Harnessing Carbon Markets Linked to Regenerative Agriculture

Farmers in arid western Rajasthan significantly contribute to the ecosystem services within and beyond the arid region. Farming systems, including parkland systems, contribute to stopping the desertification process. Therefore, ecosystem services generated by farmers with no payments need to be quantified and compensated.

A green fund could be created to support farmers' environmentally sustainable practices.

Carbon markets are being promoted to serve as platforms for trading carbon credits. They offer potential avenues for additional income and serve as incentives for embracing more sustainable practices in agriculture, agroforestry, or regenerative agriculture. These markets can compensate farmers for adopting management practices that enhance soil carbon storage. Moreover, revenue generated from the sale of carbon credits can improve the economic feasibility of climate projects, particularly in challenging markets, by providing an additional secured source of initial investment. This, in turn, encourages private sector investors and agribusinesses to engage in the long-term investments often required in this domain.

Farmers have excellent potential to take advantage of the emerging global voluntary carbon market. However, this needs benchmarking and developing an effective monitoring, reporting, and verification (MRV) system.

3.3.3 Harnessing the Emerging Innovations in Farming Systems

Protected Farming System

Greenhouses, plastic houses, lath houses, cloth houses, net houses, shade houses, hotbeds, and other types of structures all fall within the category of protected crop management. Protected farming of high-value fruits and vegetables (such as capsicum, tomato, cucumber, cabbage, ridge gourd, bottle gourd, okra, broccoli, chili, etc.) and other horticultural crops under greenhouses enable farmers to cultivate cash crops on tiny plots in marginal, water-deficient locations where traditional cropping is not practical. CAZRI Jodhpur has successfully demonstrated high-value vegetable production under different playhouses under arid conditions. To upscale protected farming in a suitable area of an arid region, the following steps may help:

- Capacity building of trainers and farmers and establishing cells in the relevant department, which can provide guidance and support technical aspects, planting material, good management practices, etc.
- Protected farming needs to be promoted based on market demand assessment.
- Training local youth to undertake repair and maintenance of the greenhouse.
- Government support on initial investment may be continued

Agri-Voltaic Farming Systems

The Agri-voltaic farming system (AFS) is a distinctive approach that enables the simultaneous generation of electricity, cultivation of crops in interspace areas, and harvesting of rainwater from the top surface of the PV module. Approximately 50% of the land area occupied by a solar PV installation system can be utilised for crop cultivation, effectively utilising what would otherwise be left as fallow land. This system allows for the cultivation of various crops, including medicinal plants such as Aloe vera, Sonamukhi, and Sankhpuspi, as well as aromatic plants like lemongrass, under the shade of solar panels. About 1,50,000 L of rainwater can be harvested on one acre. The stored water can be used as supplemental irrigation to crops. There is good scope to promote an Agri-voltaic farming system with the expanding solar infrastructure in western Rajasthan. Emphasising the following aspects will help in harnessing the potential of the Agri-voltaic farming system:

- There is a need to have a policy that mandates solar companies to work with relevant experts and departments to ensure rainwater harvesting is linked to appropriate cropping systems. However, further applied research may be needed to generate more information to develop recommendation domains for different conditions.
- Farmers and other stakeholders have very little knowledge about AFS, and hence there is a need for a structured program to increase awareness and training of farmers and other stakeholders on agro-voltaic farming systems.
- A cluster-based approach will be helpful for the small holders of the agro-voltaic farming system to enable them to participate in relevant value chains.

3.3.4 Optimising Livestock Value Chains

Livestock activity is one of rural western Rajasthan's most stable sources of income and livelihood. Special attention is needed to strengthen livestock value chains to improve the system's resilience. Optimising livestock value chains need a holistic approach to promoting:

- High-yield-adapted animal breeds like Tharparkar, and Kankrej- Both natural breeding and artificial insemination approaches employ semen sexing (sorting) to measure only female calves.
- Fodder augmentation through improved pastures, multi-purpose trees into farming systems, and creating market-led mechanisms to get surplus dry fodder from Haryana and Punjab, compressed bales doe at the source. Appropriate business models involving private entrepreneurs and FPOs would be needed.
- Animal health through prophylaxis of animals: Huge losses could be prevented, especially in small ruminants, if proper prophylaxis of animals is done. Since the livestock production system is extensive and sometimes nomadic, the conventional approach of providing livestock services is less effective. There is a need for creating and facilitating Mobile Ambulance Livestock Services (MALSs). The initial infrastructure and support may fully come from the Government, but it should be created in the business model mode so that it becomes self-sustainable in the future.
- Significant production and mortality losses, especially in small ruminants, occur due to diseases and parasites. Therefore, there is a pressing need to conduct an

awareness campaign via mass media channels (TV, newspapers, etc.), emphasising the importance of prevention programs for diseases and parasites in small ruminants.

- To eliminate poor-quality breeding males of small ruminants, the Panchayat must play an active role. Once the preferred animal breeds to be promoted at the Mandal and village levels are determined, the Panchayat can advise its members against breeding males of non-descript poor breeds. These poor breeding males can then be replaced as part of a breed improvement program.
- To enhance fodder availability for small ruminants, it is essential to not only develop common pasture resources (CPRs) through the involvement of panchayats and user groups but also to establish institutionalised linkages between small ruminant farmers in regions with limited fodder and those in areas with ample fodder resources. For example, the kids/lambs produced in drier districts of western Rajasthan could be sold at an early age (3–4 months) and reared in the districts with surplus fodder by rural women more advantageously. Such trade arrangements need to be facilitated.
- Small ruminant markets often favour brokers and intermediaries, leading to disadvantages for consumers, small ruminant keepers, and sellers of livestock by-products. Factors such as limited bargaining power, lack of awareness, high transaction costs, limited surplus for sale, low income (resulting in distress sales), and inadequate infrastructure are the primary reasons behind the lower price realisation of livestock produce for smallholders.
- There is a need to link the smallholder producers directly with the formal market (processors, marketers/retailers, and consumers). Local elected bodies and livestock-holders should be encouraged to improve the marketing of livestock. All the markets for live meat animals or even every village should have a common weighing machine. That machine may be used to weigh the meat animals at the time of their sale. This would help farmers to know the value of their meat animals and consequently increase their bargaining power.
- Linking the farmers, especially the small ruminant keepers, with the market. The basket of options will be different for different crop-livestock farm households depending on their nature of production, resources, and preferences.
- Inefficient slaughter operations and markets, along with high transportation costs, contribute significantly to consumer expenses. Poor utilisation of small ruminant by-products results in low price realisation by the goat and sheep rearers. Entry barriers for new players, well-established agents' networks, and resistance to government-led changes hinder the modernisation or relocation of abattoirs.

3.3.5 *Optimising Farm Mechanisation and Digital Innovations*

Appropriate farm mechanisation is of special significance to improve the sustainability and profitability of framing systems in arid western Rajasthan. Mechanisation on farms helps to boost efficiency and saves labour time by deploying machines across the whole agricultural value chain, beginning with production, and continuing through post-harvest processing, storage, and ultimately the sale. Significant farm mechanisation by promoting tractors, tillers, sprinklers water pumps, and other agricultural machinery. However, this region has some special needs for mechanisation mainly because of short sowing windows, harvesting needs of agroforestry trees, longer storage needs, etc. appropriate mechanisation can help improve systems productivity and profitability significantly. Lack of need-based small mechanisation, for example, machines to harvest multi-purpose trees that are thorny such as *Prosopis cineraria* (khejri), *Carissa carandas* (karonda), etc. limits their potential. The following options may be helpful:

- *Establishing farm machinery custom hiring centres (CHC)*- the CHCs could promote using different business models through FPOs, community groups, rural youth entrepreneurs, and public–private partnerships for multi-service centres.
- *Development and Promotion of Specialised Machine:* Establishing facilities for producing high-tech agricultural equipment to serve less productive agricultural regions. The development and promotion of tools and machines for sowing, harvesting, and coarse threshing crops like pearl millet are needed. Along with this, machines that improve the efficiency of processes such as sorting, grading, polishing, waxing, packaging, and transporting goods have to be developed.

Promoting Digital Agriculture

The e-commerce sector, digital advice platforms for agriculture, and crop insurance are all potential application areas for the digital agricultural ecosystem. It is possible for national and state digital advisory services to assist in increasing agricultural productivity and efficiency across the board, thereby enhancing the sector's capacity for resilience and adaptation in the face of climate-related impacts. It is possible for the increased accessibility of agronomic advice provided by digital platforms to assist in the improvement of agricultural productivity, the reduction of negative environmental footprints, and the enhancement of national resilience against shocks. Through the use of digital networks, digital agriculture has the potential to improve connectivity between farmers and enhance resilience throughout the agricultural value chain. This will enable better access to high-quality, real-time data, which can be used to adapt to the effects of climate change.

3.3.6 *Optimising Post-Harvest Management*

Post-harvest management refers to a collection of procedures that are put into place after harvesting to maintain the same quality in agricultural products from the farm to the very end of the agri-value chain. The supply chain needs to be equipped with effective procedures, channels of communication, and tracking methods to keep product quality consistent. The following approach may help to improve the post-harvest management system.

- *Establishing new warehouses* through a public–private partnership or supporting policies encouraging private investment into warehousing can help improve systems resilience and farm profitability.
- *Value addition through primary and secondary processing* units involving farmers'/women groups/FPOs can help improve profitability and employment opportunities. However, business incubation and handholding support are needed in the initial years for such new businesses led by farmers.
- *Technology-Enabled Tractability System:* Traceability systems could be implemented for agricultural products from the farm level up to the final level of the value chain (farm gate, transport, warehouse, market yard, customer). These traceability systems make it possible to conduct a quality and quantity check through the value chain, significantly reducing losses.
- *Awareness and Education Programme:* A variety of approaches, such as farmer field schools and training for those who train others, need to be utilised to implement an awareness and education program on post-harvest challenges. Zonal awareness initiatives could facilitate an increase in the value addition of food and techniques for reducing post-harvest losses. In addition to this, fostering learning alliances will also be of assistance.

3.3.7 *Mechanisms to Improve Market Linkage and Farmers' Access to Market Information, Knowledge, Network*

The farmers' access to market information, knowledge, and networks is essential to enhance the viability of agriculture. For that, at least up to the Panchayat Samiti level, all the *essential mandates (agricultural markets)* should be connected to the internet, and farmers' knowledge should be updated regularly, which can be done by village-level extension functionaries. The electronic media, mobile, newspapers, radio, TV Channels, etc., should be strengthened for faster communication to the farmers at the village level about the technologies and prices of different agricultural commodities in various *mandies*. The following can be the options for improvement in the marketing of agriculture and livestock products and by-products in the region.

- Enhancing public investment in marketing infrastructure, such as rural roads, storage godowns, cold chains, and processing units.

- Revise Agricultural Produce Marketing Acts to stimulate private investment in agricultural marketing and facilitate contract farming arrangements.
- Make it mandatory for the state government to hold regular elections of Market Committees to make them farmer-friendly committees.
- Ensure the effective implementation of price support policies in arid areas. Establish mechanisms for planning and preparation for price support operations at the state level.
- Redefine the role and functions of State Agricultural Marketing Boards and Agricultural Produce Market Committees to incorporate the promotion of marketing organisations and contract-marketing arrangements tailored to products specific to arid areas.
- Agricultural marketing system research should be prioritised under the National Agricultural Research System (NARS) for arid areas.
- The Government's recent announcement of special commodity parks/zones for arid zones should be operationalised promptly. Specific commodities such as kair, sangri, nagaurimethi, seed species, etc., should be established and operationalised at the earliest opportunity.
- Extension agencies and NGOs should prioritise training farmers in post-harvest operations, collective marketing, packaging, branding, and quality maintenance at the farm and village levels.
- Formation of farmers'/producer self-help group/CIG for marketing of agriculture products.
- Due to the greater distance between villages, establishing marketing infrastructure such as road connectivity, telephone links, storage warehouse, and processing units is comparatively less feasible in arid regions than in densely populated areas. Consequently, public investment in infrastructure in arid areas tends to be a lower priority, and private investors display less interest in such investments.
- Numerous commodities and products from the arid zone possess special geographical indications and enjoy significant demand in distant markets, including overseas markets. Examples include sojat henna, kair, sangari, bikaneri bhujia, and isabgol, among others. These products hold considerable potential to boost employment and income through branding and organised marketing efforts.
- The inter-year price fluctuations for crops like guar, cumin, coriander, fennel, and fenugreek are relatively larger, but there needs to be more prudent planning and support. Access to warehouses could be helpful.

3.4 Existing Institutions, Partnerships, and Their Effectiveness

The state's agriculture extension network is poor due to inadequate staff, lack of funds, improper capacity building, and motivation. Although Farm Science Centres (KVKs) are located in each district of Rajasthan, considering the districts' size in the desert region, they cannot cover the farmers in totality. However, they are now backed up by ATMA, IWMP, NGOs, and various programmers of international agencies like IFAD, ICRISAT, ICARDA, UNESCO, etc., but still much more is needed. Similarly, the farmers' groups organised under different public programs are either non-existent or need to be properly functioning. There is a need to develop such institutions and their federations from the village to the district level to ensure they are functioning effectively. Further, PRIs like *Jila Parishad, Panchayat Samiti, Gram Panchayat*, etc., must be strengthened. Capacity building of various functionaries at all levels is needed; besides, they should be provided enough funds and other infrastructure facilities to serve the rural communities as per their mandate, along with a proper monitoring system.

Relevant existing institutions in arid Rajasthan:

- SAUs and their regional research stations
- State Departments on agriculture, horticulture, animal husbandry, rural development, watershed management, forest, irrigation, DRDA, IDS, etc.
- Input supply institutions viz. National and State Seed Corporations, Fertiliser Association of India (FAI), IFFCO, KRIBHCO, Pesticide companies, District Seed and Fertiliser Cooperatives, etc.
- Food Corporation of India, Central, and State Warehousing Corporations.
- Krishi Vigyan Kendras
- ICAR Institutes and their regional stations
- NGOs
- Panchayati Raj Institutions (PRIs)
- NABARD, commercial banks, Bhumi Vikas Banks, and other financial institutions

However, there is a need to strengthen linkages among various institutions for better extension delivery services. Also, there is a need to develop a chain of public institutions and encourage collective action through self-help groups, technology transfer clubs, rural technology centres, livelihood resource centres, etc., to right from village to at least divisional level as building blocks for timely and adequate delivery of inputs, information, and marketing. Further, PRIs should be sensitised and educated about the spread of agricultural technologies through these agencies. To improve the public–private-partnership following points also need consideration:

- Various ongoing and planned Central and State development programs should be coordinated by a suitable agency to establish synergistic horizontal linkages and convergence.
- The various Mission Mode Projects, such as the National Horticultural Mission, must integrate and internalise the location specificities.
- NABARD, along with other banks, SFAC, relevant ministries, and departments, should facilitate public–private partnerships not only in infrastructural development but also in promoting rural entrepreneurship. Additionally, they should support the establishment and effective functioning of self-help groups (SHGs), cooperatives, other rural groups and community-based organisations, agri-clinics, and agribusiness centres to ensure access to quality inputs and fair markets.

- Panchayats should be appropriately empowered as a key authority to facilitate grassroots-level priority setting, programming, and implementation, in collaboration with suitable community-based organisations (CBOs), NGOs, and other institutions. This collaboration is especially crucial for facilitating technology transfer and promoting producer-oriented remunerative marketing.
- Encourage farmers to engage in group farming to leverage the benefits of economies of scale. These groups can establish comprehensive linkages with technology, inputs, services, credit, and other institutional support through an end-to-end approach. Additionally, they can develop forward linkages with remunerative markets, income generation, and profit sharing.

How does the Current Policy Impede or Promote the Existing and Proposed Agricultural Practices?

A few examples of the conflicting policies and their improper implementation, which adversely affected the agriculture development in arid Rajasthan, are cited below:

- The Green Revolution in India, on the one side, helped to solve the country's food problem but also adversely affected the land quality, resulted in environmental pollution, caused disparity in rural society, lack of attention to dryland farming, etc.
- Environmental degradation in the Thar desert is also the result of either wrong policies or faulty implementation. For example, lack of foresightedness during the planning and implementation of IGNP resulted in the loss of valuable *Lasiuruss indicus* based ecoregion of Thar desert, the problem of water logging, soil salinisation, etc., faulty implementation of policy on groundwater use have led to overexploitation leading to acute water scarcity for agriculture and drinking.
- Faulty mine allotment coupled with large-scale illegal mining in the Thar desert has led to irreparable loss to the area's geography, illegal encroachment on CPRs coupled with heavy and indiscriminate animal grazing resulted in the loss of valuable vegetation and animal resources of the region.

3.5 Conclusions

In the absence of adequate planning and execution of long and short-term climate adaptation mechanisms, valuable natural resources and agricultural systems are becoming degraded, leading to the distress and agony of the rural people of West Rajasthan. The problem is bound to get magnified in this region's changing scenario of changing climate. Dryland regions need an integrated approach for enabling current and future resilience in view of their fragile agro-ecosystem under climate change. An integrated farming systems approach oriented to harness market-led opportunities may be the best way to support the development of sustainable farming and livelihood systems in arid western Rajasthan. Policy planning and actions should adopt a long-term perspective, taking into account trade-offs and synergies while considering the fragility of the region and the possibility of increased shocks. Following the suggested framework and assessment of ground realities in context to location-specific needs and prospects, action plans, strategies, and policies focused on the arid zone people, of course, within the overall framework of national and state agricultural development policies, may be developed and implemented as a mission in a community participatory mode. The overarching policy framework must maintain consistency and coherence, especially concerning agricultural policy reform, trade liberalisation, and adherence to multilateral environmental agreements.

Acknowledgements The authors thank Soumitra Pramanik and Elias Khan for their support in accessing and analysing the data. Special thanks to T.K. Bhati, KVK Barmer, GRAVIS, and CAZRI team, and the communities for their support during the project implementation from which this draws several learnings. Authors acknowledge funding support from the funders who support the CGIAR's initiatives on Sustainable Intensification of Mixed Farming Systems and Agroecology through their contributions to the CGIAR Trust Fund: https://www.cgiar.org/funders/ and the Indian Council of Agricultural Research.

References

Bhati, T. K., Tewari, J. C. & Rathore, S. S. (2008). Productivity dynamics of integrated farming systems in western Rajasthan. In Pratap Narain, M. P., Singh, A., Kar, S. K. & Praveen-Kumar (Eds.), *Diversification of arid farming system* (pp. 23–30). Jodhpur, India: Arid Zone Research Association of India and Scientific Publisher (India).

Bhati, K. T., Kumar, S., Haileslassie, A. & Whitbread, A. M. (2017). *Assessment of agricultural technologies for dryland systems in South Asia: A case study of Western Rajasthan, India.* http://oar.icrisat.org/id/eprint/10837

Conroy, C. & Lobo, V. (2002). Silvi-pasture development and management on commonlands in semi-arid Rajasthan. In *Natural Resources Institute and BAIF, UK Department for International Development (DFID) study report*. https://assets.publishing.service.gov.uk/media/57a08d23e5274a27b2001663/R6953u.pdf

FAO (2014). Building a common vision for sustainable food and agriculture. Principles and approaches. *Food and Agriculture Organization of the United Nations (FAO).* https://www.fao.org/3/i3940e/i3940e.pdf

Garnett, T., Appleby, M. C., Balmford, A., Bateman, I. J., Benton, T. G., Bloomer, P., & Godfray, H. C. J. (2013). Sustainable intensification in agriculture: Premises and policies. *Science, 341*(6141), 33–34.

Gill, M. S. (2004). Methodologies for farming system approach-A case study. In *Proceedings of national symposium on alternative farming systems, held during 16–18 September 2004 at PDCSR. Modipuram, Meerut* (pp. 1–29).

Godfray, H. C. J., Beddington, J. R., Crute, I. R., Haddad, L., Lawrence, D., Muir, J. F., & Toulmin, C. (2010). Food security: The challenge of feeding 9 billion people. *Science, 327*(5967), 812–818.

Jayanthi, C., Rangasamy, A., Mythili, S., Balusamy, M., Chinnusamy, C. & Sankaran, N. (2001). Sustainable productivity and profitability to integrated farming systems in low land farms. *Extended Summaries*, 79–81.

Jodha, N. S. (2001). *Life on the edge: Sustaining agriculture and community resources in fragile environments.* Oxford University Press.

Kolarkar, A. S., Murthy, K. N. K., & Singh, N. (1983). Khadin-A method of harvesting water for agriculture in the Thar Desert. *Journal of Arid Environments, 6*(1), 59–66. https://doi.org/10.1016/S0140-1963(18)31433-2

Kumar, S., Ramilan, T., Ramarao, C. A., Rao, C. S., & Whitbread, A. (2016a). Farm level rainwater harvesting across different agro climatic regions of India: Assessing performance and its determinants. *Agricultural Water Management, 176*, 55–66. https://doi.org/10.1016/j.agwat.2016.05.013

Kumar, S., Craufurd, P., Haileslassie, A., Ramilan, T., Rathore, A., & Whitbread, A. (2019). Farm typology analysis and technology assessment: An application in an arid region of South Asia. *Land Use Policy, 88*, Article 104149.

Kumar, S., Whitbread, A. M. & Rao, K. P. C. (2016). Innovation platforms as vehicle to strengthen stakeholders' capacity to innovate for improved livelihoods in Drylands in Asia and Sub Saharan Africa. Technical Report. *International Crops Research Institute for the Semi-Arid Tropics.* http://oar.icrisat.org/10242/

Manjunath, B. L., & Itnal, C. J. (2003). Farming system options for small and marginal holdings in different topographies of Goa. *Indian Journal of Agronomy, 48*(1), 4–7.

Mishra, P. K., & Kumar, M. (2007). Institutionalising common pool resources management: Case studies of Pastureland management. *Economic and Political Weekly, 42*(36), 3644–3652.

Paramesh, V., Ravisankar, N., Behera, U., Arunachalam, V., Kumar, P., Solomon Rajkumar, R., & Rajkumar, S. (2022). Integrated farming system approaches to achieve food and nutritional security for enhancing profitability, employment, and climate resilience in India. *Food and Energy Security, 11*(2): e321. https://onlinelibrary.wiley.com/doi/10.1002/fes3.321

Rao, C. R., Raju, B. M. K., Rao, A. S., Rao, K. V., Ramachandran, V. R. K., Venkateswarlu, B. & Sikka, A. K. (2013). *Atlas on vulnerability of Indian agriculture to climate change.* http://www.nicra-icar.in/nicrarevised/images/publicaitons/Vulerability_Atlas_web.pdf

Kumar Shalander, Whitbread, A. M. & Bhati, T. K. (2016a). Integrated ecosystems approach for sustainable intensification of community based silvo-pastoral systems in arid Western Rajasthan, India: Implications for upscaling. In *World Congress Silvo-Pastoral Systems: Functions, management and Systems in a changing world, Evora, Portugal* (pp. 26–30). https://www.researchgate.net/publication/309033227

Kumar Shalander, Whitbread, A. M. & Falk, T. (2016b). *Institutionalizing governance arrangements for sustainable management of community-based Silvi-pasture systems in arid ecoregions in Rajasthan, India* (pp. 26–30). Evora, Portugal: World Congress Silvo-Pastoral Systems. https://www.researchgate.net/publication/309033072

Kumar Shalander, Whitbread, A. M. & Falk, T. (2016c). Pathways to sustainable intensification: Participatory designing of adapted farming system innovations. *Solutions*, 3281–3235.

Singh, R., Akuraju, V., Anantha, K. H., Garg, K. K., Barron, J., Whitbread, A. M., & Dixit, S. (2022). Traditional rainwater management (haveli cultivation) for building system level resilience in a fragile ecosystem of Bundelkhand region, Central India. *Frontiers in Sustainable Food Systems, 6*, Article 826722. https://doi.org/10.3389/fsufs.2022.826722

Singh, J. P., Gill, M. S. & Tripathi, D. (2007). Development of integrated farming system model for marginal and small farmers. In *Extended summaries of 3rd national symposium on integrated farming system and its role towards livelihood improvement* (pp. 51–57).

Singh, G. (2004). Farming systems options in sustainable management of national resources. In *Proceedings national symposium on alternative farming systems held at PDCSR, modipuram* (vol. 1618, pp. 80–94).

Sivakumar, M. V., Das, H. P., & Brunini, O. (2005). Impacts of present and future climate variability and change on agriculture and forestry in the arid and semi-arid tropics. *Climatic Change, 70*, 31–72.

Chapter 4
Australian Desert Ecosystems: Perspectives and Challenges for Agriculture

Robert N. Truelove, Stephen C. Lellyett, and Abul K. S. Huda

> The impression is for the writer what experiment is for scientist, with the difference that in the scientist the work of the intelligence precedes the experiment and in the writer it comes after the impression.
> Marcel Proust ***Time Regained*** p 448 Book 4 Everyman's Library edition 2001

4.1 Introduction

An ancient land, occupied for millennia by an ancient race of people, with an ancient culture centred around survival in an extremely hiarsh environment—the Australian deserts. The Australian deserts or as we describe them below, the arid and semi-arid lands of Australia, are a unique ecosystem that had been undisturbed until relatively recent times. A pressing challenge Australia faces is to adapt to the evolving state of the desert ecosystems, while at the same time maintaining the culture of its inhabitants, including those who are indigenous, without denying them the opportunity to improve their lives and participate in the benefits of the modern Australian economy, including the enjoyment of a similar standard of living to the rest of the Australian population. It should be noted, that the high per capita gross regional product cited by Rola-Rubzen and MacGregor (2009) is heavily biased towards a relatively small number of large multi-national mining organisations and their employees

R. N. Truelove
Cootamundra, Australia

S. C. Lellyett
Bureau of Meteorology, Melbourne, Australia

A. K. S. Huda (✉)
School of Science, Western Sydney University, Penrith, Australia
e-mail: s.huda@westernsydney.edu.au

M. S. Rathore et al. (eds.), *Desert Ecosystems*,
https://doi.org/10.1007/978-3-032-03237-9_4

or wealthy business owners; as against others, especially indigenous people who are often marginalised or living in poverty, even while supplemented by welfare.

The aim of this chapter is to survey the literature on the Australian desert ecosystems and the people living in these ecosystems, with a particular focus on agriculture, to gain an understanding of what the situation is now and the challenges the ecosystem and the people that live in this ecosystem face in the future. As Proust makes clear, there is a difference between the work of a scientist and the work of a writer to arrive at the truth. We often need both forms of inquiry, so this chapter will employ a mix of analysis and logical inquiry of some of the available scientific data and record the impressions of the authors to try to arrive at what the authors believe is a realistic picture of the current situation and the challenges ahead.

The chapter will also delve into how the Australian deserts are currently used by both the indigenous population and the more recent European settlers; in particular, for agriculture (which includes pastoral use) and the scope for future beneficial use. We will provide a perspective on how sustainable these practices are: given the climate, soils, and water constraints. The question of whether the area of land suitable for agriculture might change under a changing climate will also be considered.

4.2 The Australian Deserts

Desert environments can vary considerably, so it is difficult to define precisely what a desert is, (and as Tolstoy said about unhappy families, each one is different!). The most distinguishing characteristic is a shortage of available moisture for plants, generally because of high evapotranspiration and low precipitation. So, plants that adapt to desert environments, either; reduce evapotranspiration (such as cacti that have evolved spines to reduce leaf area through which evapotranspiration occurs); or store water (cacti again); or survive on modest amounts of precipitation and soil moisture, as many of Australia's desert species are able to do. For most of Australia's arid and semi-arid regions, another defining feature is the highly variable timing of rainfall events and the great variation in the amount of rainfall per event. Additionally, low atmospheric humidity, high daytime temperatures and winds, exacerbate the general "dryness" of the Australian deserts.

The Australian deserts are continental in scale (see Fig. 4.1). They stretch 2000 km from the arid coast of Western Australia, through the Northern Territory to western Queensland and NSW, and southwards to parts of western Victoria and across most of South Australia. Thus, deserts constitute a major geographic component of the Australian continent. There are a number of different ways deserts can be defined; by rainfall; by vegetation; and by other meteorological measures. The arid and semi-arid zones of Australia as measured by the Commonwealth Department of Environment are cited by Macfarlane and McConnell according to a "moisture index" (MI) developed by the Desert Knowledge Co-operative Research Centre, which measures the ratio of water lost through evaporation to moisture gained from rainfall. Where the

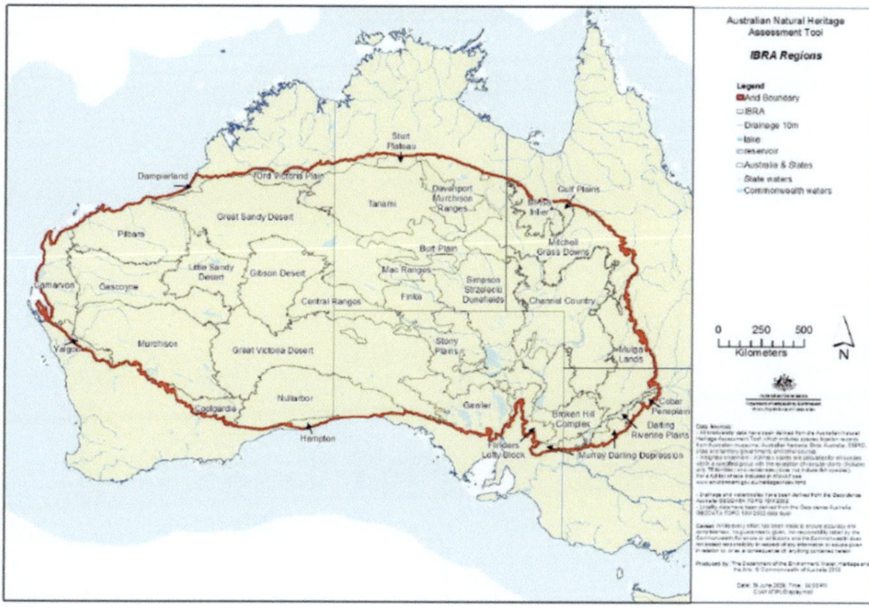

Fig. 4.1 The arid and semi-arid regions of Australia, as defined by a moisture index of below 0.4 (Macfarlane & McConnell, 2017)

MI is less than 0.2, it is considered arid.[1] Where the moisture index is between 0.2 and 0.4, the land is considered semi-arid. The geographic extent of the area under this definition is shown in Fig. 4.1 and covers approximately 70% of the Australian continental land mass.

That vast swathes of Australia are arid or semi-arid is not surprising. Australia is situated in latitudes where sub-tropical anticyclones and subsiding air usually occur. Most of the world's hot deserts occur at these latitudes. The interior of the continent is also the driest region of Australia and the major drainage basin at the centre of Australia is endorheic, where any surface run-off, which occurs very sporadically, discharges into large salt lakes from which any surface water evaporates. Australia's deserts form one of the world's major desert regions: the largest in the southern hemisphere—70% of the continent's current landmass, or 5.3 million km^2 (Smith & Hesse, 2005). They combine (excluding Antarctica) to make Australia the driest of the world's inhabited continents (Australian Bureau of Meteorology, 2008: 45 cited by Smith, 2013: 2).

The major deserts by name are shown in Fig. 4.2. Although they are not necessarily contiguous, the interlocutory areas are almost entirely either arid or semi-arid. So, in this chapter when speaking of the Australian deserts we refer more generally to the

[1] Ingereth Macfarlane and Anne McConnell Final Report March 2017, 'The Waters of Australian Deserts' Cultural Heritage Study A report to the Department of Environment and Energy and the Australian Heritage Council p 8.

Fig. 4.2 Main Australian deserts

arid and semi-arid area, which include the named deserts and the vast interconnecting lands between and around these deserts.

The average annual rainfall (AAR) of the Australian deserts ranges from 81 to 250 mm. However, caution needs to be taken when using AAR to define deserts, particularly in Australia, which we discuss in detail below. High daytime temperatures resulting in extreme evaporation counteract the perhaps higher-than-normal rainfall when compared with deserts in other parts of the world. Central Australia is arid, with the driest areas averaging only 150 mm of rainfall each year, but again this can be highly variable.

Figure 4.3 shows a Koppen-based classification of the Australian climate by the Australian Bureau of Meteorology (Stern et al., 2000). It includes two vast interconnected areas across central and western Australia, called Desert and Grasslands. These areas broadly correspond to the arid and semi-arid environmental classification of the DOE based on the MI (see Fig. 4.1). Stern et al. (2000) describe these as the desert and steppe subdivisions of what is termed the dry area of Australia. The

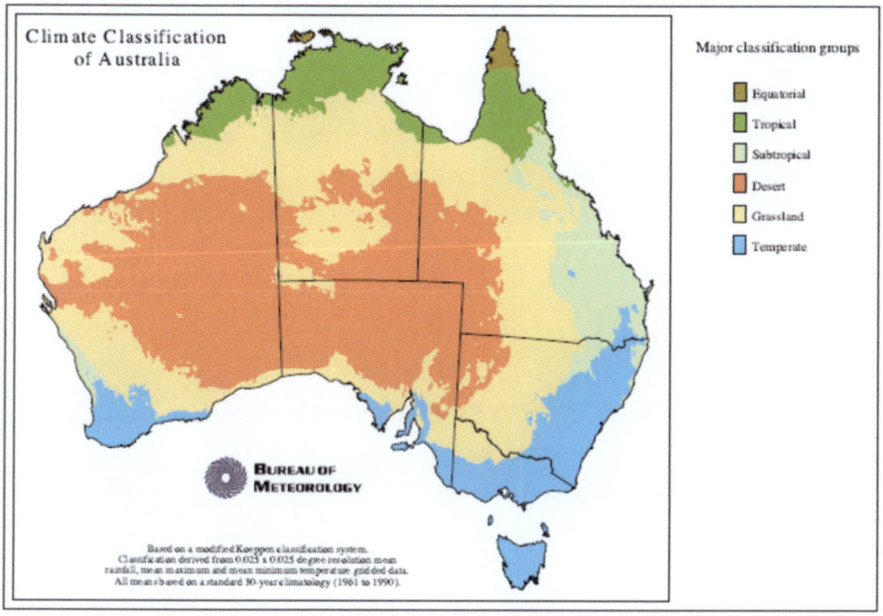

Fig. 4.3 Climate classification of Australia (Stern et al., 2000)

reasoning behind this is because of "the significant differences between the types of vegetation found in deserts and grasslands". He also noted a part of central Australia that has a grassland group of climates, despite being surrounded by desert—the yellow patch in (Fig. 4.2) in the southern part of the Northern Territory (around the town of Alice Springs), and this is a consequence of the higher rainfall due to the MacDonnell ranges of that region (Stern et al., 2000).

The desert and grassland regions of Australia can be divided into two major weather patterns: the summer rainfall dominant north and the winter rainfall dominant south, shown in Fig. 4.4, or conversely, as winter and the summer drought areas respectively. Smaller areas that are persistently dry also occur around the fringes of the winter drought and summer drought areas. The desert and grassland winter drought and summer drought areas are defined by a criterion that there is more than 30 mm of rainfall in the wettest summer month or winter month respectively. This definitional change is carried out because drought conditions may be said to prevail throughout the year in climates without at least a few relatively wet months. It should be noted that the original set of Köppen climates employed the phrases "winter drought" and "summer drought" to respectively describe seasonally dry climates. Figure 4.4 shows that the summer drought subdivisions are found in the southern half of the country, while the winter drought subdivisions are found in the northern half of the country.

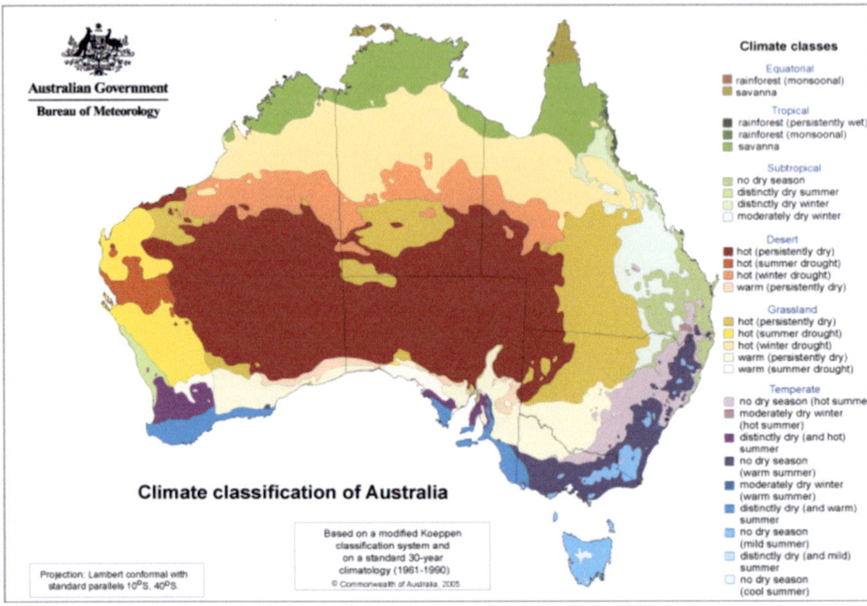

Fig. 4.4 Climate classes of Australia (Stern et al., 2000)

4.3 Socio-Economic Use of Desert Regions

The Australian deserts are home to over half a million people, or about 2.5% of the total population in 2017, but about half of this population live in major towns, primarily servicing either mining or tourism. Of the other half who live in small towns, on pastoral properties, or in remote indigenous settlements, 45% are indigenous. The actual desert regions are therefore quite sparsely populated.[2] There are only a handful of larger towns to be found within the desert regions. There are around 90 indigenous language areas within the Australian desert regions. Macfarlane and McConnel (2017) quote estimates by (Smith, 2013) of pre-colonial population estimates of the indigenous people, ranging from 1 person per 90–200 km^2 in the sandy deserts to 1 per 13 km^2 in the well-watered central ranges, 1 per 50 km^2 in other uplands. Smith (2013) also cites estimates of higher densities of 1/8 km^2 in the eastern rivers. The current population density of the desert regions of Australia is estimated to be 0.11 person per square kilometre, but as stated above around half live in major towns (Macfarlane & McConnell, 2017, p. 37).

[2] According to Macfarlane and McConnel (2017), 574,000 people lived in the arid and semi-arid zones of Australia, but about half lived in major regional service centres or mining towns such as Alice Springs (services and tourism), Mt Isa (Copper, Zinc and Lead Mining), Broken Hill (Lead and Zinc mining) or Kalgoorlie (Gold mining), with the rest scattered across 400 Aboriginal settlements and about 3000 pastoral properties. Macfarlane and McConnel (2017), p. 37.

While, what constitutes deserts depends on the definition used, as discussed above, it is safe to say over two thirds of continental Australia is arid or semi-arid. Given the size of this land mass, understanding its ecology, weather patterns, land use and related data is a matter of vital economic, social, and political importance. To address this, the Australian government in collaboration with 28, mainly indigenous partners, created the Desert Knowledge Co-operative Research Centre (CRC), in 2003. The CRC operated until 2010. It appears to have had a somewhat troubled existence, but in its final report claimed to have "produced 59 research reports, 61 working papers and nine book sections and 102 journal articles" (DKC Exit Report, 2010, p. 13). It also claimed its educational achievements were "At the DKCRC's closure, 11 Ph.D. candidates had completed their research and 18 more were in the final writing stages. Seven postgraduate students had completed and 32 more were on track to complete by mid-2010. Fourteen graduate students and trainees had found jobs, 10 of which were in desert Australia. A total of 11 Aboriginal students and trainees are being supported to completion or to other employment, of whom five had completed by 30 June 2009." DKC Exit Report 17.

The original objectives of the DKCRC were:

- Provide sustainable livelihoods for desert people that are based on natural resources and service enterprise opportunities that are environmentally and socially appropriate; and
- Encourage sustainable remote desert settlements that support the presence of desert people, particularly remote Aboriginal communities, as a result of improved and efficient governance and access to services; and
- Foster thriving desert regional economies that are based on desert competitive advantages, bringing together Indigenous and non-Indigenous communities, government and industry; and
- Apply social science insights into governance, human capacity and the design of appropriate institutions to all these outcomes.[3]

The final report lists a range of outcomes that resulted from the DKCRC research and development including:

- Creating sustainable livelihoods for desert people.
- Develop sustainable desert settlements that support the presence of desert people, particularly in remote desert communities, as a result of improved and efficient governance and access to services.
- Thriving desert regional economies that are based on desert competitive advantage, bringing together Aboriginal and Non-Aboriginal communities, government and industry.
- Increase social capital in their communities or service organisations (DKCRC, 2010, pp. 6–12).

The report goes on to list a number of Case Studies as to how and where these outcomes were successfully achieved. While no doubt during its existence, the

[3] https://www.desertknowledgecrc.com.au/desert-knowledge-australia/.

DKCRC made some contribution to improving the situation of people living in remote desert communities, the problems are by no means solved, in fact, some would say they are still quite intractable and continue to be major concern for all jurisdictions with desert communities.

One of the leaders of the DKCRC Dr. Fay Rola-Rubzen of Curtain University, in collaboration with colleague Professor Murray MacGregor, used input–output analysis to estimate the structure of the Australian desert economy.[4] Their analysis calculated an output (or gross revenue) of about $AUD90.5 billion was generated by businesses and organisations in Australia's desert regions in 2008. While now quite out dated, this is the only comprehensive assessment of the size of the economy of Australia's desert regions the authors could find. Adjusting for inter firm trade within the region, Rola-Rubzen and MacGregor estimated the gross regional product, that is the total value of final goods and services produced in the desert regions, was approximately $45 billion. They estimated the per capita annual gross regional product from the Australian desert regions was $AUD80,360. This compares to around $AUD52,700 in the broader economy at that time. The major contributors to total output in the desert regions of Australia were the mining sector (39.5%), followed by the manufacturing sector (12.6%) and agriculture, forestry, and fishing (9.3%).

The total employment in the desert areas of Australia was estimated by Rola-Rubzen and McGregor at about 238,563 jobs in 2008, with agriculture, forestry and fishing; retail trade; and mining as the main contributors to employment (Rola-Rubzen & MacGregor, 2009, p. 5).

That the mining industry was, and today remains, the predominate economic contributor to the Australian desert region's economy, is not surprising, as Australia's desert region is rich in a range of minerals that are sought after internationally. In the Pilbara region of Western Australia, the Hammersley Range has some of the largest iron ore mines in the world, including Mount Newman (BHP), Tom Price (Rio Tinto), Paraburdoo (Rio Tinto), Hope Downs (BHP), Christmas Creek (Fortescue) and Roy Hill (Hancock Prospecting). The eastern goldfields around Kalgoorlie are also one of the richest gold provinces in the world with numerous mines operating for more than a century. Western Australia also has large nickel deposits and many new economic metals, such as lithium. The northwest Gulf region of Queensland has world-class copper, zinc, and lead mining operations at Mt Isa (Glencore) and the Century zinc, lead and silver at Lawn Hill (Oz Minerals). The desert region of South Australia's Gawler Craton is also home to a world-class copper/gold mine at Olympic Dam (BHP). Besides these large, world-class, mining operations numerous smaller mines operate throughout Australia's Desert regions. The Surat Basin of northeast South Australia is a major oil and gas producing region with a major gas pipeline extending from Moomba to the east coast.

Considering all these mining operations have sales in the billions of dollars per annum, it is likely much of the output of some of these, and all the numerous other

[4] Rola-Rubzen and MacGregor (2009) Draft Paper presented to AARES 53rd Annual Conference, Cairns, QLD 10–14 February 2009.

mining operations in the desert regions of Australia, are not captured in the Rola-Rubzen and McGregor calculations and that the mining economy is larger, perhaps considerably larger, than their estimates.[5] The reason for focussing on the extent of mining in the Australian desert regions is, despite the actual mines themselves, being mere "pinpricks" in terms of the land area impacted in relation to the extent of the Australian deserts, they do open up desert regions to occupation and visitation by road, rail and air transport, which while providing much needed infrastructure, also has a broader impact on the environment. Moreover, the need for power and water to operate mines has an environmental impact. A significant change is currently underway at many mining sites as mine operators switch to renewables as a source of energy, in particular solar and wind and these projects while environmentally desirable in limiting carbon emissions, have detrimental environmental impacts on native flora and fauna.[6]

Mining developments in Australia are subject to environmental impact assessments. These vary, with different legislation in each of the different state and Commonwealth jurisdictions. Mining generates massive wealth from the desert regions of Australia, driving economic growth that is increasing the living standards of the entire population, in part via the coffers of the various state treasuries and the Federal government, through royalties and taxes. Nevertheless, pressure is continually being placed on the fragile eco-environment through further development. The need for jobs and export earnings are powerful drivers of development and without proper environmental oversight, through regulation and enforcement, the risk of environmental damage to the desert regions is high.

This puts the desert environment at risk from global challenges, such as the demand for mineral raw materials that is driving growth in developing countries like China and India, including now a new suite of rare minerals required by technologies to combat climate change. Desert residents and those relying on the desert for their livelihood, are also at risk if the developments degrade the environment. A key concern therefore

---

[5] It is not clear from the paper by Rola-Rubzen and McGregor what mining operations are included in the data and what are not. It is possible much of the sales revenue from the major mining companies is not captured in their data and the underlying statistics they use, given the practice of multi-national companies shifting sales and revenue recognition off shore for taxation purposes.

[6] Global mining company AngloGold Ashanti plans to build one of the largest renewable projects in Australia's natural resource sector. The company has entered into an agreement with power developer Pacific Energy to construct and operate 62 MW of wind and solar generation capacity at their Tropicana gold mine site. Electricity generated will be fed into the power supply for the mine site, which lays 330 km east-northeast of Kalgoorlie in Western Australia. The project, which is due to be completed in early 2025, will be integrated into the existing 54 MW gas-fired power station at the mine. The renewable infrastructure will comprise four 6 MW wind turbines, each with a hub height of 130 m and rotor diameter of 165 m, equal to the largest wind turbines in Australia. It will hold an additional 44,100 solar panels with single access tracking to make up a 24 MW PV solar plant. A 14 MW battery storage system will underpin the wind and solar renewable energy generation, which together will deliver an expected 50% reduction in overall natural gas consumption. This currently represents the largest off-grid gas-wind-solar with battery storage facility in the Australian resources sector (Cossins-Smith, 2023) AngloGold Ashanti to build renewable project in Australia. Mining Technology, June 27, 2023, viewed at https://www.mining-technology.com/news/anglogold-ashanti-renewable-project-australia-mine/).

of desert stakeholders, including local and state governments, is managing these risks so that the benefits of economic development outweigh the costs; that externalities such as the potential for environmental damage are properly included in the cost of the development and that strategies are put in place to mitigate the environmental impact of development, so that the resilience of the desert economy is maintained.

Nielsen et al. (2020) in a paper focussed more broadly on what they regard as the rangelands of Australia, state that pastoralism (non-intensive beef cattle, sheep and goat production that allows the animals to forage largely unfettered over large areas), occupies 55.4% of the rangelands of Australia. Conservation estates represent 10.5% of the area and 33.6% is under some form of exclusive Indigenous land tenure determined under the Native Title Act, which may also include additional conservation and pastoral areas. There are also small (1% by area), but nationally important, holdings by the Department of Defence, telecommunications infrastructure, and transport corridors (Foran et al., 2019 as cited by Nielsen et al., 2020).

Hacker et al., defined the southern rangelands of Australia, as those semi-arid and arid areas that lie inland of the wheat–sheep zone and south of the summer-dominant rainfall zone and occupy 4.3 million square kilometres or 57% of mainland Australia. Using data from the Australian Bureau of Agriculture Resource Economics and Sciences (ABARES), they estimated that 54% of the area is used for extensive grazing, mainly sheep, but also cattle and more recently goats. They stated "*Recent decades, however, have also seen an evolutionary transformation of the pastoral industry with a contraction of wool production and expansion of beef, goat and sheep meat production, the latter being associated with the widespread adoption of new breeds that pose specific issues for sustainable land use*" (Hacker et al., 2019, pp. 186–187). In western Queensland and in parts of South Australia and Western Australia, cattle have replaced sheep as the main enterprise, because of wild dog predation (Hacker et al., 2019, p. 187). Hacker et al. also note significant carbon sequestration in native forests has emerged alongside pastoral production, and foreshadowed renewable energy production from solar and wind, which may complement ongoing changes within the pastoral industry (Hacker et al., 2019, p. 187).

The northern pastoral industry is based around very large tracts of leasehold land, with little clearing of the native fauna. The executive summary of the Pastoral Land Board of the Northern Territory, states that pastoral stations cover around 596 091 km^2 or 45% of the area of the Northern Territory. The 221 pastoral leases vary from the smallest station of 198 km^2 to the Northern Territory's largest station which runs cattle over 12,212 km^2 (Pastoral Land Board Northern Territory, Annual Report 2013–14, viewed at https://depws.nt.gov.au/__data/assets/pdf_file/0019/254800/Pastoral-Land-Board_AR-2013-14.pdf).

Cattle mostly roam free in large paddocks and are mustered once a year to draft off marketable stock as well as carry out other animal husbandry operations. Sometimes only station boundaries are fenced. The cattle are mainly sold for live export to Southeast Asian countries (Indonesia, Malaysia, Vietnam and the Philippines). Over the past half century *Bos indicus* breeds have replaced more traditional European *Bos taurus* breeds, in Northern Australia, because they are better adapted to the hot sub-tropical and tropical climates and are more able to withstand tick infestation.

An important area of research carried out by DKCRC was to assess the involvement of indigenous people in the pastoral industry.[7] Aboriginal stockmen have been a key source of labour on pastoral stations; often in the past, working for rations and living in their traditional "*humpy*'s"—(crude bark shelters)—on creeks and billabongs, near the main stations.[8] After a walk off by Gurindji stockmen on Wave Hill station in the Northern Territory, in August 1966, (that lasted 7 years), indigenous stockmen and other station staff were able to achieve equal pay with white workers. But the long strike was not just about pay, it was also about land and the beginning of a long process of recognising Aboriginal prior settlement and land rights. Today many aborigines are still employed on white-owned stations, but there are many also that own their own stations, either through aboriginal corporations, or individual families that have accumulated capital to be able to buy properties and employ their fellow indigenous people.

4.4 Natural Capital

4.4.1 Land—Current State, Management Practices etc.

Australia's continental landscape is distinguished by tectonically stable, highly weathered, and geological ancient landforms, so that formerly rugged ranges are worn down low. Soils are often poor, skeletal, and predominantly derived from the underlying rock or sedimentary deposits from ancient streams or lakes, low in fertility and in the arid and semi-arid zone low in humus and other organic material. The desert regions are characterised by sand dunes formed by wind erosion, dry gibber plains and various other non-vegetative features. In both the arid and semi-arid regions swells and depressions called *gilgai*, are also often a feature of the landscape. This is caused by the alternate swelling and contraction following wetting and drying of clay soil. The famous "red earth" of central and western Australia (red because of the high iron continent derived from the underlying rock) is typically alkaline and interspersed with calcareous deposits. Most of the soils of the arid and semi-arid regions are leached and their infertility is a dominant feature of the ecology and potential land use (Woinarski et al., 2015: 60).

Prior to European colonisation, indigenous communities had subsisted in the arid and semi-arid regions of Australia, including in the deserts, for tens of thousands of years, living off a range of native flora and fauna. The early settlers introduced European livestock, (sheep, cattle, horses, pigs and goats), many of which

---

[7] Desert Knowledge CRC (2008). Review of Indigenous Pastoralism and Employment.

[8] The classic Australian novel, We of the Never Never, by Mrs Aeneas (Jeanie) Gunn, describes life on "Elsey" a remote Northern Territory cattle station in the Roper River district in the early 1900's. In it are vivid descriptions of the aboriginal camps on the station and interactions between the white settlers and the natives (First published in 1908, Penguin Imprint July 2008 ISBN: 9,781,741,667,622).

escaped captivity and quickly bred up feral populations, destroying native habitat and competing with both the native fauna and the indigenous communities for the scarce native flora. Another import was the camel, used by the early explorers and settlers for transport. Some escaped and many were turned loose after the introduction of the railways and later motorised transport and subsequently built-up feral populations spread across central Australia.[9] The impact of these species, both feral and pastoral was to dramatically degrade the composition of native flora and change ecological processes in the arid and semi-arid zones. In addition, the sinking of artesian bores to tap sub-terrarium water for domestic livestock allowed the populations of feral livestock to increase, as well as large native fauna like kangaroos, which enabled them to survive in areas previously too dry for them to survive. Feral European rabbits denuded the native grasslands and other vegetation, decreasing ground cover, increasing soil erosion and further altering the flora species mix. Along with rabbits, came foxes, cats and an increase in the population of native dogs (dingoes), all a result of the additional watering points, with the consequent impact of predation on small native fauna. The net result has been the extinction and near extinction of many native animals and a changing composition of the vegetative cover across much of the arid and semi-arid lands.[10]

See Kingsford et al. (2020) for a more comprehensive discussion of the degradation of the native ecology, as a result of European incursion and settlement.

In all but the driest regions, pastoral activities are carried out. From the outset of colonisation, the European settlers adopted a positive, if not confrontational approach, to utilising the dry inland for agricultural, mainly pastoral, activities. In 1930, historian Sir Keith Hancock observed;

> "there are in Central Australia frogs which have learned to store within themselves sufficient water to keep them alive through twelve or eighteen months of rain-lessness. Men, by adapting themselves to their environment and modifying their environment to suit themselves, may achieve even greater marvels." (W.K. Hancock, 1930, Australia, Beur, London as quoted by P.G. James "Economic Policy for Drought" in J.V. Lovett (1973) Drought Angus and Robertson, Chap 15, p. 245).

Perhaps nothing illustrates the pioneering attitude of those early white settlers better than the epic trek of Patsy Durack and his family. Leaving Goulburn in N.S.W., in 1879, they drove a mob of 7,250 breeding cattle and 200 horses, 3000 miles (4,800 km) to the Kimberly region of northern Western Australia.[11] The arduous

---

[9] Australian feral camels are derived from camels imported from the Indian sub-continent and Afghanistan during the nineteenth century, as beasts of burden. Camel 'trains' were the main means of transport before motor vehicles and along with the camels a significant number of 'cameleers' came, mainly from what is now Pakistan and Afghanistan. Their descendants still populate many of the small remote villages and towns in central Australia. The feral camel population consist mainly of dromedary, or one-humped, camels (*Camelus dromedarius*), widespread throughout the Middle East and India. There are also some two-humped or Bactrian camels (*C. bactrianus*) native to Central Asia.

[10] Kingsford et al. (2020) citing papers by Silcock and Fensham (2019); Woinarski et al. (2015).

[11] The trek is described in detail in Mary Durack's Australian classic novel Kings in Grass Castles. (Originally published in 1959. Current Penguin Imprint, 19 March 2018 ISBN: 978,014,379,073.

trek took three years and led to the first white settlement of the rugged Kimberly region of northern Western Australia.

The semi-arid regions of Australia have been used for extensive grazing activities from the mid 1800's. Sheep, and more recently goats, are mainly run in the south and cattle in central and northern Australia. The main sheep enterprise is wool, based on the hardy Merino breed,[12] which has been a feature of the Australian "outback", for over two centuries, and around which much Australian folklore is based, although the industry is now in decline. On many outback stations in Southern Australia, goats have replaced sheep. The goat meat industry developed from opportunistic harvesting of feral goat populations but has since become a domesticated livestock enterprise.

Recently, the American term "rangelands" has been used to describe the wide expanse of natural grazing land used for pastoral activities in Australia. The Australian Rangeland Society defines rangelands as:

> "Rangelands include all those environments where natural ecological processes predominate and where values and benefits are based primarily on natural resource areas which have not been intensively developed for primary production.
>
> "The rangelands of the semiarid and arid zones cover approximately 75 percent of the Australian continent and equate broadly with the 'Outback'" (ARS, 2019). It should be noted however that rangelands can also include some higher rainfall environments, outside of desert areas.

Rangelands play a key role in Australia's environmental, industrial and societal framework, with unique landscapes that cover a large proportion of the continent supporting diverse communities, industries, and a range of ecosystem goods and services (Nielsen et al., 2020).

There is ample evidence that Australia's rangeland ecosystems are under increasing pressure from anthropogenic impacts, both directly through inappropriate land use and indirectly through climate change (Hughes et al., 2016). Reduced vegetation cover in times of drought frequently results in soil erosion. Webb claims there is evidence that degradation is linked directly to the vulnerability of ecosystems to global change (Webb et al., 2017). Soil degradation is also a substantial concern for Australian rangelands, as it is elsewhere (Bai et al., 2008; Cherlet et al., 2018).

Projected changes by late this century to rainfall, including increased rainfall variability, and prolonged drought interrupted by floods, as well as rising temperatures and more days with heat stress and high fire risk, pose potential risks to livestock and remote communities in most of Australia's rangeland (Howden, 2017).

Diversification in the agricultural use of the dry arid and semi-arid rangelands is limited, primarily by precipitation, although some new opportunities are arising such as carbon sequestration, which can both make current livestock production more sustainable or replace existing marginal livestock enterprises with a potentially more profitable alternative. El Hassan (2019) suggests sustainable use of goats to meet growing global demand for goat meat, as we discussed above, can be a profitable

[12] Spanish merino sheep were first brought to the colony of N.S.W. in 1797, nine years after the First Fleet landed. Later an officer of the NSW Regiment, John McArthur initiated a lucrative trade in wool with his native England.

alternative enterprise to wool growing. Goats have also been particularly successful in parts of western N.S.W. that have been invaded by woody weeds, (as a consequence of the cessation of native burning) and are now unsuited to sheep grazing. Goats can control the spread of these weeds and if properly managed help with restoration of a more productive native grass-based pasture.

The woody weed problem of western N.S.W. is just one example of where the discontinuation of native cultural practices has led to a change in the ecology of arid and semi-arid regions. Nielsen et al. suggests:

> there is great value in developing modern rangeland management that centres on a careful blending of Indigenous and local knowledge with modern science and technology. Management would also benefit from an increased focus on building a sense of identity and culture to create connection with the land and a sense of belonging to a community. This should include increased sharing of knowledge and education to promote awareness of the rangelands locally and beyond (Nielsen et al., 2020, p. 360).

Arid and semi-arid rangelands around the world are experiencing change in response to various global pressures; population growth; food scarcity; fuel scarcity and climate change etc. A key aim of this chapter is to increase awareness of the challenges facing the arid and semi-arid rangelands of Australia and share that with others from parts of the world facing similar challenges. Australian rangeland management can also learn from what is happening in other parts of the world; and those responsible for rangeland management elsewhere in the world, can learn from the Australian experience (Godde et al., 2020; Stafford Smith, 2008).

Australia's rangelands face a number of challenges; declining population (both indigenous and non-indigenous); market shifts such as the decline in relative profitability of wool growing; environmental degradation; abandonment of native cultural burning and other custodial practices; and potentially climate change. It is not at all clear how these challenges will be addressed, but a starting point must include bringing together all the stakeholders to collaborate on solutions.

4.4.2 Water

Rivers and streams are dry most of the time but can revert to broad, multi-channel, "raging torrents", with very wide flood plains after high rainfall events. Lakes are also mostly dry, with surface salt pans, that only fill after extensive rains. Evaporation rates are extremely high, so the shallow lakes dry out quickly. Elevated areas are sparse and as mentioned above, worn low by erosion over the millennia, so rivers and streams meander across broad plains with little gravitational fall. This characterises the drainage patterns of central Australia where the major rivers or streams, as mentioned above, are endorheic, that is they do not flow to oceans. The two main drainage systems of central Australia, Coopers Creek and the Diamantina River, empty into Lake Eyre in northern South Australia, one of the largest salt lakes in the world. The Lake Eyre basin itself covers roughly a sixth of the Australian continental land mass.[13]

The sources of water in the arid and semi-arid zones are;

- rainfall, the northern monsoonal rains in particular, which generate high seasonal precipitation in the north of the continent and are the source of the large water flows that occur from time to time along the central Australian rivers and creeks mentioned above;
- ephemeral and permanent surface waters including rock holes and rainfall-fed non-porous claypans;
- sub-surface or sub-artesian soakages;
- wells, including native wells and bores that access deep artesian waters.

The main source of water for the introduced livestock on pastoral stations is bores tapping sub-terrarium (or artesian) water. The largest of these basins, The Great Artesian Basin (GAB), lies beneath parts of the Northern Territory, Queensland, South Australia and New South Wales. It covers 1.7 million km^2 (a sixth of the Australian mainland). It contains 65 million GL of groundwater which is enough water to fill the Sydney Harbour 130,000 times (https://www.dcceew.gov.au/water/policy/national/great-artesian-basin). The fact that there was plentiful water underground that was easy to tap, led to much wastage in earlier times. Bores were sunk and water was allowed to flow freely through open runnels to various watering points, often open dams. The loss through evaporation and soakage was extremely high, up to 95% in some instances. As a result, over the past three decades, there has been a concerted push by the various basin jurisdictions to encourage better management of water and the upgrading of infrastructure. Various government-subsidised projects aimed at controlling bore flow with fully plumbed well heads as well as improving drainage and watering points to reduce evaporation loss, particularly by piping water to troughs. Although this type of work contributed to hundreds of bores being upgraded, there was criticism that it was uncoordinated and underinvested. So, in 2018, the federal government announced a new Improving Great Artesian Basin Drought Resilience Plan aimed at speeding up the investment in water-saving infrastructure and making farms and stations more drought-resilient.[14]

Bayley (1999) provides a detailed study of Indigenous water sources and knowledge of their use.

[13] 1.2 million square kilometres.

[14] The Improving Great Artesian Basin Drought Resilience program provided basin jurisdictions with up to $27.6 million of matched funding from the Australian Government over five years from 1 July 2019 (2019–20 to 2023–24).

4.5 Sustainability Issues

The United Nations' definition of sustainability is—"meeting the needs of the present without compromising the ability of future generations to meet their own needs.[15] Essentially what this definition of sustainability means is intertemporal fairness and since this is the criteria by which the United Nations Sustainable Development Goals were determined, the authors will also adopt this meaning in the discussion of sustainability of the Australian desert regions, below.

The threats or potential threats to the sustainability of the Australian desert ecosystems are or could be:

- The impact of higher global temperatures that result from increasing atmospheric carbon due to anthropogenic pollution, on an already relatively hot environment. The second-round effect on already highly variable rainfall of potentially more extreme weather and climatic events (heatwaves, floods, cyclones, extended dry seasons);
- Invasive plants and animals;
- Habitat fragmentation and destruction;
- Land use including clearing of native vegetation;
- Altered fire regimes (in particular, possibly more frequent and severe fires, and a lengthened fire season).

4.5.1 Impact of Climate Change on Viable Agriculture in Australian Deserts

Very commonly deserts are defined in terms of low average annual rainfall (AAR). However, consideration of AAR alone can be misleading in the case of Australia's arid and semi-arid regions. Throughout much of inland Australia, rainfall is so sparse and sporadic when it does occur, that the annual probability distribution patterns are of limited utility. Similarly, AAR means and medians are of limited direct practical value in decision making. Even measures of variation will be difficult to ascertain and contextualise because of sparse and inconsistent rainfall events, exacerbated further by the problem of a lack of data due to sparse measuring points (Lellyett et al., 2022).

Eurocentric seasonal definitions (winter, spring, summer, autumn) are also largely unsuited to the climate of much of Australia's arid zones. The north has a wet season and dry season that follows an annual pattern, but much of the centre has very little in the way of seasonal variation as far as rainfall is concerned, although there is a seasonal temperature pattern. The south encompassing the Nullarbor Plain, and parts of South Australia has a winter seasonal rainfall pattern. Indigenous seasonal climate knowledge can provide some insights towards understanding the desert ecology response to climate and the likely impact of climate change. Notably,

[15] First articulated by United Nations Brundtland Commission in 1987 as cited by Tietenberg (2000).

the Bureau of Meteorology has collected and transcribed a number of aboriginal seasonal "calendars" and the associated prevailing weather patterns. (http://www.bom.gov.au/iwk/calendars/walabunnba.shtml). An example from the Wallabunnba desert community 300 km north of Alice Springs in the Northern Territory, is a two season Watangka (Summer) season with high temperatures, bush fires, and rains, and a Yurluurrp (Winter) season that is cooler and dryer. Other Indigenous communities have differing systems, such as the Nyoongar people from the arid southwest (Western Australia) that have six seasons (Bureau of Meteorology, Indigenous Weather Knowledge website).

What is important in terms of season is what the ecosystem is adapted to. Much of the arid and semi-arid regions have highly variable rainfall, not annually variable, but variable over multi-year time frames. Even if we could develop a suitable rainfall distribution curve, it would be unlikely to conform to the common slightly skewed bell curve with an extended right-hand tail reflecting low probability high rainfall events and truncated left-hand tail at zero on the horizontal axis (i.e., corresponding to zero rainfall) typical of wetter sub-tropical and coastal climates in Australia. So, framing the likely impacts of climate change calls for a different set of statistics. Because dry spells (drought) can last for many seasons, even years, and rainfall events can be clustered and vary enormously in amplitude, we need to analyse various multi-year to decadal time frames to identify emergent patterns.

Current information regarding climate change strongly suggests mean and extreme temperatures are rising and as a consequence, this will have a direct impact on some ecological systems. Second, the effect of temperature rises is to increase evaporation, impacting water availability and in a second-round effect, influencing precipitation. This second-round effect thus alters weather. However, weather systems are complex and often occur at scales below the resolution of climate models. In particular, the distribution of rainfall is a case in point, especially rainfall associated with thunderstorms. That is, delineating specifically when and where rainfall will occur as a result of changing weather patterns under climate change, carries significant uncertainty. Finally, carbon dioxide in the atmosphere is increasing. The effect of this is to increase plant growth (and quite considerably as has been known for a long time in agriculture with the use of high carbon dioxide environments in greenhouse agriculture to increase plant production). This is what led to the current climate change often being described as the "greenhouse effect". This means where water becomes less limiting plant production will increase as a result of higher atmospheric carbon dioxide. This in turn will increase yields of agricultural crops and stimulate rangeland pasture growth, both potentially beneficial as far as human food production is concerned, an often under-reported consequence of climate change. A further consequence is that the increased vegetation will sequester more carbon dioxide from the atmosphere. This is known as a feedback loop in the carbon cycle and in the present circumstances a "virtuous feedback loop". So, the issue of concern becomes to what extent this positive impact on growth will be offset by changes in evaporation and rainfall (see Mukherjee et al., (2022) for a fuller general discussion).

Firstly, on the impact of rising temperatures, the 2023 Intergovernmental Panel on Climate Change (IPPC) report says "*Human activities, principally through emissions of greenhouse gases, have unequivocally caused global warming, with global surface temperature reaching 1.1 °C above 1850–1900 in 2011–2020*" (Headline Statement 1 A Observed Warming and its Causes at https://www.ipcc.ch/report/ar6/syr/resources/spm-headline-statementsPCC). The IPCC goes on to forecast "*global warming will exceed 1.5 °C during the twenty-first century and make it harder to limit warming below 2 °C*" (Headline Statement A4 Current Mitigation Progress, Gaps and Challenges IPPC *op cit*).

The Bureau of Meteorology (BOM) estimates that Australia's national mean temperature in 2022 was 0.50 °C warmer than the 1961–1990 average.[16] Hughes et al. (2016) state that the average surface temperature of Australia has increased by 0.9 °C since 1910, a date selected as being before significant industrial emissions and they say the largest increases were across inland QLD and NT, which includes a considerable expanse of the arid and semi-arid zones of Australia. Annual average temperatures by 2030 were projected by Hughes et al. to be 0.6–1.3 °C above the 1986–2005 baseline, or 1.2–1.9 °C above pre-industrial levels. Note these differing estimates of warming, and show the importance of the base line when making estimates of temperature change. There also appears to be some evidence that warming may not be occurring at the rate of earlier predictions, so caution is required when considering absolute changes, none the less a clear and accelerating warming trend has been observed over the last century and is projected to continue in coming decades.

So, the question then is, will these hotter temperatures and longer periods of higher temperatures impact an ecosystem already accustomed to an extremely hot and dry climate? In an already extreme temperature environment will these changes have a significant impact? There will undoubtedly be an impact, but it may be marginal in terms of ecological disturbance in such extreme environments, depending upon the extent and rate of adaptation of species living within the desert ecological systems. For instance, Litcher-Marck and Baldwin (2023) found that Desert Rock Daises (Perityleae) were already pre-adapted to hot-dry conditions before the emergence of North American deserts and that they subsequently thrived under desertification. It is possible that other Australian species may be similarly pre-adapted, or otherwise already so highly adapted to hot dry climates that the impact of future climate change may be minimal, as compared to species in cooler high rainfall zones.

The second-round effect of global warming, on rainfall, is harder to analyse. Australian rainfall is influenced by four main drivers; ENSO (El-Niño Southern Oscillation), IOD (Indian Ocean Dipole), SAM (Southern Annular Mode), and MJO (Madden–Julian Oscillation) which impacts the northern monsoons. These coupled ocean–atmosphere phenomena account for the bulk of seasonal to interannual variability in Australian rainfall, especially over the eastern states. ENSO and the IOD

---

[16] Bureau of Meteorology (2023) Annual Climate Statement 2023, viewed at http://www.bom.gov.au/climate/current/annual/aus/#:~:text=Australias%20area%2Daveraged%20mean%20temperature,0.67%20%C2%B0C%20above%20average.

are the dominant drivers at interannual timescales, while at shorter timescales SAM further modulates climate over the dry southern and western arid zones, while the MJO modulates conditions in the northern tropics.

Some long-term trends in rainfall have been observed that have been attributed to climate change. Across the southwest and to a lesser extent in the southeast, a drying trend exists over the cool season months of April to October (BOM & CSIRO, 2022). Figure 4.5 depicts areas of decline for the period 2000–2021, as compared to records overall since 1900. Southern Australia receives much of its rainfall during the cooler months from cold fronts driven by low-pressure systems south of the sub-tropical high-pressure ridge. The frequency of these systems has diminished and are less likely to produce rainfall when they do occur. An observed increase in the number of high-pressure systems over southern Australia, which bring dry clear weather and little rain, is known to be a climate system response to climate change.

The north of Australia has been wetter in the last two decades than in the previous century, across all seasons, but particularly through the northern Wet season (October to April). Rain events during this period have also penetrated further south, extending all the way to Southern Ocean, so that the impact has been felt across nearly all the arid and semi-arid zones the subject to analysis in this paper as can be seen in Fig. 4.6.

Thus, either a wetter or dryer climate may result from climate change across the arid and semi-arid landscapes of Australia, depending upon the exact location and season under consideration. Dryer in the south during the cooler months and wetter

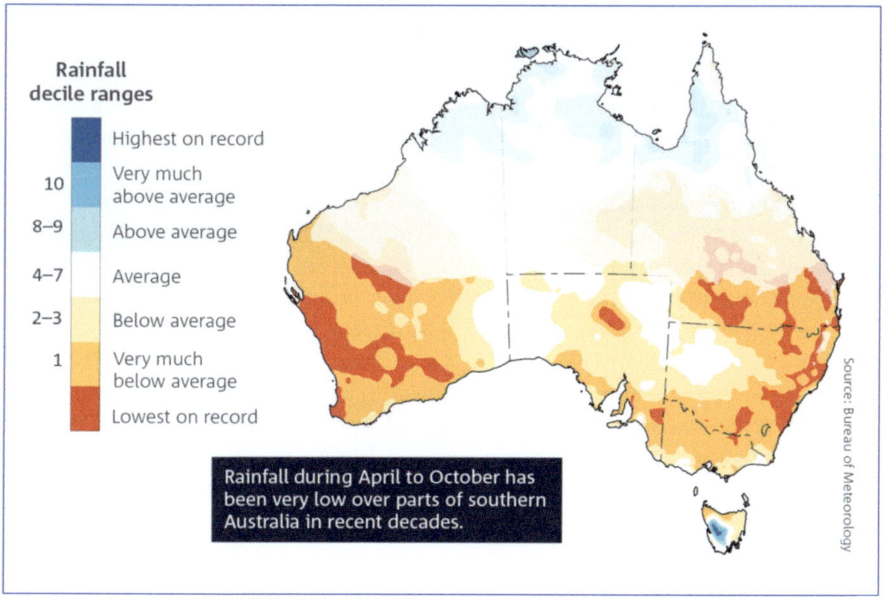

Fig. 4.5 April–October rainfall deciles for the period 2000–2021 as compared to the reference period 1900–2021. Areas across central and northern Australia receiving 40% or less of their annual rainfall during April–October are faded (BOM & CSIRO, 2022)

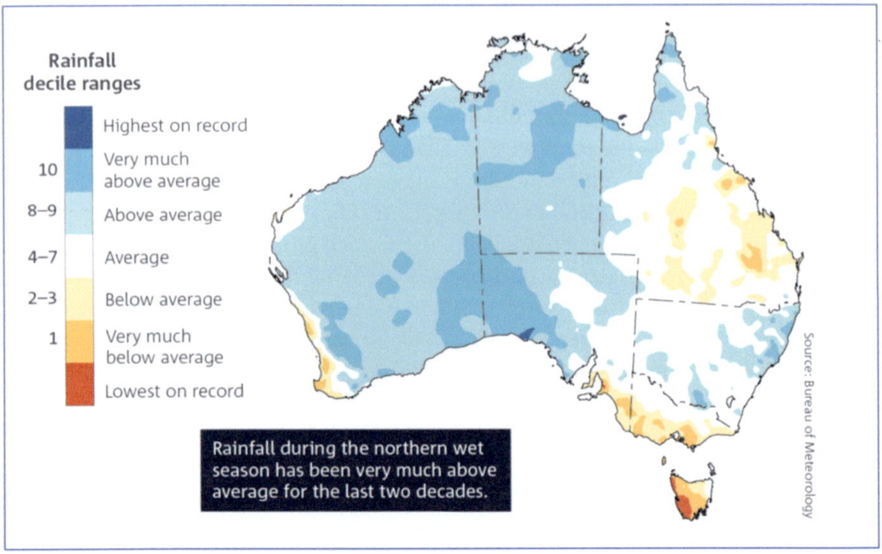

Fig. 4.6 October–March (northern wet season) rainfall deciles for the period 2000–2021 as compared to the reference period 1900–2021 (BOM & CSIRO, 2022)

in the north through the northern "wet" season, possibly extending into southern regions as well in summer. Again, it is hard to say how the ecosystems already well adapted to highly variable rainfall will be impacted by further change. Unless there is a significant shift in the rainfall distribution to the right, i.e. a much wetter climate, it seems feasible that the impact may not be all that significant on an ecosystem already adapted to long periods of drought and occasional floods. If a significant shift in rainfall to the right (i.e., an increase), does occur it is more likely through the northern "wet" season and may extend well south, increasing summer season rainfall in the temperate zone as well as the sub-tropics and tropics. This might prompt a significant increase in plant growth due to the "greenhouse" effect of higher carbon dioxide in the atmosphere and more available water. Otherwise, an incremental shift in the rainfall distribution, in situations where rainfall is sufficient to form the skewed bell curve as we discussed above, would imply the mean shifting a few percentage (%) points to the right or left depending on whether the climate is wetter or dryer, and perhaps some expansion or contraction of the standard deviation. So, the question is—and this is deserving of further research—would this be enough to alter an already highly adapted ecosystem to high temperatures, long periods of droughts and intermittent flooding?

4.5.2 *Invasive Plants and Animals*

Kingsford et al. believe much of the environmental degradation across Australia, is irreversible, some obviously so because of species extinction (Kingsford et al. 2020). Kingsford cites other papers where authors have suggested partial restoration is possible by mitigating threats including, removing introduced animals and artificial watering points, reducing populations of native herbivores and reintroducing regionally extinct species (Moseby et al., 2009; Silcock & Fensham, 2019). How practical these measures are, especially given the scale of the problem, is unclear. Measures that could improve the composition and functionality of the desert ecosystem, such as restoring vegetation cover and species mix, and increasing the abundance of fungal, invertebrate, small mammal, reptile, frog and bird populations are all obviously desirable as suggested by Kingsford who cites these sources (Contos & Letnic, 2019; Moseby et al., 2009; Rees et al., 2019), but again the practicality of this is the issue. To what extent a satisfactory new ecological balance can be achieved in conservation areas is currently being explored. One example is the Arid Recovery project in northern South Australia. This not-for-profit organisation has created a 123 km^2 fenced sanctuary for native wildlife by excluding feral predators and domestic livestock (see https://aridrecovery.org.au/about-us).

Sustainability as defined by the U.N. and as we use it in this chapter, doesn't necessarily imply conservation. Some authors seem to confuse the two concepts, but in trying to distinguish between the two concepts, Vucetich et al. (2021) make the point "*Disagreements in conservation and sustainability are more likely…, to rise from differences in how one adjudicates situations where one's own values conflict with each other, as might occur when a conflict arises between felt obligations to both humans and non-humans*". Vucetich considers sustainability as essentially viewing the world from a human self-interest perspective, while conservation may be more altruistic, preservation of other species simply because they are there and not from the perspective of how they might benefit mankind.

Species loss and other environmental damage, while generally regarded as undesirable, may not "*compromise the ability of future generations to meet their own nee*ds" and therefore not fail the human-centric test for sustainability. Ecosystems are dynamic, adapting to changes in the external environment by evolving new ecological balances. The conservationist may want to preserve a particular ecological balance, but that may not take account of background natural evolution. So, this can lead to problems if that static ecosystem balance at a particular point in time is used as a reference point.

What is important is to distinguish between anthropogenic threats and non-anthropogenic threats. This is a complex problem, because of the need to measure only a certain type of change (that which is anthropogenic causal), against a background of changing (or evolving) ecological balances. It may be difficult with such complex systems, to untangle the two types of change, however in assessing the impacts of climate change the focus needs to be on anthropogenic threats which are clearly not evolutionary, and then decide whether they are sustainable, or not. The assessment we have been tasked with here is; therefore, whether or not anthropogenic environmental disturbances are sustainable, that is "*do not compromise the ability of future generations to meet their own needs*", or unsustainable.

In the case of the Australian desert ecosystems which are already highly disturbed the issue is extremely difficult to assess. It is further complicated by the question of restoration, which in turn raises the question of restoration to what prior ecological balance? What is the reference point to use? Then, it raises the question of what is practical. In certain small, contained areas it may be feasible to eliminate feral pests and restore to some extent a prior ecological balance of native flora and in particular fauna that have been largely eradicated from the wider areas where feral predators prevail, but this approach is probably impractical on the broader scale.[17]

So, finally, the question is now, whether or not another anthropogenic threat such as global warming, as a result of greenhouse gas emissions, will cause unsustainable changes to the already highly disturbed ecosystems of the Australian desert regions. Again, the issue of what reference point to compare changes to becomes relevant. If it is to the current degraded state of the ecosystems, it is probable a modest degree of warming, say to the current IPCC forecasts to 2050, may make little difference to sustainability in areas where higher temperatures with increased rainfall and higher atmospheric CO_2 are experienced that promote plant growth. This could be the situation across much of the northern and central Australian arid regions. Where higher temperatures, with lower rainfall and increased droughts, occur, such as is projected for parts of the south west and the south east arid and semi-arid zones, then sustainability may well be threatened.

4.6 Conclusion—Policy Implications Including Climate Adaptation Planning

Australia is the driest continent on earth, apart from the Antarctic. Seventy percent of mainland Australia is classified as arid or semi-arid, and within this zone, large areas have been designated as deserts. The arid and semi-arid rangelands have undergone

[17] The Wild Deserts project, cited by Kingsford et al. (2020), which is trying to restore a desert ecosystem in Sturt National Park, New South Wales, Australia (35,000 ha), has a novel approach to the problem. Essentially, they are trying to train native fauna, decimated by feral predators in the wider environment, to become more resilient, by first breeding them up in exclosures with no feral predator pressure and gradually moving them up to more and more exposure. The abstract of the paper outlines the plan of the experiment; as, "*management treatments included two "no restoration" areas and three "restoration" areas. The latter include two exclosures (each 2,000 ha), free of introduced animals (foxes, cats, rabbits), with previously abundant kangaroos removed and regionally extinct mammals to be reintroduced. The third management treatment is a Wild Training Zone (10,400 ha), with introduced animals and kangaroos managed at low levels, using innovative methods, improving survivorship and avoidance behaviour of reintroduced mammals to introduced predators. These measures will allow populations of threatened animals to establish, initially in the exclosures, then the Wild Training Zone and potentially more widely* (Kingsford et al. December 2019 Abstract p. 1).

significant ecological disturbance due to European incursion, both in terms of settlement and exploitation of resources (chiefly pastoral agriculture and mining), much of which is irreversible. This has been exacerbated by the removal, or migration, of native populations from the land they once occupied and lived off and the cessation of traditional land management activities. However, there are prospects for small scale, partial, remediation and perhaps on a larger scale, changes to land use that could create a new more sustainable ecological balance, including the prospect of increasing carbon sequestration over vast tracts of Northern Australia.

The likely impact of anthropogenically induced climate change on the Australian deserts is difficult to assess. Many research papers discuss the threat of climate change, but while there are spatially uniform trends of increasing temperature that have low uncertainty, future projections of regional patterns of rainfall still vary significantly and hold higher uncertainty. Rainfall trends in recent decades influenced by climate change indicate declines in winter (April–October) rainfall, but these are most pronounced in the far southwest and southeast regions of the continent, and less so over southern and central parts of the desert regions. Conversely, in the north with significant overlap of desert regions an increase in the northern wet season (October–March) rainfall, has been observed since 2000.

Even if more certain projections of rainfall were available, key questions remain. Firstly, as to whether hotter and wetter conditions over extensive central-northern desert areas during the northern wet seasons might more than offset increased evaporation and drier other times of year to produce a net positive effect facilitating increased growth and significant carbon sequestration. If so, it would create, a carbon sink, potentially significant at a global scale. Secondly, the question for regions within desert areas, where rainfall is projected to decrease, as to whether or to what extent the various components of the desert ecosystem which are already highly adapted to extended periods of drought and occasional flooding, can cope. It may turn out that adaptations to date of native flora and fauna are so effective that the impacts are relatively inconsequential. Both questions are worthy of further research.

The main agricultural activities across the arid and semi-arid zones of Australia are pastoral activities, mainly cattle grazing in the north and sheep grazing in the south on natural rangelands. Under the more likely future climate scenarios described in this chapter, a hotter and wetter climate in the north extending through the centre of Australia and comprising the greater part of the arid and semi-arid zones on the one hand, and a likely hotter and dryer climate in southwest and possibly smaller areas of the southeast on the other, the question of sustainability of the pastoral industry is likely to be different for each of the two impact zones. The hotter and wetter zone is likely to lead to more plant growth with the potential for higher sustainable stocking rates, subject of course to management of the resource for a sustainable future. For the dryer and hotter southern areas, some destocking may be required to achieve sustainability in the future. Overall, though the impact on economic systems is not likely to be great, simply because the population dependent on pastoral activities is relatively small and practices are generally quite adaptive to harsh environmental conditions.

As regional scale climate projections improve with a higher spatial and temporal resolution and reduced uncertainty in highly variable parameters such as rainfall, the realistic modelling of yet more complex and uncertain parameters such as plant available soil moisture should become possible, thereby enabling more accurate estimates of plant growth. Lellyett et al. (2022) show how in the context of seasonal forecasts, improved drought forecasts could lead to probability distributions of plant available soil moisture, which in turn could help identify economically optimal management and planning decisions. With further research and investment, there is a similar potential for climate scale projections. Such improved future projections would also help to better define the changing geographic extents of the arid and semi-arid zones. Combined with animal heat stress projections to genetically breed or engineer more climate-resilient stock varieties, the refined geographic extent of suitable land plus better estimates of plant growth may enable the probabilistic quantification of projected economically sustainable stocking rates to aid with adaptation and future planning.

Regardless of climate change, remediation of the ecology of arid and semi-arid zones, which involves exclusion of feral animals, cessation of low value pastoral activities and reduction (or control) of artesian watering points to reduce the number of native macropods, would be of high value in conserving and increasing the resilience of native flora and fauna. This would not result in the restoration of the ecological balance that existed prior to European incursion but may create a new ecological balance skewed towards survival of native species, that could make a contribution to ameliorating the impact of global climate change.

Careful and considered cost–benefit analysis is suggested in order to optimise future investments to reduce uncertainty in climate projections and better quantify the likely impact of global warming on Australian desert ecosystems on the one hand, and investments in the remediation of existing ecosystem problems on the other.

References

ARS (2019). Home page. Australian Rangelands Society, Available at: https://www.austrangesoc.com.au/

Australian Bureau of Meteorology. (2023). Annual climate statement 2023, viewed at http://www.bom.gov.au/climate/current/annual/aus/

Australian Bureau of Meteorology. Indigenous weather knowledge website http://www.bom.gov.au/iwk

Australian Bureau of Meteorology and CSIRO (2022). State of the climate, 28 pp. http://www.bom.gov.au/state-of-the-climate/

Bai, Z. G., Dent, D. L., Olsson, L., & Schaepman, M. E. (2008). Proxy global assessment of land degradation. *Soil Use and Management, 24*(3), 223–234. https://doi.org/10.1111/j.1475-2743.2008.00169.x

Bayley, I. A. E. (1999). Review of how indigenous people managed for water in desert regions of Australia. *Journal of the Royal Society of Western Australia, 82*, 17–25.

Cherlet, M., Hutchinson, C., Reynolds, J., Hill, J., Sommer, S., & Von Maltitz, G. (2018). *World atlas of desertification.* Publications Office of the European Union. ISBN 978-92-79-75350-3. https://doi.org/10.2760/9205, JRC111155.

Contos, P., & Letnic, M. (2019). Top-down effects of a large mammalian carnivore in arid Australia extend to epigeic arthropod assemblages. *Journal of Arid Environments, 165*, 16–27.

Cossins-Smith, A. (2023). AngloGold Ashanti to build renewable project in Australia. Mining Technology, June 27, 2023 viewed at https://www.mining-technology.com/news/anglogold-ashanti-renewable-project-australia-mine/

Department of Climate Change, Energy, the Environment and Water. (2023). Great Artesian Basin info page, viewed at https://www.dcceew.gov.au/water/policy/national/great-artesian-basin

Desert Knowledge Cooperative Research Centre. (2008). Review of indigenous pastoralism and employment.

Desert Knowledge Cooperative Research Centre. (2010). Exit Report. https://www.desertknowledgecrc.com.au/desert-knowledge-australia/

Durack, M. (1959). *Kings in grass castles.* Penguin Imprint. ISBN: 978014379073.

El Hassan, M. (2019). *Goats, people and rangelands: Analysis of the rangeland goat industry as a complex social ecological system.* The Australian National University (Australia), ProQuest Dissertations Publishing.

Foran, B., Stafford Smith, M., Burnside, D., Andrew, M., Blesing, D., Forrest, K., & Taylor, J. (2019). Australian rangeland futures: Time now for systemic responses to interconnected challenges. *The Rangeland Journal, 41*, 271–292.

Foster, D., Mitchell, J., Ulrik, J., & R. Williams (2005). Population and mobility in the town camps of Alice springs. A report prepared by Tangentyere council research unit, Desert Knowledge Cooperative Research Centre, Alice Springs.

Goddan, D. (1993). Economics of climate change opportunities. In B. Shelley (Ed.), *Climate change-implications for natural resource conservation* (pp. 144–161). University of Western Sydney. ISBN 1863410813.

Godde, C. M., Boone, R. B., Ash, A. J., Waha, K., Sloat, L. L., Thornton, P. K., & Herrero, M. (2020). Global rangeland production systems and livelihoods at threat under climate change and variability. *Environmental Research Letters, 15*(4), Article 044021.

Gunn, A. (1908). We of the Never Never. Penguin Imprint. ISBN: 9781741667622.

Hancock, W. K. (1930). Australia, Beur, London as quoted by James, P.G., Economic policy for drought. In Lovett, J.V. (1973), woinarskiDrought, Angus and Robertson (Chap 15 p. 24).

Howden, S. M. (2017). Living with climate change. In P. Erkelenz & M. Andrew (Eds.), *Proceedings of the Australian rangeland society 19th biannual conference.* Port Augusta South Australia.

Hughes, L. H., Rickards, L, Stefan, W., Stock, P., & Rice. (2016). *On the frontline: Climate change & rural communities* (pp. 102). Published by the Climate Council of Australia Limited.

IPCC. (2007). *Climate change 2007: Working group II: Impacts, adaptation and vulnerability.* Fourth Assessment Report.

IPCC. (2013). Climate change 2013: The physical science basis. *Working group I contribution to the IPCC 5th assessment report—changes to the underlying scientific/technical assessment.* http://www.ipcc.ch/report/ar5/wg1/

IPCC. (2014). Climate change 2014: Impacts, adaptation and vulnerability. In C. Field, V. Barros, K. Mach, & M. Mastrandrea (Eds.), *Contribution of working group II to the fifth assessment report of the intergovernmental panel on climate change* (pp. 485–533). Cambridge University Press.

IPSOS. (2015). Climate change report 2015. Accessed at http://ipsos.com.au/Ipsos_docs/CC2015/Ipsos_Climate_Change_Report_2015.pdf

Lichter-Marck, I. H., & Baldwin, G. (2023). Edaphic specialization onto bare, rocky outcrops as a factor in the evolution of desert angiosperms. *PNAS, 120*(6), e2214729120. https://doi.org/10.1073/pnas.2214729120

Kingsford, R. T., West, R. S., Pedler, R. D., Keith, D. A., Moseby, K. E., Read, J. L., Letnic, M., KeLeggett, K. E. A., & Ryall, S. R. (2020). *Strategic adaptive management planning—restoring a desert ecosystem by managing introduced species and native herbivores and reintroducing mammals*. Contributed Paper. Conservation Science and Practice. Wiley.

Lellyett, S. C., Truelove, R. N., & Huda, A. K. S. (2022). Improving early warning of drought in Australia. *Climate, 10*, 91. https://doi.org/10.3390/cli10070091

Lovett, J. V. (ed.) (1973). *The environmental, economic and social significance of drought*. Angus and Robertson. ISBN 0207128057.

Proust, M. (2001). Time regained. Book 4, Everyman's Library edition, p. 448.

Macfarlane, I., & McConnell, A. (2017). The waters of Australian deserts. Final Report March 2017, Cultural heritage study report to the department of environment and energy and the Australian heritage council.

Moseby, K. E., Hill, B. M., & Read, J. L. (2009). Arid recovery—a comparison of reptile and small mammal populations inside and outside a large rabbit, cat and fox-proof enclosure in arid South Australia. *Austral Ecology, 34*, 156–169. https://doi.org/10.1111/j.1442-9993.2008.01916.x

Nielsen, U. N., Stafford-Smith, M., Metternicht, G. I., Ash, A., Baumber, A., Boer, M. M., Booth, S., Burnside, D., Churchill, A. C., El Hassan, M., Friedel, M. H., Godde, C. M., Kelly, D., Kelly, M., Leys, J. F., McDonald, S. E., Maru, Y. T., et al. (2020). Challenges, solutions and research priorities for sustainable rangelands. *The Rangeland Journal, 42*, 359–373. https://doi.org/10.1071/RJ20059

Pastoral Land Board Northern Territory, Annual Report 2013–14. Available at https://depws.nt.gov.au/__data/assets/pdf_file/0019/254800/Pastoral-Land-Board_AR-2013-14.pdf

Rees, J. D., Kingsford, R. T., & Letnic, M. (2019). Changes in desert avifauna associated with the functional extinction of a terrestrial top predator. *Ecography, 42*(1), 67–76.

Rola-Rubzen, M. F., & MacGregor, M. J. (2009). Draft paper presented to AARES 53rd annual conference, Cairns, QLD.

Silcock, J. L., & Fensham, R. J. (2019). Degraded or just dusty? Examining ecological change in arid lands. *BioScience, 69*(7), 508–522.

Smith, M. A. (2013). *The archeology of Australia's deserts*. Cambridge University Press. ISBN 978-0-521-40745-8

Smith, M. A., & Hesse, P. (2005). *23 Degrees south: Archaeology and environmental history of the southern deserts*. National Museum of Australia, ISBN 1876944307.

Stafford-Smith, M. (2008). The 'desert syndrome'—Causally-linked factors that characterise outback Australia. *The Rangeland Journal, 30*(1), 3–14. https://doi.org/10.1071/RJ07063

Stern, H., de Hoedt, G., & Ernst, J. (2000). Objective classification of Australian climates. *Australian Meteorological Magazine, 49*(2000), 87–96.

Tietenberg, T. (2000). The quest for sustainable development. In *Environmental and resource economics* (Chap 23, 5th edn, p. 547).

Vucetic, J. A., Jeremy, T. B., van Eden, L. M., & Macdonald, E. A. (2021). How scholars prioritize the competing values of conservation and sustainability. *Biological Conservation, 257*, Article 109126.

Webb, N. P., Marshall, N. A., Stringer, L. C., Reed, M. S., Chappell, A., & Herrick, J. E. (2017). Land degradation and climate change: Building climate resilience in agriculture. *Frontiers in Ecology and the Environment, 15*(8), 450–459. https://doi.org/10.1002/fee.1530

Woinarski, J. C. Z., Burbidge, A. A., & Harrison, P. L. (2015). Ongoing unravelling of a continental fauna: Decline and extinction of Australian mammals since European settlement. *Proceedings of the National Academy of Sciences, 112*(15), 4531–4540.

Mukherjee, A., Saha, S., Lellyett, S. C., & Huda, A. K. S. (2022). Impact of climate change and variability on food security in the Asia-Pacific region. *Asia-Pacific Sustainable Development Journal, 29* (1), 119–141.

Chapter 5
Phytodiversity of Thar Desert: Traditional Knowledge, Sustainable Utilisation and Conservation

Narpat Singh Shekhawat, Richhpal Singh, Smita Shekhawat, Ashok Kumar Patel, Vinod Kataria, Sumitra K. Choudhary, Kheta Ram, Manoj K. Rai, and Harish

5.1 Introduction

Humans changed the biosphere and domesticated both biotic and abiotic resources—stressing arid land/ecosystems that were the cradle of humanity, cultures, origin and decline of civilisations. Prehistoric cultures, Vedic, Harappan, Sindhu and Saraswati civilisations flourished along the margins/regions of the Indian Thar Desert.

The scientific literature on plant resources of the Indian Thar Desert is limited. Anthropogenic pressures on valuable plants like *Calligonum polygonoides* and *Ziziphus* species are threatening. Blatter and Hallberg (1918), in an introduction to Flora of Indian Desert Part-I (1918) and Part-II (1921) stated "The Indian Thar Desert is perhaps the least known part of the plains of India". The vegetation of the Rajasthan/the Thar attracted the least attention from the botanists. It was neglected as compared

N. S. Shekhawat (✉)
Jodhpur, Rajasthan, India
e-mail: biotechunit@gmail.com

R. Singh
Government College, Jodhpur, Rajasthan, India

S. Shekhawat
Jaipur, Rajasthan, India

N. S. Shekhawat · S. Shekhawat · A. K. Patel · V. Kataria · S. K. Choudhary · K. Ram
Biotechnology Unit, Department of Botany, Jai Narain Vyas University, Jodhpur, Rajasthan, India

M. K. Rai
Department of Environmental Science, Indira Gandhi National Tribal University, Amarkantak, MP, India

Harish
Department of Botany, Mohan Lal Sukhadia University, Udaipur, Rajasthan, India

M. S. Rathore et al. (eds.), *Desert Ecosystems*,
https://doi.org/10.1007/978-3-032-03237-9_5

to plants in other deserts. The biodiversity of medicinally important slow-growing Guggul (*Commiphora wightii*) and numerous rare and endemic edible plants is dwindling. Cycles of ignorance and heavy machines/earth-movers/bulldozers enabled land clearing for developmental activities together are harming the ecological landscape. Impact assessments/evaluations of the developmental project(s) in fragile ecosystems/zones for the safety of the environment and life systems are non-existent.

5.2 Missing Conservation Initiatives

Policy decisions are taken and implemented by plant-blind and eco-insensitive agencies for land use and introduction/promotion of cultivation of exotic/invasive plant species like *Prosopis juliflora* and *Acacia tortilis* endangering reserve forest areas of the Aravalli/Desert, where Babul, Dhawra/Indrokh (*Anogeissus* spp) Phog, Kumbhtia, Ber, Ker and Kankera, and Khejari (*Prosopis cineraria*; Fig. 5.1a and b) existed. *Lantana camara* and other invasive non-native plant species are causing havoc to the native flora. Promotion of cultivation of cash crops on water-starved lands need the attention of all.

5.3 Potentials of Biodiversity of Thar Desert

The Thar Desert of India provides valuable ecosystem services. The region has the potential to become the hub of organic farming/harvesting products of native plants. Fruits of Ber, Ker, Khejari, Cucurbits (Kakoda, Kachhara, Kharbuja, Kakadi, Matira/Laamp) are important for economy, ecology and health-care as sources of nutraceuticals. Bhatkania, Dodhi, Dodi, Dhaloo, Dansaria, Phogla, Dachub, Sinia (Kinsara), Khedula, Khiroli, Khimpoli, Khoka/Sangari, Khirkheenp, Kharinti, Oont Phog, Piloo, Chirpotia and Painpa/Pimpa are rich in nutrients. Nature nurture roots/tubers crops of Nagauri Ashawagandha (Indian Ginseng–the Padalsi)/Satavari/Mulhati/Safed-Musali and Ramas. Environmental conditions and edaphic factors of Rajasthan have allowed evolution of Kharchia wheat, Chaadi Bajri, NagauriMethi, Mathania Mirchi/Gajar/Podina. Nutritious drought-hardy vegetation, the ever-ready sources of fodder and nourishing grains/seeds of Bharoot/Dhaman/Kuri/Makada/Sewan/Bhankari/Bhangari/Gokhrus are assets. Daab (*Desmotachyabipinnata*), a grass considered sacred in Vedic scriptures provides biomass during droughts and famines (Minnis, 2021). The desert and the Aravalli lands are home to numerous medicinal/herbal/edible plants.

Conventional and Biotechnology-based methods have been defined for the characterisation and conservation/propagation of plants in stressed ecosystems (Shekhawat et al., 2012, 2014). Advanced Micropropagation Facilities (Greenhouse and Aeroponic System) are constructed and established. Human resources educated, mentored and trained at Jodhpur have been contributing at national and international levels.

Fig. 5.1 **a** Field photograph of *Prosopis cineraria*, a keystone species, in the Thar Desert. **b** A multipurpose tree source of nutrient-rich fodder (loong), fuel and other ecosystem services. **c** Unripe pods (Sangari), a rich protein source for human. **d** Tender pods processed, dried and stored. **e** *P. cineraria* with edible fruits and **f** *P. cineraria* mature sweet fruits (Khoka) harvested from an elite tree

Experiences, achievements, constraints faced, limitations and future prospects are shared in this chapter.

5.4 The Thar Desert—A Tale of Trail of Dispersion of Hominin/*Homo sapiens*

India is considered a geographical accident of the world. The Aravalli Mountain range (the oldest geological feature, 350 million years old) of the country, originated in the Precambrian Supereon/Proterozoic eon/Neoproterozoic era and the Himalayan Mountain ranges (the youngest) formed 50 million years ago. Indian Thar Desert marks an adaptive boundary between monsoonal Asia to the east and the Sahara-Arabian Desert to the west (Blinkhorn, 2021). Blinkhorn et al. (2021), studied palaeolithics at Singi Talav, Didwana, and the Acheulean and Middle Palaeolithic occupations horizons at Jayal formation Katoati, Chhajoli, Amarpura (Nagaur) and Budhaa Pushkar of Ajmer. Blinkhorn and Petraglia (2017) and Blinkhorn et al. (2021) suggested the role of the Thar lands as ecological and demographic frontiers to Palaeolithic human dispersion. Scientific arguments on Aryan/Vedic civilisation/Sarasvati and Indus River(s) culture and shifting of Asian monsoon/mega-droughts during the Meghalaya period (with effect from 4200 before now) of the Holocene, exist. It is suggested Monsoonal precipitation sharply increases moving from west to east thus lowering aridity to the east of the Thar Desert. Blinkhorn et al. (2020) recorded the oldest chronometrically dated evidence for fluvial activity within this region at the Nal Quarry site, Bikaner. He suggested that the landscape of the late Middle Pleistocene and early Late Pleistocene were a key threshold for modern human dispersals.

5.5 Dry Lands/Fragile Ecosystems in the Anthropocene

Drylands refer to regions with an aridity index (AI, i.e., the ratio of mean annual precipitation to mean annual potential evapotranspiration) lower than 0.65 (Liu et al., 2021). These lands cover 40–45% of the surface of the earth. Around 38% of the world's population resides in the drylands, the prominent drivers of global biogeochemical cycles. Rotenberg and Yakir (2010) suggested that forests of drylands fix carbon at rates similar to those of Pine Forest in Europe. Dryland ecosystems are highly vulnerable. Soil salinisation is a major issue on these lands (Negacz et al., 2022). Thar Desert falls under the dry land biome. The Thar Desert is the most populated arid land in the world (85 to 90 persons/square kilometre). During the last century, the population increased by 400% and the livestock by 127% (Dhir et al., 2018). Natural resources have been over-exploited and overstressed (Choudhary et al., 2022). Among the arid lands of the world Thar region is the most intensively farmed. During the 1960s there were a couple of thousands of tractors. These numbers jumped to 30,000 by 1983 and 70,000 in 1992 (Dhir et al., 2018). The mechanised operations increased the gross cropped areas drastically by 1990. During 1990–2022, natural habitats of life forms were disrupted impoverishing fertile land cover and reducing productivity. Native shrubs of *Calligonum polygonoides* (Phog), *Ziziphus* species (Ber/Jharber), *Commiphora wightii* (Guggul) and numerous other plant species were eliminated from Thar landscapes. Green cover of evergreen Phog-lands disappeared. Vegetation (woody forest) of Guggul near Balarawa village (Jodhpur) flourishing during 1999–2001 has vanished. Encroachments/unregulated cultivation of community lands (Orans/Aagors) and water shades are very common. Water bodies/water sheds used to feed rainwater to small/large tanks, ponds, ditches, Beries (temporary small shallow dug-wells)/stepwells and low lands spread all over the state. The Indigenous Rainwater Harvesting Systems (IRHS) have been disrupted. Plants species like *Caralluma edulis* (Pimpa of Pimpthali, Patel et al., 2014a), *Ceropegia* spp (Khedula), *Dipcadi erythraeum, Glossonema varians/Cynanchum registanense* (Khiroli) and *Leptadenia reticulata* (Jivanti; Patel et al., 2014b) and *Sarcostemma* species (Somlata/Khir-khimp) are dwindling. Soil-based seed/propagule(s) banks/ repertoires of genetic diversity are eliminated. The salt lakes, viz. Sambhar and saline tracts of Didwana, Degana, Kuchaaman and Pachpadra, are features of the Desert. Sambhar Lake is declared one of India's "Ramsar sites".

King (1870) in Famine Foods of Marwar, stated "for ages, the struggle for life in the plains of Marwar has been between men and cattle on the one side, and vegetation on the other. It is an unequal fight, and vegetation is now (the than during the 1870s) losing. Nothing is conserved; the leftover few indigenous trees are being cut down, and no replacements are planted".

5.6 Flora of Rajasthan

Rajasthan has the largest area of the country (869 km from west to east and 826 km from north to south); an area similar to Italy/Norway/Poland. Study of plant diversity in this state attracted little attention prior to 1918. While writing an introduction to Flora of Indian Desert Part-I (1918) and Part-II (1921) Blatter and Hallberg started with the line "the Indian Thar Desert is perhaps the least known part of the plains of India". Only 140 plant species from the Thar were enumerated. The vegetation of the Indian Thar Desert and neighbouring regions attracted the least attention from botanists. It was neglected as compared to the deserts of the rest of the world. It is difficult to comprehend the values of *Prosopis juliflora* and *P. cineraria* even *Calligonum polygonoides* and *Simmondsia chinensis* in light of recent R and D of pioneer institutions. The earliest recorded information on botanical exploration in Rajasthan is about Jacquemont's journey in 1832 from Delhi to Bombay, via Ajmer and Neemuch (cited by Jain, 1970). George King in 1869 studied plants consumed in the western Rajasthan. He recorded "*Notes on vegetable products used as food during the late famine in Rajpootana*" and also "*Notes on the famine foods of Marwar*". In a paper entitled "Sketch of the Flora of Rajputana" (1878, 1879), he expressed "as might be expected from its geographical position and limited rainfall, the flora of Rajputana is not a rich one. The number of indigenous species is, but small, a few of these are attractive in appearance". The botany of Rajputana was a report of a tour by Duthie (1886). Kanodia and Gupta (1968) and Gupta and Kanodia (1968) enlisted some useful and interesting supplementary food plants of arid regions, and plants used during scarcity and famine periods in the dry regions, respectively. Jain (1970) described the history of studies on the flora of Rajasthan. Professor Bhandari (1974) reviewed famine foods of the Desert. Bhandari (1978, revised in 1990) published Flora of the Indian Desert. Shetty and Singh (1987, 1991, 1993) of Botanical Survey of India, Calcutta published "Flora of Rajasthan" in 3 volumes. Ethnobotany of Rajasthan, India was published by Singh and Pandey (1998).

Water of Himalayan origin flows on the sandy surfaces of the Thar Desert initially through Gang Nahar (Ganga Canal) and now through the Indira Gandhi Canal Project (IGCP), a major ecological change (one of the biggest canal projects of the country). The canal starts from the Harike Barrage below the confluence of the Sutlej, runs parts of Punjab and Haryana, and runs about 650 km in Rajasthan. This project has changed socio-economy and environment/ecology drastically (Shekhawat et al., 2012). Intensive cropping is practiced with inputs of chemicals/fertilisers that cause pollution including drinking water. Many plant species in the desert have disappeared. Pandey et al. (2012) identified 106 plant species that have been introduced in Rajasthan through canal water.

Presently, the Thar Desert is witnessing the exploration of minerals and hydrocarbons, harvesting solar and wind energy. The Bharatmala Pariyojana (Indian Garland Highways Project), is a Government of India-sponsored mega National Highways construction project that runs through the Thar Desert. Cultivation of groundnut, pomegranate, oranges, Jojoba, Jatropha and Olive (Jatoon) needs caution as some of these crops have water footprints equal to potatoes. The state contributes to oilseed Brassica(s) production; about 40–45% of the oil seed of India, 45.5% in acreage and 48.6% in production. These over-exploits all types of water resources.

India is the source of about 80% of the world's production of guar gum. Rajasthan leads in Guar (*Cyamopsis tetragonoloba*) cultivation. The state contributes to the production of Sesame (*Sesamum indicum*). Pearl-millet (Bajra) and minor millets, and a number of pulses including moth-bean (*Vigna aconitifolia*) are cultivated in the state. In addition, varieties of bamboo, Katha (*Acacia catechu/Senegalia catechu*) mangos, Sal (*Shorea* species), Shisham(*Dalbergia* species), Mahua (*Madhuca longifolia* var. *latifolia*) even sandalwood grow in this region. The Teak of Marwar (*Tecomella undulata*) is part of the wealth of the state. Piloo (*Salvadora persica* and *Salvadora oleoides*) are drought and salt-resistant trees providing fruits and non-edible oils. Indrayan/tumba (*Citrullus colocynthis*) is a desert vine and numerous other plants inhabiting arid soils are repertoires of phytochemicals, genes, unique biochemical pathways and pharmaceuticals for bioprospecting. Sewan grass (*Lasiurus sindicus*), species of *Cenchrus*, the Daab grass (*Desmostachya bipinnata*) and Kurichinki (*Urochloa panicoides*) are nutritious perennial sources of fodder for arid lands. These also harbour unidentified microbiomes, not yet been explored. A small forest of salt-tolerant *Tamarix* species appeared during the last 15 years at the Kawas region of the Barmer district, and flooded catastrophically during August 19–22, 2006; an ecological succession seems to have taken place post-disturbance.

Pandey et al. (2012) described the flora of Rajasthan comprising 2090 species of 819 genera under 159 families of the angiosperms; 66 species of wild plants present in the state may have been relatives of several cultivated plants/crops. A solitary species of Gymnosperms, *Ephedra foliata/Ephedra ciliata* (Oont Phog), that was abundant but is now threatened. Pandey et al. (2012) while commenting on Red Taxa and their conservation strategies reviewed 65 depleting plant taxa including rare, endangered and threatened taxa, their present status as per the latest IUCN categories of "threat". Thirty-six taxa of Angiosperm are endemic to the state of Rajasthan. Among the endemics, 20 taxa are of specific level, while 16 are of infra-specific level. Out of these, 14 taxa alone are present in the Rajasthan desert.

5.7 Trees' Culture in Rajasthan

Photosynthetic canopies of trees are connected with the essential belowground activities for water and nutrient acquisition through woody bodies. These cover about 27% of the terrestrial surface and contribute 90% of biomass on the planet; range from 60,000 to 100,000 taxa, and 15–25% of the 350,000–450,000 vascular plants. Gatti et al. (2022) estimated about 73,000 tree species globally, among which approximately 9,000 tree species are yet to be discovered. Nearly one-third of trees, more than 17,500 species are threatened with extinction (Irwin, 2022). Thomas Crowther, formerly a researcher at Yale University, estimated that "there are 3.04 trillion trees on earth (bringing around 422 trees per person"; Crowther et al., 2015). For India guesstimate is that there are 35 billion trees which comes to 24–28 trees per Indian. Rajasthan may have still a lower number of trees per citizen!

Late Prof. Swaminathan in 2004–2005, asked a senior author (NSS) at ICRISAT, Hyderabad, during a presentation on bioresources of Pokaran; "why the ever-stressed farmers/people do not commit suicide in Rajasthan? The answer is that humans live in harmony with nature and the organic environment. Farmers practice agriculture allowing much number of species of *Acacia, Anogeissus, Capparis, Maytenus, Moringa, Prosopis cineraria* (Khejari), *Salvadora,, Tecomella undulata, Tamarind, Tamarix* and woody shrubs of Borti, Bui, Khinp, Muragali, Phog and others in and around farms. More trees grow in fields in Marwar as compared to agricultural fields/farms in the rest of India. (Shekhawat et al., 2014). Agroforestry is a land-use practice that integrates multipurpose woody plants/trees on farm land (khet) optimising socioecological and economic advantages as stated by Brandt et al. (2024). Accordingly, 56% of India is covered by farmland, and only 20% is covered by forest. Separation of forests and tree plantations is not always clear, these classifications exclude a large part of India's trees, within/around farm-a key to sustain and survival in these challenged environments. Much worrying recent report on the severe decline (11% decline; eliminating more than 5 million trees) in farmlands trees in India (from 2010/2011 to 2020/2022), is a warning. Studies in clinical nutrition revealed the absence of severe grade protein-calorie malnutrition (PCM) (Gupta et al., 1974) in socio-economically stressed communities. The people consume unconventional food of *Prosospis cineraria*, a tree, Khejari (Fig. 5.1c and d) in the form of unripe (Fig. 5.1c) and ripened dry fruits (Fig. 5.1e and f). *P. cineraria*, the Shami, provider of ecosystem services in the desert, is the subject of global research deeply entrenched with local culture, history and ethnobotanical values (Baibout et al., 2022). Uses of *Prosopis cineraria* are recorded in Ayurvedic literature. Puppo and Felker (2022) edited a book on *Prosopis*, devoting most of the text to praise of *P. juliflora* and, a part to *P. cineraria*! Shami is a promising tree for a warming/hot world.

Traditional concepts of Mool (keeping root stock/stump (seed bank) of perennials unharmed but gently pruned in khet/farm), Seenva/Baad (wide persistent common boundary kept undisturbed for years around and in-between farms-a sacred repertoire of germplasm), Goas/Daras (the Kaccha/unpaved pathway/roads/highways, kilometres long protected by permanent organic fences-cum-boundaries), Vanies (sacred mini-forests), community grazing lands (Talab/Talai/Nadi/Nada/Water bodies/Orans/Grasslands) and Beehar/Beeds (mini-forests/Conservatory). These were and are conservatories of rare/threatened/endemic plant species. Reckless encroachments have fragmented/destroyed habitat supporting and eliminating different life forms. Value of native varieties/landraces like Chhadi/Desi Bajri (pearl millet) were under-estimated and hence vanished. Rain-fed vegetables/cucurbits/wild plants/fruit trees are great losers. The conservatories/the sacred forests (even individual trees) are dedicated to one or more folk heroes/deities venerated by the local communities; the sacred groves are home to such diversity of plants. *A. tortilis*, *Prosospis juliflora*, Jojoba, biodiesel crops, date palm, olive-oil (Jaitun) are invasive species in this region. A good example is the introduction of *P. juliflora* in the Taal Chappar sanctuary during the 1960s which endangered wildlife and invaded traditional water bodies. Ultimately, the Rajasthan government implemented a project to up-root and eliminate Vilayati Babul for the safety of wildlife and nature.

5.8 Foods, Dietary Supplements and Nutraceuticals from Nature

5.8.1 Seed Spices

Rajasthan is a major producer, consumer and exporter of spices. Seed spices are, cumin or "jeera" (*Cuminum cyminum*), coriander or "dhania" (*Coriandrum sativum*), fennel or "saunf" (*Foeniculum vulgare*), fenugreek or "methi", (*Trigonella foenum-graecum*), carom seeds or "ajwain" (*Trachyspermum ammi*), celery (*Apium graveolens*), dill (*Anethum graveolens*), caraway or "shahijeera" (*Carum carvi*) and black caraway or "kalonji" (*Nigella sativa*). These are considered immunity modulators. Indian Council of Agricultural Research has established ICAR—National Research Centre on Seed Spices (ICAR-NRCSS) Tabiji, Ajmer.

5.8.2 *Edible Gums and Resins of Rajasthan*

Resins and gums with traces of essential oils are valuable non-wood Forest Produce (NWFP). Plants of arid regions produce gums. Guggul, a product of *Commiphora wightii* is an important oleoresin. Bhainsa (buffalo eye like-guggul/Gum Salaya/ Salaki is derived from *Boswellia serrata*. Gum Arabic of *Acacia senegal* (Fig. 5.2a and b), is an edible dried gummy exudate utilised in food industry as a stabiliser, emulsifier, flavouring agent, thickener and surface-finishing agent. *Sterculia urens* (Gum Kadaya), when ripe red covered outside with many bristles, is used as a food additive and as an aphrodisiac. The other gum-yielding species are; *Anogeissus pendula* (Dhav/Dhokra), *Anogeissus latifolia* (Safed Dhav, Gum Ghati), *Acacia nilotica* (Babul Gum), *Acacia jacquemonti* (Banwali), *Aegle marmelos* (Bael), *Azadirachta indica*, *Balanites aegyptiaca* (Hingota), *Butea monosperma* (Palaas), *Bauhinia retusa*, *Commiphora agallocha/Commiphora madagascariensis, Lannea coromandelica*, *Moringa concanensis* (JangaliSargua) (Fig. 5.2c and d), *Moringa oleifera* (Sahajan/Sajana) gum (Fig. 5.2e), *Prosopis cineraria* (Janti/Khejari/Mesquite) gum. Gums are also extracted from seeds of certain plants like guar, tamarind, *Cassia tora/Senna tora*, etc. Dhaincha (*Sesbania aculeata/Sesbania bispinosa)* gum and Fenugreek/Methi (*Trigonella foenum-graecum*) gum are seed-derived. Gum(s) are pounded, powdered and fried in butter (Ghee) mixed with separately fried atta (cereal flour) and jaggery/sugar and other ingredients that provide flavour/colouring and finally fried in enough butter and converted into Laddus. These are named after sources of gum/spice used as ingredients like Haldi (turmeric), Saunth (ginger), Methi, Ajwain, Khariti (*Sida* species), Ghritkumari (*Aloe vera*) and Shatavar/ Shatavari (*Asparagus*).

5.8.3 *Roots, Tubers and Bulbs as Food Additives/ Nutraceuticals*

Withania somnifera (Nagauri Ashavagandha/the Indian Ginseng), a wonderful immunomodulatory herb. Roots (Fig. 5.3a) of wild and cultivated Padalsi are consumed in the form of a variety of preparations. *Asparagus racemosus* (Shatavari/ Nahar Kanta) fasciculated roots (Fig. 5.3b and c) are potent herbs for human health particularly for women. Roots/tubers of *Butea monosperma* (Dhàk/Palâas), *Blyttia spiralis/Vincetoxicum spirale* (Patel et al., 2016), *Ceropegia bulbosa*, *Ceropegia tuberosa/Ceropegia candelabrum* var. *biflora* (Khedula/Khapparkddu; Fig. 5.3d), *Chlorophytum borivilianum* (Safed musali), *Curculigo orchioides* (Kali musli), *Costus speciosus/Hellenia speciosa*, *Curcuma amada*, *Cyperus tuberosus/Cyperus esculentus*, *Cyperus rotundus*, *Cyperus triceps* (the Mothia/Motha; popularly the Mothia Kheer of Tal Chhaper Sanctuary, known home of the blackbuck and a variety of birds that feed on tubers of *Cyperus* spp). Bulbs of *Dipcadi erythraeum*

Fig. 5.2 **a**, **b** Gum Arabica from *Acacia senegal* harvested by a local family of Jaisalmer. **c** A tree of *Moringa concanensis* in bloom. **d** Gum of *M. concanensis*. **e** Gum of *M. oleifera*

(Askanda, Junglee Pyazi; Fig. 5.3e) and *Eulophia ochreata* (Salam misri) and products of *Hemidesmus indicus* (Anantmul), *Leptadenia reticulata* (Dodisakha/Jivanti/Shakshresth), *L.pyrotechnica* (Swarna Jiwanti/Khimp/Prasarani), *Maerua oblongifolia* are consumed in different forms for the promotion of health. So are tubers of *Momordica* and other cucurbits and *Pueraria tuberosa* (GhoraBel/Bilavakand) and *Urginea indica* (JungliPiaz, Safedkando). Many of these herbal products are considered immunomodulatory. These are candidates too for bioprospecting.

5.8.4 Leaf and Tender Stems as a Vegetable

Leaves and tender stems of *Abutilon indicum* (Kanghi), *Acalypha indica/Acalypha lanceolata* var. *lanceolata* (Khokali), *Achyranthes aspera* (Lathjira/Undakanta), *Alhagi maurorum, Alysicarpus rugosus, Amaranthus gracilis/A. viridis* (Cholayee), *Amaranthus hybridus* (Lalchulâyee), *Amaranthus spinosus*

Fig. 5.3 **a** Roots of *Withania somnifera* (Indian Ginseng). **b** *Asparagus racemosus* (Shatavari) plant. **c** Fasciculated roots of *A. racemosus*. **d** Tubers of cultivated *Ceropegia* species. **e** Bulbs of *Dipcadi erythraeum* (Junglee Pyazi)

(Jangli-chulayee), *Bauhinia* species, *Boerhavia diffusa/Boerhavia coccinea* (Saata/Punarnuva), *Caralluma edulis* (Painpa, Patel et al., 2014a), *Cassia tora/Senna tora* (Chakundâ), *Cassia occidentalis/Senna occidentalis* (Kasondi), *Cocculus hirsutus* (Jamti-ki-bél), *Cleome viscosa* (Hulhul), *Cardiospermum halicacabum* (Kanphuti), *Commelina benghalensis* (Boknâ/Moria-bati), *Chenopodium album* (Bâthua/Sâg), *Chenopodium murale, Celosia argentea* (Safèd-murgâ), *Corchorus depressus* (Kaga-Bati), *Calligonum polygonoides, Crotalaria burhia* (Kinsara/Sinia), *Cyperus arenarius/Cyperus stoloniferus* (Dachubari/Jucchabari), *Dendrocalamus strictus, Digéra muricata, Euphorbia caducifolia* (Danda Thor), *Leucas cephalotes* (Halkurâ), *Grewia tenax* (Gangârân), *Gisekia pharnaceoides* (Saleri), *Gymnema sylvestre* (Gur-mar), *Indigofera linifolia, Indigofera cordifolia, Launaea pinnatifida/Launaeas armentosa* (Pathri) *Leptadenia reticulata* (Jeewanti/Doodhi, Patel et al., 2014b), *Moringa species, Oxalis corniculata* (Amrul), *Ocimum americanum* (Bavachi, Vala-tulsi), *Pedalium murex* (Bada Gokharau), *Phoenix sylvestris* (Khajur; leaf buds and inner part of the stem), *Portulaca oleracea* (Khatta), *Portulaca quadrifida, Sesbania grandiflora, Talinum portulacifolium, Trianthemamonogynum/Trianthema portulacastrum, Tamarindus indica* (Imli), *Tinospora cordifolia, Vigna trilobata* (Chidi moth) and *Zaleya govindia* (Gadbâni) are edible and consume as food additives. Fleshy leaves of *Aloe vera/A. barbadensis* (Guarpatha/Ghritkumari) leaves, peeled off and cooked as vegetable after washing 3–4 times. Fleshy pulp of the leaves is also mixed with wheat flour and made into Chapattis and Pakora/Chilra during winter as nourishing and antirheumatic diet.

5.8.5 Edible Fruit of Native Plants of Thar

Cultivated/semi-cultivated and wild plant species of Thar Desert are sources of energy and nutrients. At times the graziers/nomads feed and survive entirely on these diets for days together. The eatable species are *Acacia leucophloea* (Safed-kikar), *Aegle marmelos* (Bael/bilwa), *Annona squamosa* (Sitaphal/Custard Apple, Sharifa), *Azadirachta indica* (Nimboli), *Caralluma edulis* (Painpa), *Ceropegia bulbosa, Ceropegia tuberosa/C. candelabrum* var. *biflora* (Khedula), *Citrullus fistulosus* (Tindsiand Phophalia), *Benincasa fistulosa* (Petha) *Citrullus lanatus* (Matira/Laamp), *Coccinia grandis* (Kanduri, Tondli), *Cordia gharaf/Cordia sinensis* (Goondi), *Cordia dichotoma/Varronia dichotoma* (Barigoondi), *Cucumis melo* var. *callosus* (Kacchari), *Cucumis melo* var. *momordica* (Phootkakri), *Cucumis melo* var. *utilissimum* (Kakadi) (Fig. 5.4a and b), *Cucumis melo* subsp. *melo* (Kharbuja), *Cucumis prophetarum* (Khat kachhario), *Momordica dioica* (Fig. 5.4c), *Cyamopsis tetragonoloba* (Guar), *Diospyros melanoxylon, Dregea volubilis/Wattakaka volubilis, Ephedra foliata/E.ciliata* (Oont Phog; Fig. 5.5a) are edible. Roasted seeds of *Euryale ferox* (Talmakhana/Makhana)/Gorgon nut or fox nut are prized healthy diets. The Government of India has taken various initiatives to boost makhana production. Makhana is low in calorific value and rich in fibre and carbohydrates (Kumar & Kumar, 2022).

Fig. 5.4 **a** Fruits of *Cucumis* species (Kachara, Matira and Kakdi). **b** *Cucumis melo* var. *callosus* (Kacchari), fruits and skinned fruits ready for making curry. **c** Fruits of *Momordica dioica* (Kakoda), a valuable nutritious wild cucurbit

Ripe figs/fruits of *Ficus glomerata/Ficus racemosa* (Gular), *F. religiosa/ F. populifolia* (Pipal) and *F. benghalensis* (Bargad) are edible. Fruits/pods of *Feronia limonia/Limoniaacidissima* (Wood Apple), *Flacourtia indica* (Ghar-googar/kankrol), *Flueggea leucopyrus/Securinega leucopyrus;* Fig. 5.5f), *Glossonema varians/Cynanchum registanense* (Khiroli/Dudia; Fig. 5.5b), *Grewia tenax* (Gangeran), *Leptadenia pyrotechnica/L. spartum* (Khimpoli; Fig. 5.5c), *Leptadenia reticulata* (Jeewanti/Duddi; Fig. 5.5d), *Lycium barbarum*, *Maerua oblongifolia* (Fig. 5.5e), *Manilkara hexandra* (Khirni), *Mangifera indica* (wild mango), *Moringa oleifera* (Sahajana, Drumstick), *Opuntia elatior* (Thor/Nagphani), *Parkinsonia aculeata, Blyttia spiralis/Vincetoxicum spirale* (Aakari Bel, Patel et al., 2016), *Phyllanthus emblica* (Amla/amalika), *Physalis minima* (Patari/Chirpotia), *Rhus mysorensis* (Dansaria), *Salvadora oleoides* (Meethi Jaal, Piloo), *Salvadora persica* (Khari Jaal), *Sesbania grandiflora, Syzygiumcumini* (Jamun/Jambuphala) are edible health foods. Likewise, fruits of *Phoenix sylvestris* (Taar, Khajoor), Fruits of *Trapanatans* var. *bispinosa* (Singhara/singoda) or dry seed/seed powder are healthy diet and widely consumed. Pods of *Wattakaka volubilis have good food values. Berries of Ziziphus nummularia* (Jhaad Ber), *Z. xylopyrus* (Ber), *Z. mauritiana*, (Ghat-ber) are extensively utilised. De-pulped fruits of *Ziziphus nummularia* with

Fig. 5.5 **a** Edible fruits of Oont phog (*Ephedra foliata/E. ciliata*). **b** Edible fruits of Khiroli (*Glossonema varians*) from Phalodi. **c** Tender follicles of Khimp (*Leptadenia pyrotechnica*), the edible Khimpoli. **d** Edible follicles of Jeewanti (*Leptadenia reticulata*), the Doodhi. **e** Sweet fruits of *Maeruaob longifolia*. **f** Edible fruits of *Flueggea leucopyrus*

stony endocarp (seeds) are pounded, powdered and eaten, even stored as a preparation called KHODI, a remedy for stomach disorders.

5.8.6 Unripe/Tender Fruits/Pods Processed and Pickled

Unripe Fruits of *Capparis decidua* (Ker), *Cordia myxa*, *Cordia dichotoma*, *C. sinensis* (Lasora/Gunda), *Carissa macrocarpa*, *Carissa carandas* (Karonda), *Grewia subinaequalis* (Phalsa), *Mangifera indica* (Mango/Kairi) and *Prosopis cineraria* (Khejari) processed, pickled and regularly eaten. Unripe dried fruits of *Capparis* are stored and marketed as food/medicine/nutraceuticals (Ram et al., 2022).

5.8.7 Wild Plants—That Are Source of Edible Seeds

Acacia senegal (Kumbhat/Bhatkaniya), *Aerva javanica* (Safed-bui), *Blepharis scindica* (Bhangari) seeds (Fig. 5.6a and b) are boiled in milk and used as a tonic. Seeds of *Citrullus lanatus* (Matira/Laamp), *C. colocynthis* (Indrâyan, tumba), *Abutilon indicum*, (Kanghi), *Sida* spp. (Khareti), *Tamarindus indica* (Imli), *Mimosa hamata* (Alaay), *Sarcostemmaacidum/Cynanchum viminale* subsp. *viminale*, *Mollugo* spp. are edible. Seed flour of *Brachiaria ramosa/Urochloa ramosa*, *Cenchrus biflorus* (Bhurat), *C. setigerus* (Kala Dhaman), *Eleusine compressa*, *Lasiurus scindicus*, *Panicum antidotale* (Gramna), *Panicum turgidum*, *Urochloa panicoides* (Kurichhinki) are used for making (Chapattis/Roti/Dhokla) during scarcity period. Seed oil of *Celastrus paniculatus* (Black oilplant/Jyotismati/Malkanguni), the "elixir of intellect" or the "elixir of life" is used in traditional Ayurvedic/Herbal medicines.

5.8.8 Pounded and Powdered Bark/Split Barks (Chhoda and Chhaal of Trees)

As reported by King (1869) bread is made from the pounded barks of *Acacia nilotica*, *A. senegal*, *A. leucophloea* (Safed-kikar) and split bark of *Prosopis cineraria* (Khejari Chhoda) with or without the addition of grain flour, which has an astringent, bitter taste and causes a good deal of gripping and discomfort in consequence but ultimately stomach gets accustomed. The tales/stories about such uses of tree barks/Chhodas powdered and mixed with flours of seeds of wild grasses are known since the 1860s great famine.

5.8.9 Unopened Flower Buds/Inflorescence

Flower buds and flowers of *Tamarindus indica* (Imli) are acrid in taste and are taken locally. Flower buds of *Thevetia neriifolia/Cascabela thevetia* (Kaner) are sucked

Fig. 5.6 **a**, **b** Fruits and seeds of *Blephariss indica* (Bhangari), an herbal medicine. **c** *Calligonum polygonoidies* plants raised through stem cuttings (after 11 months). **d** Immature dried flower buds (Phogla) and fruits (Ghantiyal)

for the nectar. *Leptadenia* species, *Moringa oleifera* and *Moringa concanensis* are edible. Flower buds of Phog (*C. polygonoides* plants raised from a stem cutting, after 11 months; Fig. 5.6c) are harvested, sun-dried (Phogla/Badia) and consumed as Raita, a native recipe of anti-oxidants/nutraceuticals against summer heat.

5.8.10 *Sajji-Khaar*

The barilla is the soda ash obtained by burning air-dried foliage of halophytic chenopod shrubs, *Haloxylon salicornicum*, *Haloxylon recurvum/Soda stocksii*, *Salsola baryosma/Caroxylon imbricatum* var. *imbricatum* and *Suaeda fruticosa*. Khaar, *Haloxylon recurvum/Soda stocksii* is the main and preferred species for making quality Sajji.

5.8.11 Miscellaneous

Rajasthan is a potential producer of carrots, chilies, *Plantago ovata* (Psyllium/Isabgol) and pomegranate (*Punica granatum* cv. Jalore seedless; Dinesh et al., 2019). Date palm (*Phoenix dactylifera*), Mint (*Mentha* spp.) of Mathania, Wild Tulsi (*Ocimum* spp.), Marua (*Origanum vulgare*) and *Psoralea corylifolia/Cullen corylifolium* cultivation is promoted. During the rainy season, protein-rich nutritious Bhunga (Mushrooms) belonging to *Phellorinia, Podaxis* spp grow on sandy/organic soils. These are harvested, eaten in the form of curry-sabji and marketed.

5.9 End Notes/Cautions

Drought-/heat-induced drying, die-back (Oobh-Sookh) and mortality of *Acacia* spp., *P. cineraria, C. decidua, A. indica*) and *T. undulata* are threatening. Tree mortality increases with rising climate-induced stresses and normal functioning is strained.

The Thar lands offer theatre for the growth, evolution and genetic enrichment of a diversity of animals and microbes (nitrogen-fixing bacteria, plant growth promoting root-associated Mycorrhizal fungi). *Piriformospora indica* the novel endophytic root-colonising fungus of plants of the Thar was isolated by Verma et al. (1998). The organism becomes a model for basic and applied research published in reputed journals like Nature and PNAS, an example of wealth. *Arnebia hispidissima, Asparagus recemosus, Boerhavia diffusa, Caralluma edulis, Crotalaria burhia* (Kinsara/Sinia), *C. polygonoides* (Phog), *Cenchrus* spp, *Dipcadi erythraeum, Fagonia indica, L. pyrotechnica, L. reticulata, Lyciumbarbarum* and *Pedalium murex* may have potential metabolites/genes for prospecting.

Phog (*Calligonum polygonoides*) and the number of threatened plants can be propagated by traditional techniques of stem cutting. Applications of aeroponics/hydroponics does help induce rootings in short periods in most woody plants and even herbals. At the field gene bank of ICAR-Central Institute for Arid Horticulture, Bikaner, buds of *P. cineraria* trees that yielded sweet pods were, grafted by in situ budding on wild rootstocks. A rapid-growing, thornless genotype was developed and released as the cultivar “Thar Shobha” which has longer racemes, a higher number of pods per raceme, longer tender pods with higher value pulp and more seeds per pod (Sivalingam et al., 2016).

5.10 Conclusion (Recommendations)

Thar ecosystem and its natural resources are vulnerable. Conservation and promotion of life-sustaining (Jeev/Jivaari) systems must be given priority.

1. Data collection, documentation and curation of plant-, microbe- and animal diversity and traditional knowledge of nature are necessary for all ecological zones. Endemic, Rare and threatened plants may be conserved in situ. Rich traditions of watersheds/traditional water harvesting systems must be part of heritage data. Besides increasing agricultural productivity, healthy watersheds have the potential to minimise land degradation, mitigating the adverse impact of climate change and generating several other ecosystem services (Meena et al., 2022). Records of Vanies/talabs/orans/thans/beheads and other forms of bioresources/biodiversity repositories like organic Darras/Goas can be collected from revenues records/ and curators of bodies of ancient literature.
2. Environmental and biodiversity impact assessment of land-use-based projects for social welfare and developments must be prerequisite.
3. Nurseries of native plants in the premises of schools/colleges/institutions be created; Botanical/herbal gardens/conservatories/Seed banks be established; Plant-blindness be eradicated.
4. The project on the Desert Biosphere Reserve has been pending since 1988. Accelerated efforts must be made to convince all parties for the establishment of Desert Biosphere Reserve.
5. The United Nations General Assembly (UNGA) at its 75th session in 2021 declared 2023 as the International Year of Millets (IYM 2023). Rajasthan is a major producer of pearl millet and related grasses. A National Research Institute should be established in this ecozone on grasses/legumes/Khejri. Pastoral lands allow rearing animals. Milk & products of camel, goat and jenny/jennet are prized health products. Vegetarianism is the demand of time as it is a climate and water-friendly human practice. There is a need for the integration of food safety regulations, certification, packaging and labelling.

References

Baibout, M., Corcket, E., Kothari, S. L., & Fievet, V. (2022). Ecosystem services provided by *Prosopis cineraria* (L.) Druce in the drylands of Southern and Western Asia. *Botany Letters, 169*, 30–42.

Bhandari, M. M. (1974). Famine foods in the Rajasthan desert. *Economic Botany, 28*, 73–81.

Bhandari, M. M. (1978). *Flora of the Indian desert*. Scientific Publishers.

Bhandari, M. M. (1990). Flora of the Indian desert. MPS Repros.

Blatter, E. A., & F. Hallberg. (1918, 1921). The flora of the Indian Desert (Jodhpur and Jaisalmer). *Journal of the Bombay Natural History Society, 26*, 418-&, 525–551, 811–818, 962–967; W7; 270–279 & 506–519.

Blinkhorn, J. (2021). The gateway to the oriental zone: Environmental change and palaeolithic behaviour in the Thar Desert. *Quaternary International, 596*, 79–92.

Blinkhorn, J., & Petraglia, M. D. (2017). Environments and cultural change in the Indian Subcontinent: Implications for the dispersal of *Homo sapiens* in the Late Pleistocene. *Current Anthropology, 58*, S463–S479.

Blinkhorn, J., Achyuthan, H., Jaiswal, M., & Singh, A. K. (2020). The first dated evidence for Middle-Late Pleistocene fluvial activity in the central Thar Desert. *Quaternary Science Reviews, 250*, Article 106656.

Blinkhorn, J., Achyuthan, H., Durcan, J., Roberts, P., & Ilgner, J. (2021). Constraining the chronology and ecology of Late Acheulean and Middle Palaeolithic occupations at the margins of the monsoon. *Scientific Reports, 11*, 19665.

Brandt, M., Gominski, D., Reiner, F., et al. (2024). Severe decline in large farmland trees in India over the past decade. *Nature Sustainability*. https://doi.org/10.1038/s41893-024-01356-0

Cazzolla Gatti, R., Reich, P. B., Gamarra, J. G., Crowther, T., Hui, C., Morera, A., Bastin, J. F., De-Miguel, S., Nabuurs, G. J., Svenning, J. C., Serra-Diaz, J. M., et al. (2022). The number of tree species on Earth. *Proceedings of the National Academy of Sciences, 119*, Article e2115329119.

Choudhary, S. K., Patel, A. K., Ram, K., & Shekhawat, N. S. (2022). Thar desert bioresources: Significance, conservation and sustainable management in anthropocene. *International Journal of Ecology and Environmental Sciences, 48*, 267–280.

Crowther, T. W., Glick, H. B., Covey, K. R., Bettigole, C., Maynard, D. S., Thomas, S. M., Smith, J. R., Hintler, G., Duguid, M. C., Amatulli, G., & Tuanmu, M. N. (2015). Mapping tree density at a global scale. *Nature, 525*, 201–205.

Dhir, R. P., Joshi, D. C., & Kathju, S. (2018). *Thar desert in retrospect and prospect*. Scientific Publishers.

Dinesh, R. M., Patel, A. K., Vibha, J. B., Shekhawat, S., & Shekhawat, N. S. (2019). Cloning of mature pomegranate (*Punica granatum*) cv. Jalore seedless via in vitro shoot production and ex vitro rooting. *Vegetos, 32*, 181–189.

Gupta, R., & Kanodia, K. (1968). Plants used during scarcity and famine periods in the dry regions of India. *Journal D'agriculture Traditionnelle Et De Botanique Appliquée, 15*, 265–285.

Gupta, M. C., Gandhi, B. M., & Tandon, B. N. (1974). An unconventional legume-*Prosopis cineraria. The American Journal of Clinical Nutrition, 27*, 1035–1036.

Irwin, A. (2022). The loneliest trees: Can science save these threatened species from extinction? *Nature, 609*, 24–27.

Jain, S. K. (1970). Floral composition of Rajasthan—A review. *Nelumbo, 12*, 176–187.

Kanodia, K. C., & Gupta, R. (1968). Some useful and interesting supplementary food plants of the arid regions. *Journal D'agriculture Traditionnelle Et De Botanique Appliquée, 15*, 71–74.

King, G. (1869). Famine foods of Marwar. *Proceeding Asian Society Bengal, 38*, 116–122.

King, G. (1870). IV. Notes of vegetable products used as food during the late famine in Rajpootana. *Transactions of the Botanical Society of Edinburgh, 10*, 198.

Kumar, S., & Kumar, T. (2022). Indian intervention to boost makhana in the domestic market and overseas. *Current Science, 123*, 1294–1295.

Liu, Z., Chen, Z., Yu, G., Zhang, T., & Yang, M. (2021). A bibliometric analysis of carbon exchange in global drylands. *Journal of Arid Land, 13*, 1089–1102.

Meena, D. C., Pal, S., & Chand, P. (2022). Assessment of watershed management ecosystem services in India: A meta-analysis. *Current Science, 123*, 1352–1358.

Minnis, P. E. (2021). *Famine foods: Plants we eat to survive*. University of Arizona Press.

Negacz, K., Malek, Ž, de Vos, A., & Vellinga, P. (2022). Saline soils worldwide: Identifying the most promising areas for saline agriculture. *Journal of Arid Environments, 203*, Article 104775.

Pandey, R. P., Meena, S. L., Padhye, P. M., & Singhadiya, M. K. (2012). A review of depleting plant resources, their present status and conservation in Rajasthan, India. *Biological Forum, 4*, 213–230.

Patel, A. K., Agarwal, T., Phulwaria, M., Kataria, V., & Shekhawat, N. S. (2014a). An efficient in vitro plant regeneration system from leaf of mature plant of *Leptadenia reticulata* (Jeewanti): A life giving endangered woody climber. *Industrial Crops and Products, 52*, 499–505.

Patel, A. K., Phulwaria, M., Rai, M. K., Gupta, A. K., Shekhawat, S., & Shekhawat, N. S. (2014b). In vitro propagation and ex vitro rooting of *Caralluma edulis* (Edgew.) Benth. & Hook. f.: An endemic and endangered edible plant species of the Thar Desert. *Scientia Horticulturae, 165*, 175–180.

Patel, A. K., Lodha, D., Ram, K., Shekhawat, S., & Shekhawat, N. S. (2016). Evaluation of physiochemical factors affecting high-frequency plant regeneration of *Blyttia spiralis* (synonym: *Pentatropis spiralis*), a threatened climber of medicinal value. *In Vitro Cellular & Developmental Biology-Plant, 52*, 10–19.

Puppo, M. C., & Felker, P. (Eds.) (2022). Prosopis as a heat-tolerant nitrogen fixing desert food legume: Prospects for economic development in arid lands. Academic Press. Academic Press.

Ram, K., Patel, A. K., Choudhary, S. K., & Shekhawat, N. S. (2022). Synergetic effects of TDZ with various phytohormones on high-frequency plant regeneration from mature nodal explants of *Capparis decidua* and their ex vivo implications. *Plant Cell, Tissue and Organ Culture, 149*, 621–633.

Rotenberg, E., & Yakir, D. (2010). Contribution of semi-arid forests to the climate system. *Science, 327*, 451–454.

Shekhawat, N. S., Rai, M. K., Phulwaria, M., Rathore, J. S., Gupta, A. K., Purohit, M., Patel, A. K., Kataria, V., & Shekhawat, S. (2014). Tree biotechnology with special reference to species of fragile ecosystems and arid environments. In K. G. Ramawat, J. M. Mérillon, & M. R. Ahuja (Eds.), *Tree Biotechnology* (pp. 187–222). CRC Press.

Shekhawat, N. S., Phulwaria, M., Harish, Rai, M. K., Kataria, V., Shekhawat, S., Gupta, A. K., Rathore, N. S., Vyas, M., Rathore, N., Vibha, J. B., Choudhary, S. K., Patel, A. K., Lodha, D., & Modi, R. (2012). Bioresearches of fragile ecosystem/desert. *Proceedings of the National Academy of Sciences, India Section B: Biological Sciences, 82*, 319–334.

Shetty, B. V., & Singh, V. (1987, 1991, 1993). Flora of Rajasthan (Vol. I, II & III). Botanical Survey of India.

Singh, V., & Pandey, R. P. (1998). *Ethnobotany of Rajasthan*. Scientific Publishers, Jodhpur, India.

Sivalingam, P. N., Samadia, D. K., Singh, D., & Chauhan, S. (2016). Molecular markers to distinguish 'Thar Shoba', a variety of khejri [*Prosopis cineraria* (L.) Druce], from trees in natural populations. *The Journal of Horticultural Science and Biotechnology, 91*, 353–361.

Verma, S., Varma, A., Rexer, K. H., Hassel, A., Kost, G., Sarbhoy, A., Bisen, P., Bütehorn, B., & Franken, P. (1998). *Piriformospora indica*, gen. et sp. nov., a new root-colonizing fungus. *Mycologia, 90*, 896–903.

Chapter 6
Status of Temporal Changes in Agriculture in Western Rajasthan

S. S. Burark, M. S. Rathore, and G. L. Meena

6.1 Introduction

Desert ecosystems are part of the larger earth's ecosystem. Deserts act as carbon sink, i.e., bacteria in the sand facilitate storing carbon dioxide and restrict it from entering into the atmosphere. Minerals, natural gas and oil deposits are also the endowments of desert regions. They have a niche to produce special medicinal and aromatic plants. The ecosystems of Indian desert regions have been severely disturbed because of many natural and anthropogenic reasons and there is an immediate necessity to take reversal measures to save them from devastation.

Climate change erodes and damages soils through heat and droughts, sand and dust storms and heavy rainfalls that ultimately degrade land in desert regions. Consequently, it intensifies biodiversity loss, which in turn erodes soils further and favours desertification. The problems of desert regions are further aggravated by human activities leading to desertification. It is mainly because of the factors such as; deforestation, over exploitation and contamination of water and other natural resources, unsustainable farm practices (overgrazing, the excessive use of chemicals and mechanisation), introduction of industry and mining, urbanisation and increase in artificial surfaces, etc. More specifically, changes in land use because of urbanisation, industrialisation and the use of fertile land for development needs are the reasons responsible for the imbalance created in the ecosystem. Further, with the fast increase in human and livestock population resulting in higher water demand for human, livestock, urban and other development needs depleted ground water to a drastic level.

S. S. Burark
MPUAT, Udaipur, India

M. S. Rathore (✉)
Centre for Environment and Development Studies, Jaipur, India
e-mail: msr@cedsj.org

G. L. Meena
Department of Agricultural Economics, MPUAT, Udaipur, India

M. S. Rathore et al. (eds.), *Desert Ecosystems*,
https://doi.org/10.1007/978-3-032-03237-9_6

Historically, the traditional arid farming systems provided sustenance to both land owners and land less dependent on arid farming, common property resources and traditional enterprises. However, recently the focus has shifted from sustenance to remunerative farming. It is mainly the increased population pressure that is affecting food, fodder and livelihood security. In order to deal with the negative impact of these natural and human interventions there is a need for evolving sustainable management of natural resources and farming practices facilitating adaptation to such changes. The plant and animal genomes of the arid lands and their synergy have tremendous potential which need to be identified and harnessed for further economic and environmental benefits.

Western Rajasthan is primarily a desert that gets 299.2 mm average annual rainfall but in recent years western Rajasthan has witnessed 84% more rain than eastern parts. Multiple factors may be responsible for the occurring changes in weather patterns during monsoon. However, the changes in land use patterns and land cover, deforestation, urbanisation, global warming, rise in normal temperature and pollution are some of the relevant factors affecting the weather pattern in desert ecosystems. Thar desert ecosystem is changing mainly because of changes in the increasing density of the human population, land use patterns caused by the invention of the Indira Gandhi Irrigation Canal and Narmada Irrigation Canal, road network, industries, mining activity, green energy projects, etc. Interestingly, there is a lack of research on the changing pattern of the southwest monsoon and its impact on the desert ecosystem and consequently on Rajasthan's economy. Most of the studies available on the subject relate to the physical features, geomorphology of landforms, rainfed agriculture, livestock, climatic variability and change, etc., and are mostly independent in nature. Though local traditional knowledge somehow supports the Thar desert communities to adopt to the vulnerable situations and are already taking a number of initiatives. The wind of modern economic development is forcing them to ignore age old robust traditional wisdom and land into the trap of modern-technology-based solutions. The need is to have studies with a holistic approach or integrated approach to guide policy makers in the sustainable use of desert resources which can lead to sustainable development of the region.

People knowingly or unknowingly have been exploiting and destroying Rajasthan's desert ecosystem causing the loss of natural habitats of flora and fauna and carbon sinks, resulting in intensification of climatic crises. Communities that depend on natural resources for their livelihoods are the most affected by the climate crisis. Agriculture being the main occupation in the western region of Rajasthan suffers mostly from changing climatic conditions, despite the people having adopted a robust farming system to absorb the shocks of climate variability. The changing paradigm of economic development is also responsible for creating unsustainability in the desert ecosystem. Land resources and the land use pattern play an important role in shaping desert ecosystems.

Desert ecosystems are generally found in arid, inhospitable and ecologically fragile areas. High day temperatures, strong winds during summers, low and uncertain rainfall, large unculturable areas, moving sediments, grazing lands and limited vegetation are the prominent features of the western desert region. Rainfed cropping and livestock rearing form the desert farming system. Harsh environmental conditions make the germination of seeds and survival of trees, shrubs, and grasslands difficult. Land use planning is central to addressing sustainability issues in dry lands, including biodiversity conservation, climate change, food security, poverty alleviation and sustainable energy. The emerging serious environmental and social problems associated with the degradation of natural resources in dry lands deserve increasing attention. They provide ecological services and sustain the livelihoods of people dependent on the natural resource base.

Indian Council of Agricultural Research (ICAR), based on the climatic, edaphic and terrain characteristics, has demarcated the Thar Desert as Western Dry region zone XIV, which is further subdivided into three sub-zones: ACZ-IA-Arid Western, ACZ-IB-Irrigated North Western Plain and ACZ-IC-Hyper Arid Partial Irrigated Zone. In this region, the soils are desert soils, red desert soils and sand dunes aeolian soils, coarse to loamy course in texture. Soils are an integral part of any ecosystem which nurtures and supports its biological components. This study includes the semi-arid and arid districts; namely Jodhpur, Barmer, Bikaner, Nagaur, Churu, Jaisalmer, Pali, Jalore, Sikar, Jhunjhunu, Hanumangarh and Sri Ganganagar.

Frequent droughts and high variability in terms of quantity and pattern of rainfall are major features of the Thar Desert. Rainfed farming has been practiced in the western region. Only one-fifth of the area (21.3%) is irrigated. Despite surface water availability through the Indra Gandhi Canal Irrigation system irrigated crop cultivation is primarily dependent on ground water resources. Groundwater is saline and the quality varies from highly brackish to potable. Pearl millet, moth bean, sesame and cluster bean in the rainy season and wheat, rapeseed and mustard, gram and cumin in the winter season are the major crops of the area wherever irrigation water is available. Livestock rearing is practiced as a complimentary and supplementary enterprise because of the climatic risk associated with farming. The natural resource base in this region is fragile. Almost 60% of the geographical area is categorised as degraded and/or wasteland. Given the poor natural resource base, it is imperative to pay attention to the sustainability, efficiency and health aspects of the ecosystem of the western region of Rajasthan. Thus, the present study was undertaken with the following specific objectives.

6.1.1 *Objectives of the Study*

The broad objective of the study is to document the Agro-ecological changes over the last 50 years in the desert region of Rajasthan and the emerging challenges for the future economic development of the region. The specific objectives are as follows:

i. To analyse the changes in land and water resources in the western desert region of Rajasthan.
ii. Study the changes in agricultural practices in the region.

iii. Analyse the changes in the composition of livestock in the region.
iv. To assess the impact of sectoral changes on the desert environment.

Demographics of Arid Districts of Western Rajasthan

Human population and their behaviour, livestock population and their composition, economic development policies and their implementation, etc. are the factors that facilitate planning for sustainable development, usage and management of natural resources in a region. Demography plays a significant role in the management and use of natural resources for the economic development of a region. The human and livestock populations and their density and growth rates in the desert districts of western Rajasthan are reported in Table 6.1. It shows that the 12 arid districts of Rajasthan comprising of Barmer, Bikaner, Sri Ganganagar, Hanumangarh, Jaisalmer, Jodhpur, Jalore, Pali, Nagaur, Sikar, Churu and Jhunjhunu account for a total population of around 27 million (Census of India 2011). The highest population was in Jodhpur district followed by Nagaur district, whereas the highest population density was in Jhunjhunu district (361) followed by Sikar district (346) and the lowest in Jaisalmer (17). The density of human and livestock population are much higher in the Thar desert region than the deserts in other parts of the world.

Table 6.1 Demographics of Western Rajasthan (Census of India 2011)

District	Area[1] (000, Km^2)	Population[2] (000')	Literacy[2] (in %)	Sex ratio[2]	Livestock density (per Km^2)	Human density (per Km^2)	Human population growth rate (2001–2011) (%)
Barmer	27.76	2603.8	56.53	902	204	92	2.86
Bikaner	22.61	2363.9	65.13	903	123	78	2.20
Churu	16.81	2039.5	67.46	938	110	148	1.86
Shri Ganganagar	10.98	1969.2	69.64	887	144	179	0.96
Hanumangarh	9.66	1774.7	67.13	906	138	184	1.57
Jaisalmer	43.58	669.9	58.04	849	73	17	2.80
Jalore	12.21	1828.7	71.08	921	134	172	2.36
Jodhpur	17.66	3687.2	79.38	916	203	161	2.48
Jhunjhunu	5.88	2137.0	86.9	950	218	361	1.11
Nagaur	11.13	3307.7	70.64	951	283	187	1.77
Pali	12.39	2037.6	68.2	934	186	164	1.13
Sikar	7.77	2677.3	72.98	947	273	346	1.59
Western Region	198.42	27,096.6	–	–	154	114	1.87
Rajasthan	342.24	68,548.4	66.11	928	169	200	1.95

Source Locust Control & Research (2016)[1] Census of India (2011)[2]
Note *Census 2001 figures for Bikaner & Churu districts have changed as tehsil Dungargarh has been transferred from Churu district to Bikaner

Table 6.2 District wise human population growth rate in western region of Rajasthan

District	Growth rate 1901–1961%	Growth rate 1961–2011%
Barmer	1.25	2.81
Bikaner	1.47	3.11
Churu	1.59	2.50
Shri Ganganagar	3.41	2.39
Hanumangarh	3.44	2.86
Jaisalmer	1.07	3.16
Jalore	1.20	2.44
Jhunjhunu	1.27	2.20
Jodhpur	1.26	2.90
Nagaur	0.99	2.56
Pali	1.22	1.87
Sikar	0.96	2.39
Western Region	1.38	2.56
Rajasthan	1.15	2.48

Source Census of India

Table 6.2 shows district-wise human population growth rate for the periods pre-independence (1901–1961) and post-independence (1961–2011). The population growth rate of the western region was higher than that of the Rajasthan state during both periods.[1] It is interesting to observe that population growth was higher in the five districts namely, Barmer, Bikaner, Jaisalmer, Jalore and Jodhpur. The higher growth rate in Ganganagar district and Hanumangarh district is mainly because of the new settlements after the construction of the Gang Canal and Indira Gandhi Nahar Project. However, the population in the desert region is growing much faster than the state average despite the increase in literacy rate, health facilities and transport system. The increasing population pressure will have a detrimental impact on the fragile desert ecosystem. It is only in the decade 2001–2011 the growth rate of the western region was less than the state (Table 6.1). The challenge is to assess the carrying capacity of the region and check the fast-growing human population.

[1] The decadal growth in human population since 1901 is given in Appendix Table 1.

6.2 Natural Resources

6.2.1 Water Resources

Rainfall is the main source of surface and ground water used for irrigation purposes. High variability in the occurrence of rainfall during the monsoon season directly affects the availability of surface and ground water and consequently agricultural production system. This further adds to uncertainty in cropping patterns and agricultural production. The decadal rainfall pattern in the western region is depicted in Table 6.3. The normal rainfall of the state is 554.6 mm and the state's actual rainfall varies from 727.3 mm in the year 1990 to 416.4 mm in the year 2000. Similarly, the average rainfall of the western region varies from 481.4 mm (1990) to 200.8 mm (1980). District wise and western region rainfall pattern figures high inter-annual variability. Maximum variability in rainfall was observed in Jalore district i.e., from 189.2 mm in the year 1980 to 1039.4 mm in the year 1990. This high variability is because of the occurrence of frequent droughts in the State leading to climatic uncertainty. All the western districts received less precipitation compared to state precipitation which directly affects the production and productivity of agricultural crops and livestock in this region.

The district-wise rainfall pattern of the Western Region of Rajasthan vis-a-vis Rajasthan is depicted in Figs. 6.1, 6.2, 6.3, 6.4, 6.5, 6.6, 6.7, 6.8, 6.9, 6.10, 6.11, 6.12, 6.13 and 6.14 showing wide variations in different years. Minimum variation in rainfall was observed in Nagaur district and maximum in Bikaner district.

Table 6.3 Rainfall pattern in western region of Rajasthan (in mm)

District/year	1970	1980	1990	2000	2010	2020
Barmer	162.9	88.0	731.6	402.5	504.8	302.0
Bikaner	280.9	171.8	257.0	402.4	400.1	220.1
Churu	536.8	394.6	324.0	450.4	598.5	395.7
Sri Ganganagar	237	208.2	270.0	247.3	319.7	140.6
Hanumangarh	NR	NR	NR	437.4	413.1	221.9
Jaisalmer	254.5	146	87.1	383.4	361.4	309.0
Jalore	476.1	189.2	1039.4	312.8	658.5	558.1
Jodhpur	643.0	263.0	789.5	384.0	415.8	368.6
Jhunjhunu	539.1	341.0	363.0	522.0	705.1	343.6
Nagaur	502.1	202.3	432.6	425.7	509.2	434.1
Pali	593.1	267.4	1014.0	564.9	551.6	521.9
Sikar	710.7	562.5	373.0	674.7	751.6	434.1
Western Region	380	200.8	481.4	391.1	451.4	347.2
Rajasthan	617.5	423.9	727.3	416.4	606.3	520.8
Average Rainfall Rajasthan	554.6	554.6	554.6	554.6	554.6	521.0

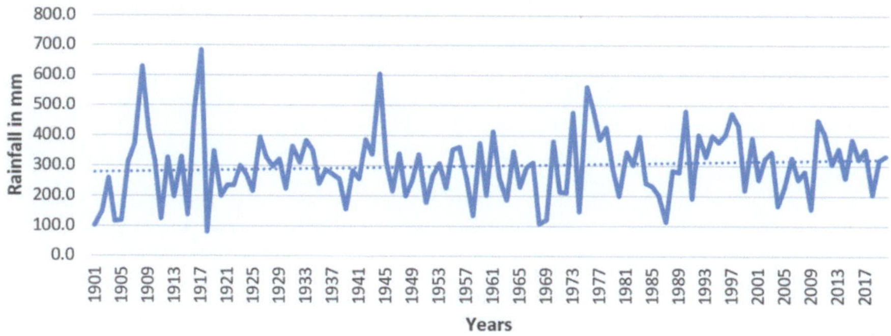

Fig. 6.1 Rainfall pattern of western region of Rajasthan

Fig. 6.2 Rainfall pattern of District Barmer

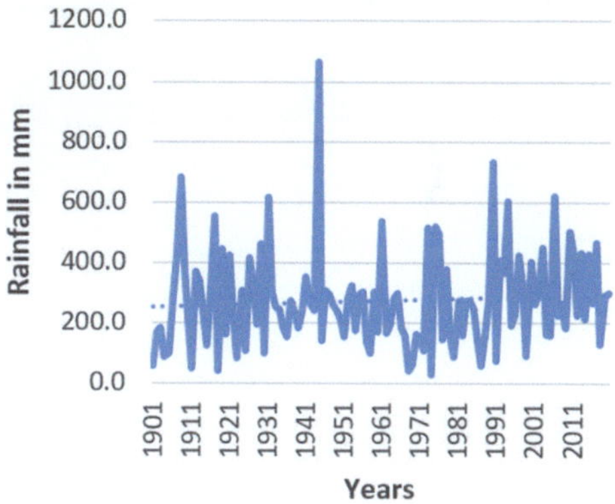

Fig. 6.3 Rainfall pattern of District Bikaner

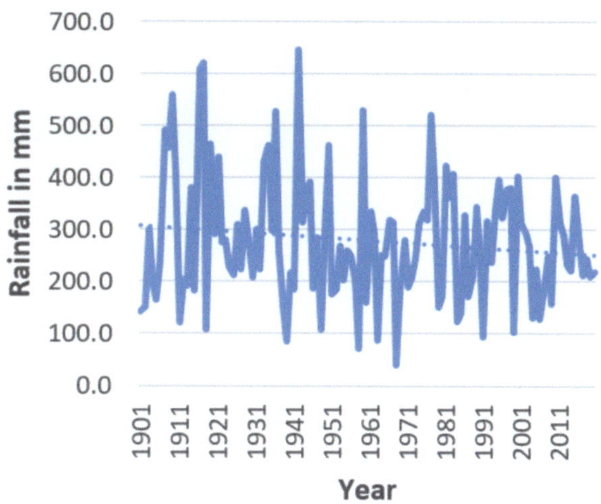

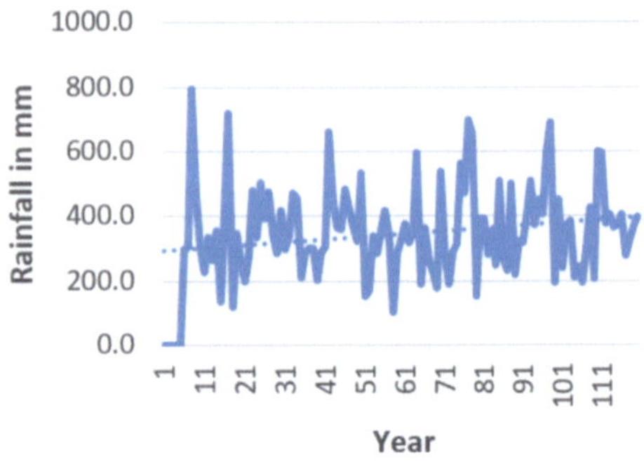

Fig. 6.4 Rainfall pattern of District Churu

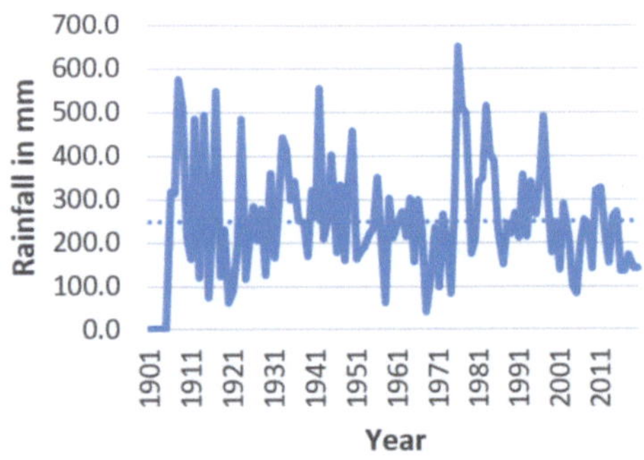

Fig. 6.5 Rainfall pattern of District Shri Ganganagar

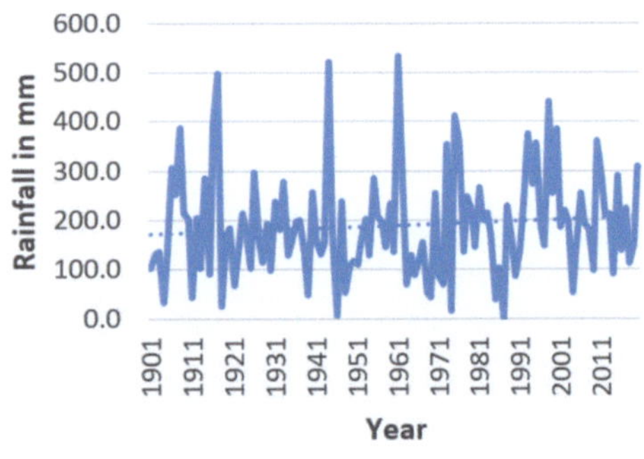

Fig. 6.6 Rainfall pattern of District Jaisalmer

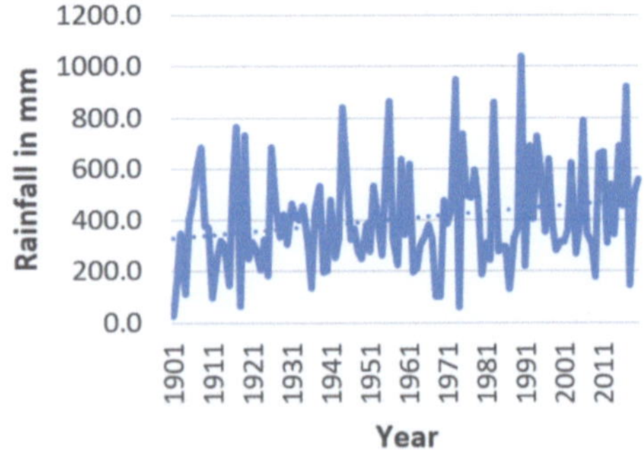

Fig. 6.7 Rainfall pattern of District Jalore

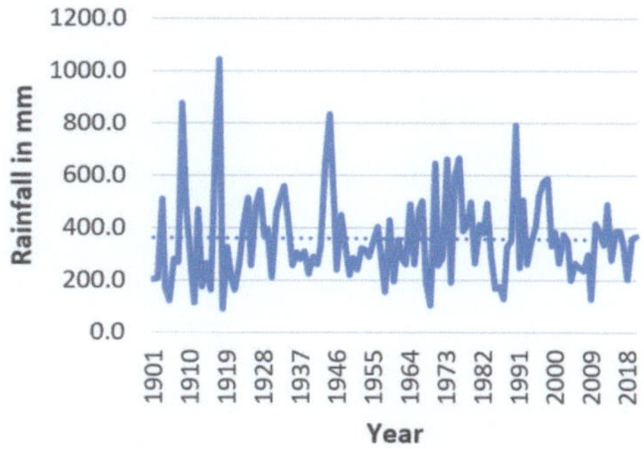

Fig. 6.8 Rainfall pattern of District Jodhpur

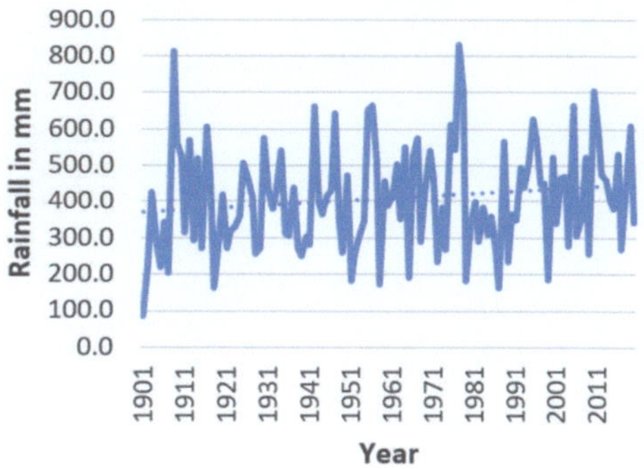

Fig. 6.9 Rainfall pattern in District Jhunjhunu

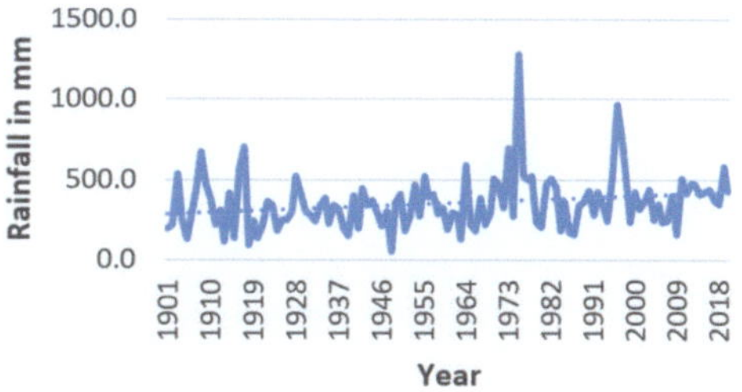

Fig. 6.10 Rainfall pattern of District Nagaur

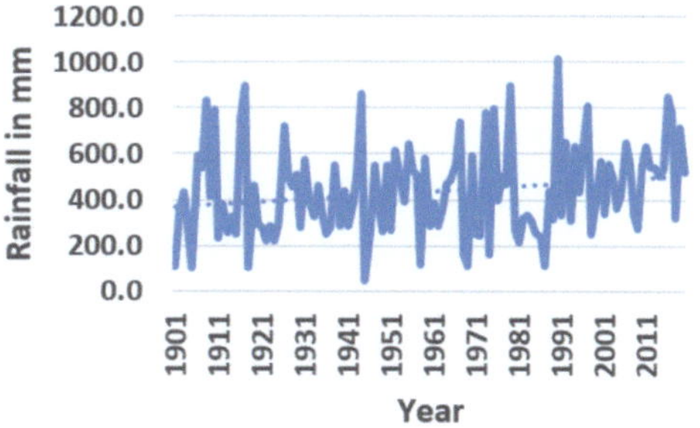

Fig. 6.11 Rainfall pattern of District Pali

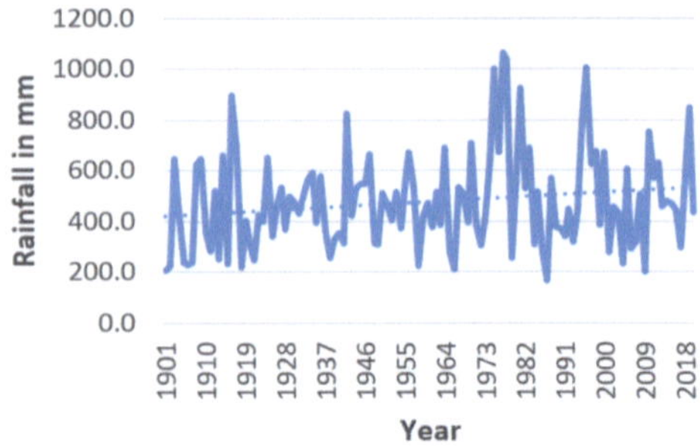

Fig. 6.12 Rainfall pattern in District Sikar

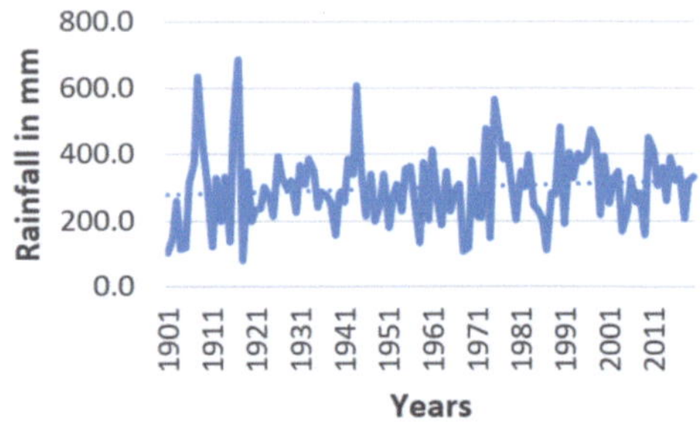

Fig. 6.13 Rainfall pattern of western region of Rajasthan

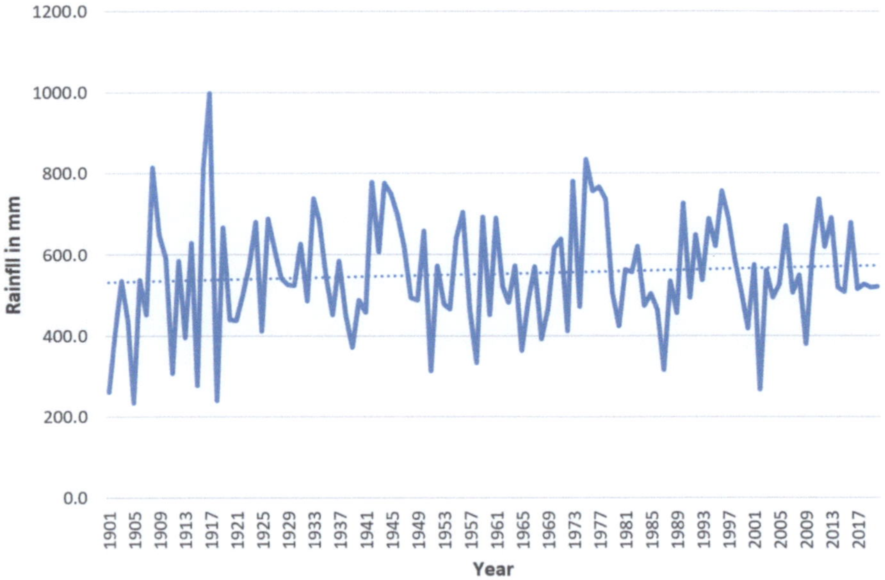

Fig. 6.14 Rainfall pattern of Rajasthan

The surface and ground water in the western region is a precious and valued environmental resource. Water scarcity plays an important role in shaping the life of the human and livestock population in the region. The western region is highly vulnerable to water resources because of its geographical location, low rainfall, saline and fluoride-contaminated groundwater and recurring droughts. It is a big challenge to ensure drinking water availability round the year for the animal and human population, except in areas served by canal systems i.e., Ganganagar, Hanumangarh and part of Bikaner districts. Increasing demand for water and a shortfall in the availability of surface water resources has put tremendous pressure on groundwater resources in the region.

Groundwater and traditional surface water bodies are the principal sources of drinking water. In most parts of the Thar desert, the groundwater is brackish or saline. People have to walk many kilometers to fetch drinking water to meet domestic demand (Navada, 1999). The status of groundwater in the region is shown in Table 6.4. The current rate of annual withdrawal of ground water is higher than its recharge to the extent of 164.5%. The problem is more severe in the case of Jaisalmer and Jodhpur districts as the abstraction rate is 318.6% and 254.1%, respectively. The groundwater utilization pattern in the region shows that irrigation use is about 87.04 and domestic use is only 12.96%. The higher water use in agriculture is because of an increase in the number of tubewells in the region. As evident in Table 6.5, the growth in the number of tubewells particularly in the four most severe desert districts, namely, Barmer, Jaisalmer, Bikaner and Churu was around 40%. Available technology has helped in identifying the location of good source of groundwater and deep drilling. The hydrogeological research helped identify the buried paleo channels in the region, banks providing financial support, easy availability of deep drilling machines in the area and poor groundwater governance are the main factors responsible for over exploitation of groundwater in the region.

As the rainfall in the area is low the only chance of enhancing groundwater is the reverse flow of water in the canal-irrigated areas. However, a large part of the IGNP command area has a hardpan layer below ground at a depth ranging between 10 and 40 m, restricting water from trickling down to recharge groundwater. The emerging challenges are to check the increasing number of tubewells and area under water-intensive irrigated crops, such as cotton and groundnut and come up with a suitable agricultural policy for the desert region.

6.2.2 Forest Resources

The common conventional understanding of forest and forest areas does not fit in for the desert region of Rajasthan. It is because of the land scape, i.e., sand dunes that shifting sand dunes and type of vegetation, high temperatures and high intensity of winds in the region, determine the location and type of vegetation that can survive in the area. Forest area in this region means the area under the ownership of the Forest

Table 6.4 Status of groundwater in western region of Rajasthan

S. No	Districts	Extractable groundwater after recharge (MCM)	Groundwater extraction (MCM)	Groundwater use		*Stage of groundwater development (%)*
				Irrigation (%)	Domestic (%)	
1	Barmer	339.74	426.01	84.89	15.11	125.39
2	Bikaner	344.84	486.40	83.55	16.45	141.05
3	Churu	116.11	134.10	80.19	19.81	115.49
4	Sri Ganganagar	446.12	173.86	99.81	0.19	38.97
5	Hanumangarh	200.81	127.09	93.31	6.69	63.29
6	Jaisalmer	85.04	270.99	89.88	10.12	318.63
7	Jalore	539.34	950.76	95.60	4.40	188.04
8	Jhunjhunu	223.47	470.51	75.60	24.40	210.54
9	Jodhpur	343.03	871.54	82.04	17.96	254.07
10	Nagaur	488.24	962.08	85.52	14.48	197.05
11	Pali	296.12	443.64	92.92	7.08	149.81
12	Sikar	353.28	607.09	81.20	18.80	171.84
13	Western Region	314.68	493.67	87.04	12.96	164.51
14	Rajasthan	11,073.63	16,634.72	86.76	13.24	150.22

Source Central Ground Water Board, Government of India, 2020

Department in the village revenue records, irrespective of the type and density of vegetation.

Ministry of Rural Development, Govt. of India provides funds for the Desert Development Programme (DDP), Drought Prone Area Development Programme (DPAP), and for Integrated Wasteland Development Programme (IWDP) and Combating Desertification Project (CDP) implemented these special area developmental programmes for the purpose of water harvesting in drought-prone, desert and rainfed areas. DDP—Started in 1977–78, DPAP—Started in 1973–74, IWDP—Started in 1989, and CDP-Started in 1999–2000.

After independence two major area specific programs for the western region, namely, Desert Development Program (DDP) and Drought Prone Area Development Program (DPAP) and also two research institutes, namely Central Arid Zone Research Institute (CAZRI) and Arid Forest Research Institute (AFRI) were established at Jodhpur. Various efforts were made to improve the productivity of different types of land categories by increasing the production of firewood, fodder, and other plants, conserving biodiversity and protecting the development of infrastructures such as canals, roadways, and farmland, etc., and the land through planting trees. The type of plantations was: (1) Canal Side Plantation, (2) Road Side Plantation, (3)

Table 6.5 Growth in number of tube wells in western region of Rajasthan

District	Number of tubewells			Growth rate (%) 2006–2010	Growth rate (%) 2010–2019	Growth rate (%) 2006–2019
	2006–07	2010–011	2019–20			
Barmer	2469	4295	12,963	11.71	24.72	39.33
Bikaner	5465	16,232	29,739	24.32	12.87	40.33
Churu	1670	3254	11,641	14.27	29.04	47.45
Shri Ganganagar	11,458	15,819	36,030	6.66	17.90	25.75
Hanumangarh	16,782	35,918	49,159	16.44	6.48	23.98
Jaisalmer	1243	3054	7277	19.70	18.96	42.40
Jalore	5390	11,753	13,506	16.87	2.82	20.17
Jhunjhunu	1406	7706	13,706	40.53	12.21	57.68
Jodhpur	25,223	28,115	57,630	2.19	15.44	17.97
Nagaur	13,684	16,642	18,484	3.99	2.12	6.20
Pali	2265	2653	3296	3.21	4.44	7.79
Sikar	5679	12,512	27,499	17.11	17.06	37.09
Western Region	92,734	157,953	280,930	11.24	12.21	24.82
Rajasthan	260,748	374,511	632,898	7.51	11.06	19.41

Source Agricultural Statistic of Rajasthan for the years 2006–07, 2010–11 and 2019–20

Block Plantation/ Forest Preserves. (4) Sand Dune Stabilization, (5) Pasture Development, (6) Agricultural Forest trees, (7) Environmental Plantation, and (8) Farm Forestry.

CAZRI and AFRI the two lead institutions of Rajasthan in the last 50 years have studied the soils, plants, natural vegetation, water, landforms, land use, socio-economic aspects, etc. of the entire arid and semi-arid areas of Rajasthan. CAZRI developed techniques like sand dune stabilization, shelterbelt plantations, increasing productivity of agricultural lands and rangelands by improved seeds and management practices like fencing, contour bunding, reseeding, etc. Their research and technology played an important role in the overall development of the Thar Desert. Despite that because of the changes in the State's sectoral policies, economic development, socio-economic changes, and environmental changes, the present status of natural resources and their management systems seems to be unsustainable. The challenge is to evolve a management system that restores the ecosystem, which will be sustainable.

As per the revenue records the change in total forest cover has been depicted in Table 6.6. Total forest cover showed a positive change in Jaisalmer district (252.77 km^2) followed by Barmer (251.79 km^2) Bikaner (217.61 km^2), Ganganagar (176.88 km^2), Jhunjhunu (145.77 km^2), Jalore (136.07 km^2), Nagaur (116.04 km^2), Churu (79.00 km^2), Pali (67.85 km^2), Sikar (99.06 km^2), and Jodhpur (40.78 km^2) all

Table 6.6 Change in total forest cover in western region of Rajasthan: 1995–2019 (Area km^2)

District	Total forest cover		Absolute change (km^2)
	1995	2019	
Barmer	38.00	289.79	251.79
Bikaner	38.00	255.61	217.61
Churu	3.00	82.00	79.00
Shri Ganganagar	26.00	202.88	176.88
Jaisalmer	73.00	325.77	252.77
Jalore	132.00	268.07	136.07
Jhunjhunu	55.00	200.77	145.77
Jodhpur	67.00	107.78	40.78
Nagaur	31.00	147.04	116.04
Pali	607.00	674.85	67.85
Sikar	94.00	193.06	99.06

Source State of Forest Report, FSI, 1995 and 2019

because of the above listed initiatives of the state government. Maximum open forest cover increased in the western districts of Rajasthan.

6.2.3 Expansion of Area Under Invasive Species

As discussed above the forest cover increased but what is important is the type of plant species selected under different programs, were those native plants or alien species and how far they survived? In the field one can find a bundle of mix plantation species dominated by invasive species. Invasive plants significantly impact the natural vegetation communities. They cause displacement and hence result in an imbalance in the natural and agricultural ecosystems. It ultimately leads to the formation of a large monoculture of invasive plants in the alien environment. These plants affect the species diversity of the native areas and their ecological integrity. They are successful invaders of non-native habitats because of their fast growth, greater reproductive potential, competitive ability, high plasticity and allelopathy. One such plant namely *Prosopis Juliflora* was introduced in the Thar Desert in 1917 by the former ruler of the princely state Marwar, and *Acacia tortillas* later on by the State Forest Department, which played 'a leading role in the afforestation of arid lands.[2] The invading species could survive on little water but are good for nitrogen fixation and stabilization of sand dunes in the desert area. Invasive species are easy to plant in a desert where forestation usually is an arduous task. On the other hand, shrubs capable of living and multiplying in extremely drought-prone environments, inexpensive during propagation, rapidly spreading and able to recover soil minerals are preferred plants. However, both these

[2] For details see Rathore (2009).

species are not compatible to some extent with the wildlife of the desert. These species can be grown on degraded lands. Being a multipurpose tree, *Prosopis juliflora* controls soil erosion, stabilise sand dunes, improve soil fertility, reduce soil salinity, provide fuel energy, supply feed and forage for grazing animals. It also provides construction timber and furniture wood, supplement foods for humans, and promotes honey production. Further, it also produces high-quality charcoal. Its branches are widely used as fencing posts, while its pods, which are high in protein and sugars are important fodder for livestock, and/or food for humans. *Prosopis juliflora* has been used to shelter agricultural crops from wind by reducing the movement of soil and sand.

Unfortunately, the uncontrolled growth of *Prosopis juliflora* has become a threat to the pastoral production system in the Thar Desert. Now people feel that the *Prosopis juliflora* is responsible for a significant reduction in the production potential of the common lands. As it has been widely spreading over the area at an alarming rate it has now exerted a substantial amount of impact on the biodiversity of common and arable lands. The main reasons for this changed perception of people about the usefulness of this plant in the present context are: the availability of alternatives to fuel wood, fast economic development, creation of infrastructure, roads, transport, increase in irrigated and commercial agriculture, alternative livelihood options, etc. People have now realized that planting Indigenous trees is a better option. That would increase land fertility and provide options for planting fruit-bearing trees, as it would provide families dependence on common lands with nutrition. Simply planting trees and complete forestation of the landscape are not sustainable options in desert areas. However, any comprehensive model in the desert agro-ecology space needs to be regenerative in nature.

The challenge is how to make a balance to protect the biodiversity of the area and regulate the growth and expansion of this plant on a scientific basis by educating people and making appropriate land use policies for the district. There is also a need to have a fresh look in the changed context, i.e., get research/ survey done on the present status of biodiversity of crops, biomass (plants, trees and grasses), livestock, etc. in the district and can think of propagation of other native species providing food, fodder and medicinal usage. What has been done to the desert land and ecosystem through human beings cannot be undone in the name of development. The alternative is informing and motivating people to work towards a positive change in the desert environment.

6.3 Temporal Changes in Agriculture of Western Region

6.3.1 Land Resources

Meaningful planning for the economic development of an area requires a thorough understanding of the geography and land use pattern. The land is the most important resource for the survival of human and livestock populations. A study of land utilization is the first step in planning of any development activity or analyzing the impact of past land based developmental activities. The changes in the proportion of different categories of lands in western Rajasthan, presented in Table 6.7 reveal that there has been an increase in forest lands from 1.13% in 1976 to 1.96% in 2019 mainly due to the afforestation efforts, discussed earlier, by the state government and other agencies. The share of non-agricultural uses has increased from 3.02 to 5.27% during the same period. The land under the two categories namely; not available for cultivation and other uncultivated excluding fallow land, are the village common lands and are unfortunately called 'Waste Lands', though we know that not a single inch of land on the earth is waste land. Table 6.7 shows that these two categories of land are further divided into five sub-categories. These lands determine the nature of the ecosystem in the region. In fact, this nine-fold land use classification was inherited from the British Raj. The British Raj had an interest in collecting land revenue and at that time there were no industrial townships or large urban areas, therefore, no land use category was specified as Industrial lands or Urban lands category. In this paper, an attempt has been made to reclassify this Waste Land category from an ecological perspective. Lands that provide biodiversity through plants, trees, flora and fauna or various types of biomass species are put under the term/category of land called 'Ecological Land' and land used for agriculture is termed as Non-Ecological Lands, though they also provide biodiversity and livelihoods.

Commons lands are the lifeline of poor people particularly in the western region of Rajasthan. Mostly, the common lands are the source of biodiversity, catchment areas of water bodies, mines and mineral areas. They are the source of livelihood for poor households in villages, provide ecosystem services and are put to many other usages. Presently these lands are under serious pressure to be used for numerous economic development activities. They are also subjected to private and public encroachment and acquisition under different names or programmes/projects by rich and government agencies. Encroachment of this land will convert people from 'Rural Ecosystem Owner' to 'Ecological Refugees' and force them to migrate out of rural areas (Gadgil & Guha, 2000). The future challenge is how to protect these lands and improve their health and productivity so that they can better serve the poor and marginal section of the population. The prime beneficiary of commons development will be the village service-providing communities residing on the periphery of the village and mostly dependent on these lands. Also, it will ensure drinking water security to human and livestock population as common lands are the areas of rainwater harvesting and ground water recharge.

Table 6.7 Percentage change in land use pattern of western region of Rajasthan (percentage area)

Land use	1976	1983	2001	2011	2019
1. Geographical area according to village papers (area in thousand hectares)	100 (17,937)	100 (17,988)	100 (17,980)	100 (17,986)	100 (17,997)
2. Forest	1.13	1.38	1.70	1.97	1.96
3. Land not available for cultivation					
(A) Land put to non-agricultural use	3.02	3.24	4.37	4.92	5.27
(B) Barren and uncultivated	5.80	5.42	5.47	5.13	5.24
Total (A + B)	8.81	8.66	9.85	10.05	10.51
4. Other uncultivated excluding fallow land					
(C) Permanent pasture and other grazing land	4.23	4.46	4.27	4.26	4.16
(D) Land under misc. tree crops and groves not included in net area sown	0.00	0.03	0.00	0.02	0.04
(E) Culturable waste	27.85	24.71	20.50	18.16	16.17
Total (C + D + E)	32.08	29.20	24.78	22.45	20.37
5. Fallow land					
(F) Old fallow land	9.49	7.33	8.02	6.69	8.07
(G) Current fallow land	6.92	7.28	6.01	5.85	6.27
Total (F + G)	16.41	14.62	14.03	12.54	14.35
6. Net area sown	41.56	46.15	49.64	52.99	52.82
7. Total cropped area	43.91	48.50	55.22	63.35	73.23
8. Area sown more than once	2.35	2.35	5.58	10.36	20.41

In the absence of Land Use Policy rampant mining and quarrying activities (both legal and unauthorised), industrial areas and even solar energy projects are coming up on these lands despite failing to meet the required environmental conditions, therefore, the challenge is to mobilise people and politicians to make State Land Use Policy document as early as possible to save the rural ecosystem.

The land use category 'Net Sown Area' is a very important category for the survival of humans and livestock. Sustainable use of agriculture lands is necessary for ensuring food security for the fast-growing population in the country. There is an increase in net sown area from 41.56% in 1976 to 52.82% in 2019. It shows that the western arid and semi-arid region of Rajasthan has more than half of its land under net sown area (NSA). The decline in old fallow lands from 9.49 to 8.07% was observed between the years 1976 and 2019. On the other hand, increasing demand for land for agricultural purposes has largely been met by bringing in cultivable wastelands under the plough. There have been land use shifts from permanent pastures and grazing land, barren and uncultivable land towards areas under non-agriculture uses and to a smaller extent towards forests. As a result of such practices, the present share of culturable waste land declined from 27.85 to 16.17% over the decades.

The land records of land use within India were so organised as to maximise the revenue to the British Raj. The land ownership was mainly of two types; first, land owned by the state and community, and second, agricultural lands. Forest lands belong to the State. Ecological health was never an agenda in land use planning till the 1980s. It was after 80 s environmental issues were brought to lime light in public discourse.

6.3.2 *Measurement of Temporal Variations in Land Use Categories*

As the land is a fixed resource, therefore, the total land of any region (Western Region) is constant (i.e., R = constant), it was assumed that land use patterns can change within and between different classes of land use. Hence, land use changes overtime can be expressed in linearly additive form as follows

$$R = (\Delta E + \Delta NA + \Delta A) \tag{6.1}$$

where, R = Total reporting area

ΔE = Net change in ecological land use category

ΔNA = Change in non-agricultural land use category

ΔA = Net change in agricultural land use category

The study of the change in land use pattern can either measure shift under different categories of land as reported in the land records or can be looked at from the environment point of view. In this paper we have tried to group the revenue-based land use categories into three broad categories namely; (1) Ecological (E), (2) Non-agricultural (NA) and (3) Agricultural (A). Ecological land use (E) comprising of area under forest (F), permanent pastures and other grazing lands (P), land under miscellaneous tree crops and groves (M) and barren and un-culturable land (BU). Further, the ecological category can be divided into two sub-categories, viz., (a) Desirable Ecology (E1), comprising of area under Forest (F), Permanent Pastures and other grazing lands (P), and Land under Miscellaneous tree crops and groves (M) and (b) Undesirable Ecology category (E2), comprising of area under barren and un-culturable land (BU). Non-agricultural land use (NA) comprising of area Not Available for Cultivation (NC) and agricultural sector (A) comprising of area under culturable waste land (C), Old Fallow lands (OF), Current Fallows (CF) and Net Area Sown (S).

$$R = (\Delta F + \Delta P + \Delta M + \Delta BU) + (\Delta NC) + (\Delta C + \Delta OF + \Delta CF + \Delta S) \tag{6.2}$$

The shifting of land within the desirable Ecology category (E1) will not pose any adverse ecological implications. The shifting of Barren and Un-culturable land

(BU) to Forest (F), Permanent Pastures and Other Grazing Lands (P) and land under Miscellaneous Tree Crops and Groves (M) will have favourable ecological impact. However, the shifting of Forest (F), Permanent Pastures and Other Grazing lands (P) and land under Miscellaneous tree crops and Groves (M) to Barren and Un-culturable land (BU) and other uses, such as, industrial, school, roads, etc. will have serious adverse ecological consequences. Thus, net change in ecological category (E) can be expressed as;

$$\Delta E = (\Delta E1 + \Delta E2) \tag{6.3}$$

where, $\Delta E1 = \Delta F + \Delta P + \Delta M$ and $\Delta E2 = \Delta BU$.

6.3.3 Land Use Dynamics of Western Rajasthan

In order to study the land use dynamics in the arid and semi-arid region of Rajasthan the total land endowment is divided into three broad categories:

Ecological Land Use

The changes in the land use categories are analysed and reported in Table 6.8. It indicates the shift in land use towards ecological category. It also shows that out of 12 districts of Western Rajasthan, positive change is seen in 8 districts and negative change in 4 districts, that is, towards the desirable ecological category. This change can be associated with the fact that Government has implemented several afforestation programmes in the state to increase biomass cover and also to protect the forest areas from encroachments.

The extent of increase in desirable ecology category of land use was observed highest for Bikaner district followed by Barmer, Jodhpur, Hanumangarh, Jalore, Pali, Nagaur, and Jhunjhunu districts. Whereas negative change was observed in Jaisalmer, Ganganagar, Sikar, and Churu districts. This is how that area under desirable ecology has increased in two-third of districts in Western Rajasthan.

Table 6.8 also shows that only 6 districts of western Rajasthan have shown positive change towards undesirable ecological sector while rest of the 6 districts have registered a negative change which reflects the fact that land under undesirable sector has decreased due to its diversion to other sectors to some extent. The positive shift was found in the districts of Jaisalmer, Bikaner, Sikar, Jhunjhunu, Hanumangarh, and Sri Ganganagar, while the negative shift in land use of undesirable ecological category was observed in the districts of Jodhpur, Jalore, Barmer, Pali, Nagaur, and Churu. It clearly indicates that area under undesirable ecology category has been increasing in half of the districts of western Rajasthan.

Non-Agricultural Land Use

The results furnished in Table 6.8 show that out of 12 districts only 3 districts had negative change in non-agricultural land use while the rest of the 9 districts have

Table 6.8 District wise intersect oral shift in land use of Western Rajasthan (Area in '000Ha)

Districts	Ecological sector (E)		Non-agricultural sector (NA)	Agricultural sector (A)	Overall net change = E1 + E2 + NA + A
	Desirable ecology (E1)	Undesirable ecology (E2)			
Barmer	7.229	−1.135	8.200	−14.194	0.100
Bikaner	10.759	16.301	−21.407	−5.240	0.413
Churu	−0.786	−0.028	2.148	−1.327	0.007
Ganganagar	−4.539	0.068	2.583	1.826	−0.062
Hanumangarh	1.738	0.110	1.713	−3.541	0.020
Jaisalmer	−32.742	117.484	−45.402	−39.340	0.00
Jalore	0.987	−3.658	3.977	−1.297	0.009
Jhunjhunu	0.198	0.261	−0.278	−0.181	0.00
Jodhpur	2.709	−3.362	3.081	−2.428	0.00
Nagaur	0.371	−0.353	13.727	−2.602	11.143
Pali	0.712	−0.598	0.934	−1.048	0.00
Sikar	−1.214	0.869	2.477	−1.864	0.268

Source Statistical Abstract Rajasthan (2010); Agricultural Statistics at a Glance (2020)

shown positive change ranging from 13.727 thousand hectares in Nagaur district to 934 ha in Pali district suggesting that more land has been diverted to non-agricultural sector from other sectors. At the same time Jaisalmer district (−45.402 thousand hectares) has maximum negative change followed by Bikaner (−21.407 thousand hectares), and Jhunjhunu district (−0.278 thousand hectares). Therefore, it shows that the area under the non-agricultural sector has decreased in the 25% districts of western Rajasthan while it has increased in the remaining 75% districts. Sharma and Pandey (1992), Takle et al. (2007), Wani et al. (2009), Bardhan and Tewari (2010), Sharma (2014), Ram and Chauhan (2020) observed similar pattern of increase in land under non-agricultural use in Barmer district of Rajasthan.

Agricultural Sector

As per Table 6.8, only one district i.e., Sri Ganganagar out of 12 districts of the western region has demonstrated positive change in the agricultural sector. Whereas rest of the 11 districts have shown negative change implying that the shift has taken place away from the agricultural sector. Sharma and Pandey (1992), Bardhan and Tewari (2010), Pushpa and Akashraj (2014), Sharma (2014), Pandey and Ranganathan (2018) also observed similar results of decreasing area under agricultural sector. Maximum negative change was registered in Jaisalmer district (−39.340 thousand hectares) followed by Barmer (−14.194 thousand hectares), Bikaner (−5.240 thousand hectares), Hanumangarh (−3.541 thousand hectares), Nagaur (−2.602 thousand hectares), Jodhpur (−2.428 thousand hectares), Sikar (−1.864 thousand hectares), Churu (−1.327 thousand hectares), Jalore (−1.297 thousand hectares), Pali (−1.048 thousand hectares), and Jhunjhunu (−0.181 thousand hectares). Thus, it can be inferred that area under agricultural sector increased only in one district i.e., Sri Ganganagar district of western Rajasthan. The results obtained in this study concurred that the shifting of land has taken place from agricultural use towards ecological land use category and non-agricultural category. The overall net change was positive in all the districts except Sri Ganganagar district.

Intra Category Shift in Agricultural Sector

The shift in intra land use categories is shown in Table 6.9. Out of the 12 districts of western region of Rajasthan only 3 districts showed positive change by increase in the area under the category Culturable Waste Lands and in the rest of 9 districts there have been negative change. While in 8 districts positive changes have been observed in favour of Fallow Lands Other than Current Fallows category, while in the remaining 4 districts negative change observed by shift in the area of the Fallow Lands Other than Current Fallows category. Out of 12 districts of western Rajasthan only 2 districts showed positive shift towards Current Fallow Lands category whereas 10 districts have registered negative change in the Current Fallows category. Net Area Sown has decreased in 4 districts viz., Barmer, Jhunjhunu, Sikar, and Churu, whereas 8 districts have shown positive change in this category viz. Jaisalmer, Bikaner, Jodhpur, Hanumangarh, Sri Ganganagar, Nagaur, Pali, and Jalore district, implying that land under agricultural production has increased in the western region.

Table 6.9 District wise intrasectoral shift in agricultural sector (area in "000 ha)

Districts	Culturable wasteland (C)	Fallow lands other than current fallows (OF)	Current fallow land (CF)	Net sown area (S)
Barmer	−7.190	32.326	31.473	−70.803
Bikaner	−258.727	54.916	−19.041	217.612
Churu	13.015	0.174	−13.206	−1.310
Sri Ganganagar	−23.064	−26.421	−60.000	111.311
Hanumangarh	−1.876	−13.951	−101.282	113.568
Jaisalmer	−345.682	69.969	−21.473	257.846
Jalore	−13.826	16.455	−9.638	5.712
Jhunjhunu	1.124	15.170	1.529	−18.004
Jodhpur	−4.133	−6.939	−132.036	140.680
Nagaur	−6.678	0.028	−57.311	61.359
Pali	−2.388	−6.203	−23.444	30.987
Sikar	1.963	11.310	−4.752	−10.385

Source Statistical Abstract Rajasthan (2010); Agricultural Statistics at a Glance (2020)

Table 6.10 Degradation of range lands in western region of Rajasthan (area in "000 ha)

Districts	Rangeland		Percent of rangeland to geographical area	
	1997	2019	1997	2019
Barmer	1239	1048	43.99	37.19
Bikaner	1272	1117	46.47	36.72
Churu	239	173	14.17	12.51
Sri Ganganagar	241	232	22.01	21.22
Hanumangarh	68	61	6.99	6.33
Jaisalmer	3393	2787	88.36	72.60
Jalore	390	332	36.91	31.37
Jhunjhunu	117	154	19.75	25.96
Jodhpur	917	773	40.64	34.24
Nagaur	451	406	25.59	22.87
Pali	527	482	42.70	39.06
Sikar	189	200	24.37	25.88
Western Region	9041	7764	43.43	37.27
Rajasthan	14,642	12,894	42.76	37.61

Source Author's own computation from secondary data

6.3.4 Degradation of Rangelands

The extent of rangeland in western region of Rajasthan is presented in Table 6.10. It is evident from the table that the total rangeland has decreased from 43.43% (9041 thousand hectares) of the total geographical area in 1997 to 37.27% (7764 thousand hectares) in 2019 in western region and from 42.76% (14,642 thousand hectares) to 37.61% (12,894 thousand hectares) in Rajasthan state during the same period. Further, rangelands as a percentage of total geographical area were highest in Jaisalmer district and lowest in Hanumangarh district during both the periods.

6.3.5 Economic Loss Estimation Caused by Land Degradation

The following estimation model calculates economic losses on account of land degradation (EL) by multiplying the extent of degraded land in each district with per hectare total value of crop output standardised by district average yields and farm harvest prices of crops (Vashist et al., 2003).

$$EL = \Sigma X_{dk} * [(\Sigma Y_{ik} * P_{ik})/N_K]$$

EL = Economic losses as a result of land degradation,

Y_{ik} = quantity of ith crop output in the kth district,

P_{ik} = farm harvest price of ith crop in the kth district,

N_k = net sown area in kth district,

X_{dk} = area under dth class of degraded land in the kth district.

Degraded and Wastelands status in Western Region

The extent of various types of degraded land in the western region is depicted in Table 6.11. The total area of degraded lands in the western region and Rajasthan state was 11,729 and 20,424 thousand hectares, respectively.[3] Wind erosion is the most common cause of erosion in typical western region as well as in Rajasthan state as a whole. The extent of degraded land varied from 27 thousand hectares in the Pali district to 2772 thousand hectares in the Jaisalmer district. Wind erosion affects all twelve districts of the Thar-desert region, while water erosion affects only five districts, salinity affects seven districts, and sodicity affects six districts. The primary driver of land degradation is only due to wind erosion in Jhunjhunu and Sikar, wind erosion particularly severe in Jaisalmer, Bikaner, Barmer, Churu, and Jodhpur, water erosion in Pali, salinity in Hanumangarh, Churu, and Nagaur, and sodicity in Sri Ganganagar districts of the western region. The degraded land in relation to the geographical area was found to be approximately 56.06% in the western region, and 59.68% in the Rajasthan state. A significant chunk of the geographic area is affected by various types of land degradation, posing a significant challenge to crop and livestock production in the western region as well as in the state.

Economic Losses due to Land Degradation in Western Region

The losses due to land degradation were calculated for western region districts and are shown in Table 6.12. As per the table, actual annual losses of ₹17.93 billion were observed in Western districts, accounting for 32.46% of total losses in Rajasthan (₹55.24 billion). The severity of annual economic losses in terms of money was observed to be highest in Bikaner district (₹4.50 billion), followed by Jaisalmer (₹3.31 billion), Jodhpur (₹2.30 billion), and lowest in Pali district (₹0.05 billion) of the western region. The highest economic loss was estimated in Bikaner district, despite the fact that the total degraded land was highest in the Jaisalmer district. The economic losses differed across districts due to differences in the nature and intensity of land degradation, climatic factors, as well as cropping patterns.

[3] The area under the land use category Waste Lands is also estimated by ISRO and published in the Waste Land Atlas and as per that report the waste lands in the Western Region is given in Appendix Table 2.

Table 6.11 Degraded and wasteland statistics of western Region of Rajasthan (Area in 10^3 ha)

Districts	Water erosion	Wind erosion	Saline	Sodic	Total degraded land	Others	Total geographical area
1	2	3	4	5	6	7	8
Barmer	0 (0.00)	1908 (99.27)	10 (0.52)	4 (0.21)	1922 (67.39)	930	2852
Bikaner	1 (0.05)	2119 (99.05)	0 (0.00)	0 (0.00)	2120 (77.20)	626	2746
Churu	0 (0.00)	1346 (97.47)	35 (2.53)	0 (0.00)	1381 (81.47)	314	1695
Hanumangarh	0 (0.00)	320 (80.00)	78 (19.50)	2 (0.50)	400 (31.75)	860	1260
Jaisalmer	0 (0.00)	2753 (99.31)	11 (0.40)	8 (0.29)	2772 (71.76)	1091	3863
Jalore	4 (1.55)	244 (94.57)	2 (0.78)	8 (3.10)	258 (24.36)	801	1059
Jhunjhunu	0 (0.00)	149 (100.00)	0 (0.00)	0 (0.00)	149 (25.21)	442	591
Jodhpur	0 (0.00)	1235 (99.52)	6 (0.48)	0 (0.00)	1241 (54.36)	1042	2283
Nagaur	7 (0.91)	735 (95.08)	31 (4.01)	0 (0.00)	773 (43.72)	995	1768
Pali	24 (88.89)	1 (3.70)	0 (0.00)	2 (7.41)	27 (2.20)	1202	1229
Sikar	0 (0.00)	414 (100.00)	0 (0.00)	0 (0.00)	414 (53.42)	361	775
Sri Ganganagar	1 (0.37)	194 (71.32)	0 (0.00)	77 (28.31)	272 (33.96)	529	801
Western Region	37 (0.32)	11,418 (97.35)	173 (1.47)	101 (0.86)	11,729 (56.06)	9193	20,922
Rajasthan	8632 (42.26)	11,419 (55.91)	192 (0.94)	181 (0.89)	20,424 (59.68)	13,794	34,224

Figures in parentheses under column numbers 2, 3, 4, and 5 indicate percentage to a total of column number 6

Figures in parentheses under column number 6 indicate the percentage of total geographical area (column number 8)

Source NBSS & LUP

6.3.6 *Land Distribution*

As the population grows subdivision of land holdings takes place in all size classes of private land holdings. The resultant outcome is an increase in the number of small and marginal holdings. In the desert regions, the soils are of poor quality, low rainfall,

Table 6.12 Economic losses due to land degradation in western region of Rajasthan

S. no	Districts	Degraded land (hectares)	Value of output (₹ ha^{-1})	Economic losses (billion rupees)
1	Barmer	1922	240.02	0.46
2	Bikaner	2120	2124.69	4.50
3	Churu	1381	1034.58	1.43
4	Hanumangarh	400	3658.87	1.46
5	Jaisalmer	2772	1193.19	3.31
6	Jalore	258	1685.58	0.43
7	Jhunjhunu	149	3824.78	0.57
8	Jodhpur	1241	1853.93	2.30
9	Nagaur	773	1048.63	0.81
10	Pali	27	1700.87	0.05
11	Sikar	414	3145.36	1.30
12	Sri Ganganagar	272	4825.15	1.31
13	Western Region	11,729	26,335.65	17.93
14	Rajasthan	20,424	2704.60	55.24

low productivity of land and above all if the size of the landholdings are small, i.e., economically unviable in that case agriculture may not be the best livelihood option.

The percentage change in the number of landholdings in Western Rajasthan from 2000–01 to 2015–16 is reported in Table 6.13. In the year 2000–01, the highest percentage of operational holdings were in Medium size class (33.60) followed by the Large (23.22), Semi Medium (22.08), Small (12.17) and Marginal (8.93) classes. But in the year 2015–16, it changed to 33.71% in Medium size class followed by 25.08% in Semi Medium, 16.19% in Small, 11.36% in the Marginal class and 13.66% in the Large class of the total holdings in Western Rajasthan. It shows that in the 15 years subdivision of holdings took place and the number of marginal, small, semi-medium and medium size class increased largely because of a decrease in the number of large size holdings. With the increase in population subdivision of holdings is likely the outcome and more prominent in the case of large holdings. Increase in the number of marginal and small holdings has direct impact on the overall agriculture production and the farming system adopted in the region. An increase in small unviable economic size of holdings with poor soil conditions and increasing climate variability brings new challenges to the sustainability of agriculture in the region.

The total area under all sizes of holdings decreased from 146.11 lakh hectares to 128.04 lakh hectares during the period from 2000–01 to 2015–16 as evident from Table 6.14. The area under all sizes of holdings increased during 2000–2016 except for large sizes of holdings. This is due to the subdivision area under marginal and small categories of farms had almost doubled during the period under study. The overall status is that there was a 10% decline in the number and 16% decline in the area of large holdings. From an equity point of view, one may find a positive change

Table 6.13 Number of holdings by size group in western districts of Rajasthan

S. no	Size of holding	Number of holdings		Percent of holdings	
		2000–01	2015–16	2000–01	2015–16
1	Marginal	148,860	263,837	8.93	11.36
2	Small	202,954	376,173	12.17	16.19
3	Semi medium	368,104	582,741	22.08	25.08
4	Medium	560,265	783,105	33.60	33.71
5	Large	387,310	317,465	23.22	13.66
6	Total	1,719,991	2,323,321	100.00	100.00

Table 6.14 Area of holdings by size group (western district)

S. no	Size of holding	Area of holding (in ha.)		Percent area of total holdings	
		2000–01	2015–16	2000–01	2015–16
1	Marginal	79,563	153,660	0.65	1.20
2	Small	302,902	556,941	2.46	4.35
3	Semi medium	1,078,307	1,692,703	8.77	13.22
4	Medium	3,571,328	4,903,943	29.04	38.30
5	Large	7,265,669	5,497,106	59.08	42.93
6	Total	14,611,191	12,804,355	100.00	100.00

but from an economic point of view, this change may affect agriculture production in the region therefore, increasing agriculture production and productivity of land and for sustainability of agriculture a new strategy has to be evolved for the region.

6.3.7 Cropping Pattern

Natural resources of semi-arid and arid areas, i.e., land, water and vegetation, are very fragile, weak, and partly non-resilient, hence prone to irreversible degradation and desertification under the excessive pressure of human and livestock population. Nature has provided arid zones with valuable vegetation resources and animal wealth for sustainable livelihood. Surface and groundwater resources are very scarce due to scanty rainfall. Four major production systems are prevalent in the hot desert western districts of Rajasthan (Bhati, 2014) and these are;

i. Small ruminant based production system
ii. Large ruminant based production system
iii. Rainfed crop production system and
iv. Partly irrigated crop production system

These systems vary across districts. At a limited scale, the other production systems prevalent in the region are based on the availability of canal water (IGNP, Gang Canal and Narmada Canal) and moisture-based cultivation in Khadin's in the form of the irrigated farming system. Irrigated agriculture is also practiced in isolated areas where ground water is available, tubewell based irrigation is to produce commercial crops namely, Cumin and Isabgol (*Plantago ovata*).

The area under Rabi and Kharif crops in the Western Rajasthan has observed a significant change during the past few decades. The changes since the early 1970s are presented in Table 6.15. The cropping pattern was directly or indirectly influenced by the state policy to meet food security in the state. A remarkable increase was found in the share of Kharif cultivated area from 2.21 lakh hectares in 1976–77 to 7.28 lakh hectares in 2019–20. The outstanding increase was found in the share of Rabi cultivated area from 0.02 lakh hectares in 1976–77 to 4.57 lakh hectares in 2019–20. The increased kharif cultivated area was more pronounced in the Barmer, Nagaur and Jodhpur districts in comparison to the Bikaner, Churu, Jaisalmer, Jalore, Jhunjhunu and Pali districts. The increased Rabi cultivated area was more pronounced in Bikaner, Nagaur and Jodhpur districts in comparison to Churu, Barmer, Jaisalmer, Jalore, Jhunjhunu and Pali districts. These changes occurred because of canal irrigation in the Bikaner Region and tube well irrigation in the Nagaur and Jodhpur districts.

The absolute increase in the area under Kharif crops was mainly because of the tractorisation of agriculture practices facilitating efficient use of soil moisture in the arid and semi-arid regions. In the Rabi season, the increase in area was because of the advent of canal and energisation of wells and tubewells thereby increasing the

Table 6.15 Change in area under Rabi and Kharif crops in western region of Rajasthan

District	Change in cultivated area-kharif (Lakh hectares)		Absolute change (L. hectares)	Change in cultivated area-rabi (Lakh hectares)		Absolute change (L. hectares)
	1976–77	2019–20	Difference	1976–77	2019–20	Difference
Barmer	1.7	11.5	9.80	0.00079	3.98	3.98
Bikaner	3.4	7.6	4.20	0.00168	7.42	7.42
Churu	4.7	8.8	4.09	0.00209	2.85	2.85
Jaisalmer	0.003	2.4	2.40	0.00008	4.54	4.54
Jalore	0.7	5.5	4.78	0.00002	3.30	3.30
Jhunjhunu	1.7	3.1	1.40	0.15000	2.54	2.39
Jodhpur	2.8	10.7	7.90	0.00489	6.26	6.26
Nagaur	3.1	11.3	8.20	0.00007	6.41	6.41
Pali	1.7	4.6	2.90	0.00002	3.83	3.83
Western Region	2.21	7.28	5.07	0.02	4.57	4.55

irrigated area. The number of tubewells has increased mainly because of easy access to drilling technology and subsidised finance by the banks.

The cropping pattern adopted by the farmers from 1976 to 2019 in the western region of Rajasthan is shown in Table 6.16. The gross cropped area has increased from 7013.27 thousand hectares in 1976 to 11,169.57 thousand hectares in 2019. Overall, the cereals occupied 41.63% of the gross cultivated area (GCA) in 1976 which declined to 30.93% during 2019. The increase was found in the share of wheat in GCA from 4.12% in 1976 to 5.07% in 2019. Bajra has sharply decreased as the area under bajra in 1976 was 34.76% of GCA, which decreased to 23.50% in 2019. The share of cereals namely, sorghum and maize had increased while the share of barely declined.

The share of pulses in the gross cropped area increased from 28.17% in 1976 to 35.16% in 2019. The share of gram had remarkably increased while the share of kharif pulses had decreased under total pulses. The area under food grain production had declined from 69.80% in 1976 to 66.09% in 2019.

The area under oilseeds has recorded a sharp increase. The share of oilseeds area in GCA increased from 4.86% in 1976 to 14.43% in 2019. The share of rapeseed and mustard, groundnut and caster seed had remarkably increased while the share of sesame had decreased under total oilseeds. Soybean is not grown by farmers in the region. This change can be attributed to the development in the field of new high yielding seed varieties, crop wise improved cultural practices, the government's price policy and easy access to inputs.

The per cent share of area under cash crops/other crops has shown decreasing trend over the years. The share of Guar crop had remarkably decreased mainly because of market price fluctuations. While area under cotton significantly increased during the study period due to increased productivity and favourable prices.

The main changes observed in the desert region are: (i) an increase in Net Sown area, (ii) Culturable waste lands are brought under plough, (iii) technology, particularly tractors, made possible to bring more area under cultivation, (iv) Animal draft power is replaced by tractors facilitating farmers in timely sowing of crops in larger areas by better use of soil moisture, and (v) sharp increase in irrigated rabi crops is mainly because of canal and tubewell irrigation in the region. In the past, the entire area was rainfed and only kharif crops were grown.

As irrigated areas increased in the desert region the biodiversity in crops and other biomass declined leading to ecosystem changes. There is a need to have a deep holistic study of socio-economic and economic development factors triggering the ecosystem changes to facilitate evolving new strategies to lead a sustainable ecosystem in the hot desert region of Rajasthan.

6.3.8 Droughts in Western Region of Rajasthan

Drought is a recurring phenomenon of desert ecosystems causing serious water, food and fodder crises for human and livestock populations. The cropping pattern

Table 6.16 Changing cropping pattern in Western Region of Rajasthan (%)

Crops	1976	1983	2001	2011	2019
Bajra	34.76	47.62	54.99	32.14	23.50
Sorghum	1.11	1.90	2.69	1.93	1.83
Maize	0.41	0.44	0.37	0.17	0.10
Wheat	4.12	3.80	5.00	4.01	5.06
Barley	1.23	0.34	0.29	0.35	0.44
Total cereals	41.63	54.1	63.34	38.6	30.93
Gram	2.11	3.77	3.70	6.61	10.77
Rabi pulses	0.00	0.00	0.02	0.00	0.01
Kharif Pulses	26.06	16.94	23.71	20.03	24.38
Total pulses	28.17	20.71	27.43	26.64	35.16
Total food grains	69.80	74.81	90.77	65.24	66.09
Sesame	3.81	2.39	2.75	2.41	1.11
Rapeseed and mustard	0.86	2.83	4.40	4.04	6.91
Taramira	0.00	0.23	0.00	0.00	0.00
Linseed	0.00	0.01	0.01	0.00	0.00
Groundnut	0.16	0.15	0.69	2.26	5.04
Castor seed	0.03	0.05	0.41	2.17	1.37
Total oilseeds	4.86	5.66	8.26	10.88	14.43
Sugarcane	0.00	0.00	0.01	0.00	0.00
Dry chilly	0.20	0.15	0.23	0.03	0.02
Cotton	0.19	0.28	0.73	0.69	1.58
Tobacco	0.02	0.01	0.00	0.00	0.00
Guar seed	24.92	19.08	0.00	23.16	17.87
Coriander	0.01	0.01	0.00	0.00	0.01
Others total	25.34	19.53	0.97	23.88	19.48
Total area (area in 000' ha)	100.00 (7013.27)	100.00 (7957.01)	100.00 (6781.99)	100.00 (10,469.21)	100.00 (11,169.57)

and crop productivity had a great dependence on rainfall and its distribution across the western region of Rajasthan. The frequency of meteorological and agricultural droughts in western region areas of Rajasthan during the last 120 years (1901–2020) is shown in Table 6.17. The western region experienced droughts in 89 years and 31 normal years. Normal years ranged from 23% to 29% during the last 120 years showing drought dominance in the western region of the state, the special feature of the desert ecosystem. The occurrence of drought in Rajasthan has a pattern that every 3 years of drought is followed by one normal year in most of the districts. Though drought equally affects human and livestock populations but re-occurrence of droughts in consecutive years has had a greater impact on animal populations. The last severe drought occurred in the year 2002. It was considered unique in terms of magnitude, spacing, dispersion and duration, as all records of past years since 1918 were broken. Previously, it was estimated that about 10 million hectares of land in the arid region of Rajasthan was available for grazing to about 11.28 million adult cattle units (ACUs) (Narain et al., 2006). The estimated dry fodder requirement was about 28 million tonnes. Under normal situations, about 11 million metric tonnes of fodder was estimated to be available, having a deficit of 60% even during the normal years. The deficit under drought conditions increases by about 80–95%. To cope with the drought situation in the year 2002–03 about 1.39 million metric tonnes of fodder was distributed at subsidised rates through the opening of 2807 fodder depots in the western region of Rajasthan benefiting around 250,312 animals.

6.4 Livestock Resources

Rajasthan is endowed with a huge livestock population that is at present 56.94 million (Livestock Census 2012) and almost equal to the human population. Although the western region of Rajasthan harbours 28.02 million animal heads. Rajasthan had 13.94 million cattle, 13.69 million buffaloes, 7.90 million sheep, 20.84 million goats, and 0.56 million other animals including horses, ponies, mules, donkeys, pigs and camels. The western region of Rajasthan had 7.08 million cattle, 3.83 million buffaloes, 5.77 million sheep, 10.99 million goats, and 0.36 million other animals. Small ruminants' sheep and goats dominate livestock production systems in Rajasthan.

The status of major livestock species across the western region is reported in Table 6.18. The composition of livestock is changing over time. Some important changes have emerged in the changing dynamics of livestock population in the western region and the entire Rajasthan. The share of the western region in total livestock population in the state has increased between the Livestock Census 1997 and 2019 from 16.30 to 25.27%, respectively. Livestock species-wise changes showed that in the case of cattle, it has changed from 11.04 to 13.67%, buffaloes (except Sri Ganganagar, Hanumangarh and Sikar districts) from 33.35 to 39.22%, goats barring a few exceptions in case of Bikaner, Churu, Hanumangarh, Jalore, and Jhunjhunu districts the share increased from 33.2 to 39.2%. Despite sheep being a prominent ruminant of Rajasthan, their share sharply declined from 36.72% to 20.60% across the districts in the whole of the western region. The drastic decline in sheep population may be because of the severe drought during the years 2002–03 and 2008–09, the increase in irrigated agriculture, declining access to grazing lands/common lands and also several issues of within-state and inter-state migration. In the past two decades, the number of bovines (cattle and buffaloes) particularly buffalo and goats has increased, while the population of sheep and other animals, especially camels declined. This change can be linked to state livestock policy, particularly dairy policy and increasing meat demand within and outside the state. The other major change

Table 6.17 Drought and normal years in western districts of Rajasthan (1901 to 2020)

Districts	Years of normal rainfall	Normal years	Drought years	% of normal years
Barmer	73, 75, 76, 78, 81, 83, 84, 85, 89, 90, 92, 93, 94, 97, 98, 00, 01, 02, 03, 06, 08, 10, 11, 13, 15, 16, 17, 18, 19, 20	30	90	25
Jaisalmer	70, 73, 75, 76, 78, 79, 81, 82, 83, 84, 88, 92, 93,94, 95, 96, 98, 99, 00, 01, 02, 03, 06, 07, 08, 10, 11, 12, 13, 15, 17, 20	32	88	27
Bikaner	70, 74, 75, 76, 77, 78, 81, 82, 83, 86, 89, 92, 94, 95, 96, 97, 98, 00, 01, 02, 03, 09, 10, 11, 12, 15, 16	27	93	23
Churu	74, 75, 76, 77, 78, 80, 81, 83, 85, 88, 90, 91, 92, 93, 94, 95, 96, 97, 98, 00, 02, 08, 10, 11, 12, 13, 14, 15, 16, 18, 19, 20	32	88	27
Nagaur	70, 71, 72, 73, 75, 76, 77, 78, 81, 82, 83, 85, 88, 89, 90, 92, 93, 95, 96, 97, 98, 00, 02, 08, 10, 11, 12, 13, 14, 15, 16, 17, 18, 19, 20	35	85	29
Jodhpur	70, 73, 75, 76, 77, 78, 79, 81, 82, 83, 89, 90, 92, 94, 95, 96, 97, 98, 00, 03, 10, 11, 13, 15, 16, 19, 20	27	93	23
Pali	70, 73, 75, 76, 77, 78, 79, 88, 90, 92, 94, 95, 96, 97, 03, 05, 06, 07, 10, 11, 12, 13, 14, 15, 16, 17, 19, 20	28	92	23
Jalore	70, 71, 72, 73, 75, 76, 77, 78, 79, 83, 89, 90, 92, 93, 94, 95, 97, 98, 03, 06, 10, 11, 13, 15, 16, 17, 19, 20	28	92	23
Jhunjhunu	70, 71, 73, 75, 76, 77, 78, 81, 83, 88, 92, 93, 94, 95, 96, 97, 98, 01, 02, 03, 05, 06, 08, 10, 11, 12, 13, 14, 15, 16, 18, 19	32	88	27
Western Region	70, 73, 75, 76, 77, 78, 79, 81, 82, 83, 88, 89, 90, 92, 93, 94,95, 96, 97, 98, 00, 08, 10, 11, 12, 13, 15, 16, 17, 19, 20	31	89	26

was noticed in the case of the Camel population as the number sharply declined. An animal known as 'Desert Ship' the only transport animal in the Thar Desert districts is on the verge of extinction and to protect them the state government has declared Camel as a protected state animal. The one important factor responsible for this status of camel is automobiles replacing camel carts as the primary mode of rural transportation.

The rapid increase in cattle and buffalo populations may indicate a paradigm shift in farmers' perceptions of the role of the bovine population in the rural farm economy. In terms of average milk yield and saleable quality, supported by tremendous growth

Table 6.18 Temporal changes in live stock population in western region (% share)

Districts	Cattle		Buffalo		Sheep		Goat		Others	
	1997	2019	1997	2019	1997	2019	1997	2019	1997	2019
Barmer	12.98	17.58	2.23	4.33	36.20	19.68	44.73	57.23	3.86	1.18
Bikaner	22.11	43.45	4.24	7.57	45.35	24.08	25.38	22.80	2.92	2.09
Churu	15.65	26.95	9.88	17.04	30.59	17.31	39.81	36.51	4.07	2.18
Sri Ganganagar	28.81	45.76	22.83	14.38	24.58	16.81	21.12	21.81	2.66	1.24
Hanumangarh	25.01	44.21	26.69	24.55	23.39	13.81	19.48	14.66	5.43	2.77
Jaisalmer	12.52	16.99	0.06	0.19	48.75	34.93	36.05	46.08	2.63	1.81
Jalore	15.85	17.96	17.41	35.27	39.15	24.99	25.74	21.14	1.84	0.63
Jhunjhunu	9.81	21.89	28.25	28.23	17.85	12.44	40.97	35.96	3.13	1.47
Jodhpur	17.64	29.12	5.53	8.66	41.18	16.87	34.21	44.69	1.44	0.66
Nagaur	15.08	20.66	13.55	19.63	36.30	16.08	33.74	43.05	1.33	0.58
Pali	15.19	16.76	11.45	15.32	50.00	35.53	22.12	31.70	1.24	0.68
Sikar	11.91	19.35	25.53	24.67	17.05	8.85	43.23	46.45	2.27	0.68
Western Region	16.30	25.27	11.04	13.67	36.72	20.60	33.35	39.22	2.57	1.24
Rajasthan	22.37	24.48	17.95	24.05	26.33	13.88	31.16	36.60	2.18	0.99

Source Department of Animal Husbandry and Dairying, Livestock Census of India, 1997 and 2019

in the dairy sector and better transportation facilities, new technology and infrastructure and favourable milk price policy leading to the introduction of buffaloes in the desert ecosystem. It is evident from the above-narrated facts that farmers are interested in rearing those species of livestock, which do not need to be taken for long distances in search of fodder, i.e., for grazing and can be stall fed. Sheep rearing is rapidly losing share as a result of shrinking grazing/common lands, issues in seasonal migration and poor quality of wool and stiff competition from synthetic wool/yarn. Sheep rearing is losing importance because sheep herders do not find it economically viable. Also, they do not want to continue the migratory system in order to provide education to their children. The additional factor is that sheep are more susceptible to environmental stress, whereas goats, which are browsers, can adapt to a wide range of adverse conditions, fast growing in number, and meat demand is increasing and easy to sell.

6.4.1 *Feed and Fodder*

The annual feed and fodder demand for livestock in the western region was estimated to be 34.49 million tonnes (on a dry matter basis based on adult cattle units and per unit intake capacity) as depicted in Table 6.19. Feed and fodder demand varied because of the variations in the livestock population and its composition across districts in the western region. The highest demand for feed and fodder was in Barmer, followed by Jodhpur, Bikaner, Nagaur, and Sikar districts due to the higher population of dairy animals. The percentage gap between feed and fodder demand and supply was estimated to be 18.9%. In the western region of Rajasthan, the shortage is less severe than in the rest of the state. The gap was not distributed evenly across the western region districts. It was highest in the district of Barmer (62.3%) followed by Jalore (54.1%), Jaisalmer (50.0%), Pali (37.5%) and lowest in Jhunjhunu (7.1%) district. The supply of livestock feed and fodder exceeded demand in the Hanumangarh, Sri Ganganagar, and Bikaner districts of the western region. It might be either due to the availability of well-irrigated facilities as a result of the Indira Gandhi Canal in these districts or having low livestock density as compared to other western region districts.

Table 6.19 shows that the state's total annual potential supply of dry matter is 51.1 million tonnes. The total annual demand for dry matter is 80.3 million tonnes. The annual potential supply of dry matter from all sources, including crop residues, green roughage, and concentrates, was only 63.6% of total dry matter demand in 2017–18, implying a net deficit of 36.4%. In Rajasthan, feed and fodder were recorded 39.8, 24.9 and 21% deficits against its demand for the year 2007, 2009 and TE 2013–14, respectively (Chand et al., 2013; Meena et al., 2013, 2018). Historically, 62% forage

Table 6.19 Supply, demand and gap of livestock feed and fodder (dry matter)

Particulars	Feed and fodder demand				Livestock population pressure		
	Adult cattle unit no. 000'	Demand (million tons)	Supply (million tons)	Gap %	Grazing lands (ACU/ha)	Net area sown (ACU/ha)	Grazing lands and NSA (ACU/ha)
Barmer	1838.39	4.70	1.77	−62.31	9.06	1.10	0.98
Bikaner	1546.67	3.95	4.03	+1.88	31.14	0.97	0.94
Churu	856.52	2.19	1.91	−12.81	23.04	0.75	0.73
Sri Ganganagar	867.46	2.22	3.50	+57.21	6196.14	1.14	1.14
Hanumangarh	866.56	2.21	3.05	+37.84	187.57	1.03	1.02
Jaisalmer	784.79	2.01	1.00	−50.01	9.11	0.94	0.85
Jalore	990.82	2.53	1.16	−54.08	20.92	1.48	1.38
Jhunjhunu	702.95	1.80	1.67	−7.09	17.91	1.78	1.62
Jodhpur	1709.42	4.37	3.20	−26.69	13.71	1.22	1.12
Nagaur	1367.81	3.49	2.59	−25.82	19.27	1.09	1.03
Pali	932.13	2.38	1.49	−37.52	10.21	1.44	1.26
Sikar	1034.90	2.64	1.93	−26.97	25.93	2.04	1.89
Western Region	13,498.42	34.49	27.98	−18.88	17.00	1.15	1.08
Rajasthan	31,424.78	80.29	51.09	−36.37	18.80	1.74	1.59

Source Author's own computation

supply deficit was experienced in the Thar Region, ranging from 30.9% in Zone IV to 71% in Zone III (Venkateswarlu et al., 1992). Similar shortages due to an increase in animal populations were recorded in field surveys from 1997 to 2003 (Rahmani & Islam, 1997). The fodder demand–supply gap is widening as a result of competing land pressures, acreage shifts from cereals to other crops, and the diversified use of agricultural residues.

The overall pressure on livestock feed and fodder resources because of changes in land resources, including grazing lands and net area sown, was calculated to be 1.59 ACU per hectare for the state and 1.08 ACU per hectare for the western region (Table 6.19). The livestock pressure of stocking rate in adult cattle unit of animals was calculated to be 18.80 ACU per hectare of grazing land for Rajasthan state and 17 ACU/hectare for the western region. In comparison to overall state figures, there was a wide inter-district variability. Whereas, livestock population pressure was comparatively lower in the western region. The pressure was highest in Sikar and Jhunjhunu districts as compared to Churu and Bikaner districts. There is severe stress because of the higher stocking rate in terms of ACU per hectare of pasture and grazing lands.

6.5 Sectoral Policies

For the development of crops and livestock in the western region of Rajasthan the state and central government came out with a number of policies from time to time. The latest program for the development of agricultural products the Ministry of Food Processing Industries has launched 'One District, One Product' (ODOP) programme to assist western districts of Rajasthan in reaching their full potential, fostering economic and socio-cultural progress, and creating employment possibilities, particularly in rural areas. The list of 12 products from 12 districts of the western region is as follows:

- Barmer—Pomegranate
- Bikaner—Moth (Bhujia, Namkeen, Papad)
- Churu—Groundnut Product
- Hanumangarh—Wheat Grain Based Products—Noodles, Pasta
- Jaisalmer—Nutritious xerophytic fruits Ker, Sangri
- Jalore—Isabgol (*Plantago ovata*)
- Jhunjhunu—Fruit based product, Lemon
- Jodhpur—Cumin—(Cleaning, Grading, Classifying, Packing, Roasting, Cumin based drink)
- Nagaur—Fenugreek, poly-milk products
- Sikar—Onion
- Sri Ganganagar—Kinnu

6.6 Emerging Challenges

Looking at special agricultural and allied field characteristics of the western region of Rajasthan, the following challenges emerged:

1. Degradation of natural resources
2. Fast depleting ground water
3. Loss of ecological services
4. Reduction in biodiversity of food crops, grasses and medicinal plants
5. Extending agriculture on marginal lands
6. Changing composition of livestock and loss of descript breeds of animals (Camel in the extinction phase, endangered species list)
7. Introduction of capital-intensive high input commercial crops
8. Sustainability of newly introduced horticultural crops
9. Negative impact of green energy projects on livestock and wild animals
10. Changed rainfall pattern and the energy farming system
11. Bleak future of the livestock sector

Looking at the foregoing discussion on agriculture and livestock the important emerging question is "**How far extensive and intensive agriculture is justified in protecting the ecosystem of the desert region of Rajasthan?**".

6.7 Future Perspectives

As discussed above, agriculture including livestock is very important for the livelihood and economy of western Rajasthan. There are so many factors that directly and indirectly affect the growth of agriculture. The factors are declining size of land holding, change in cropping pattern especially towards horticulture and cash crops, depletion of ground water for irrigation, changing composition and population of livestock, etc. The western region of Rajasthan is characterised by changing prospects of agriculture due to high variability in rainfall, shrinking agricultural land and diversification towards plantation and other commercial crops. The national movement to address climate change and sustainability, the intervention in the area of the energy sector by installing solar and wind energy structures in common and grazing lands affecting water and composition/productivity of livestock. The population of sheep and camel is declining, the gap in demand and supply of fodder is increasing and so forth consequently affecting the ecosystem of the region.

The developmental activities being taken by the State and Central Government for the immediate benefit of the stakeholders are detrimental to the ecosystem of the western region. Therefore, future plans and policies should be prepared after considering the carrying capacity and sustainability of the desert ecosystem. There are frequent changes in the cropping pattern guided by policies and market systems affecting the land and water resources in the region. More research having a comprehensive and integrated view of the region be undertaken before any plan of action is prepared for the region. The livestock sector is being neglected or is at low priority despite playing an important role in providing livelihood to a large number of households. The change in the composition of livestock, mostly guided by the market and lack of policy perspective will lead to the extinction of camel and a significant reduction in the number of sheep in the region. The other important factors likely to significantly affect the desert ecosystem is the present development model of exploitation of natural resources, for example, rampant mining activity (sand stone, granite and China clay), encroachment of common lands and water bodies, green energy projects, etc. There is an urgent need to re-examine the policies and programs from a desert perspective.

Appendix 1: Decadal Growth Rate of Human Population in Western Region of Rajasthan

Table A.1 Decadal growth rate of human population in western region of Rajasthan

District	1901–1911	1921	1931	1941	1951
Barmer	−4.34	−6.04	9.53	27.98	20.98
Bikaner	9.35	−3.44	18.76	32.98	9.56
Churu	15.05	−0.09	22.26	24.56	14.98
Shri Ganganagar	43.66	−17.22	102.49	54.58	18.01
Hanumangarh	43.66	−17.21	102.49	54.58	18.01
Jaisalmer	15.37	−20.58	13.61	23.28	13.10
Jalore	10.69	−7.15	15.77	14.08	14.99
Jhunjhunu	8.05	−4.32	14.83	21.08	19.90
Jodhpur	2.47	−12.05	15.97	25.86	20.37
Nagaur	6.66	−12.81	16.61	15.47	16.37
Pali	15.94	−11.67	17.29	17.44	18.95
Sikar	0.60	−1.46	12.66	17.93	10.04
Western Region	13.93	−9.50	30.19	27.49	16.27
Rajasthan	6.70	−6.29	14.14	18.01	15.20

District	1961	1971	1981	1991	2001	2011
Barmer	36.14	19.24	44.41	28.27	36.90	32.52
Bikaner	29.30	30.14	46.34	41.84	37.71	24.28
Churu	25.66	32.07	34.81	30.23	23.51	20.25

(continued)

(continued)

District	1961	1971	1981	1991	2001	2011
Shri Ganganagar	63.38	21.04	48.08	29.36	27.59	10.04
Hanumangarh	66.42	53.02	42.90	29.03	24.39	16.91
Jaisalmer	28.80	18.82	44.84	41.73	47.52	31.81
Jalore	29.16	22.10	35.20	26.52	26.81	26.21
Jhunjhunu	22.24	29.12	30.39	30.61	20.93	11.67
Jodhpur	31.64	30.17	44.82	29.12	34.04	27.74
Nagaur	22.40	35.00	29.04	31.69	29.38	19.20
Pali	21.91	20.40	31.39	16.63	22.46	11.94
Sikar	21.29	27.11	32.09	33.81	24.14	17.03
Western region	33.20	28.18	38.69	30.74	29.62	20.80
Rajasthan	26.20	27.83	32.97	28.44	28.41	21.31

Appendix 2: Type and Change in the Area of Waste Lands in Western Region (Area Km^2)

Table A.2 Type and change in the area of waste lands in western region

Wasteland classes	Western region		
	2008–09	2015–16	Change
Gullied and/or ravinous land (medium)	282.84	269.04	13.81
Gullied and/or ravinous land (deep)	85.56	80.21	5.37
Land with dense scrub	11,213.04	9625.38	1814.66
Land with open scrub	7934.58	8815.64	1291.89
Waterlogged and Marshy land (permanent)	17.15	16.01	1.24
Waterlogged and Marshy land (seasonal)	77.5	70.11	8.11
Land affected by salinity/alkalinity (medium)	378.34	384.15	8.34
Land affected by salinity/alkalinity (strong)	282.64	299.86	19.13
Shifting cultivation—current Jhum	0	0	0
Shifting cultivation—abandoned Jhum	0	0	0
Under-utilised/degraded forest (scrub dominated)	1246.16	1149.37	132.89
Under-utilised/degraded forest (agriculture)	5.24	8,894,401	16.06
Degraded pastures/grazing land	2684.94	2431.2	253.73
Degraded land under plantation crop	0.81	0.81	0
Sands-Riverine	48.21	46.38	1.83
Sands-Coastal	0	0	0

(continued)

(continued)

Wasteland classes	Western region		
	2008–09	2015–16	Change
Sands-Desertic	2970.28	2434.73	535.54
Sands-Semi Stab. –Stab > 40 m	9486.91	9344.56	142.35
Sands-Semi Stab. –Stab 15–40 m	12,964.51	11,721.34	1243.19
Mining Wastelands	0.63	6.62	5.99
Industrial wastelands	7.56	11.34	3.78
Barren Rocky/Stony waste	2684.33	2676.62	7.72
Snow covered /Glacial area	0	0.24	0.24
Total	52,303.58	49,373.83	2929.73
TGA—total geographical area	208,751		

Source Wasteland Atlas of India 2019

References

Agricultural Statistics at a Glance. (2020). Government of India, Ministry of Agriculture & Farmers Welfare, New Delhi.

Ahmad, N., Sinha, D. K., & Singh, K. M. (2018). Changes in land use pattern and factors responsible for variations in current fallow land in Bihar, India. *Indian Journal of Agricultural Research, 52*(3), 236–242.

Bardhan, D., & Tewari, S. K. (2010). An investigation into land use dynamics in India and land underutilization. *Indian Journal of Agricultural Economics, 65*(1), 658–676.

Bhati, T. K. (2014). Techonology assessment for dryland production system in South Asia with special reference to Western Rajasthan, ICRISAT, Patancheru, A. P., India, 2014.

Census of India. (2011). Government of India, Registrar General & Census Commissioner, India, New Delhi, July 2015.

Chand, P., Sirohi, S., & Sirohi, S. K. (2013). Production and demand estimates of livestock feed and fodder in Rajasthan. *Indian Journal of Animal Nutrition, 30*(2), 149–156.

Darling, W. G., et al. (1987). Isotope techniques in water resources development. Proceedings of IAEA Symposium, Vienna.

Gadgil, M., & Guha, R. (2000). Ecology and equity: The use and abuse of nature in contemporary India. Kar, A. (2011). Caring for the Thar, geography and you (pp. 60–69).

Livestock Census of India. (2012). Ministry of Agriculture, Department of Animal Husbandry, Dairying and Fisheries, Krishi Bhawan, New Delhi.

Meena, G. L., Mishra, S., & Chand, P. (2018). Assessment of demand and supply of livestock feed and fodder resources in Rajasthan. *Indian Journal of Animal Nutrition, 35*(2), 166–173.

Meena, G. L., Burark, S. S., & Pant, D. C. (2013). Assessment of availability and requirement of animal feed & fodder in Rajasthan. Paper presented at the National Seminar on prospects of livestock and dairying in India under changing global economic era. Anand Agricultural University, Anand.

Naveda, S. V. (1999). Groundwater problems studies in the Thar Desert, India using isotope. *Water and Environment News, Quarterly, 7*, 470–475. https://doi.org/10.5455/Vet

Niranjan, P., Rathore, L. S., Singh, R. S., & Rao, A. S. (2006). Drought assessment and management in arid Rajasthan. Central Arid Zone Research Institute, Jodhpur (pp. 1–71)

Pandey, G., & Ranganathan, T. (2018). Changing land-use pattern in India: Has there been an expansion of fallow lands? *Agricultural Economics Research Review, 31*(1), 113–122. https://doi.org/10.5958/0974-0279.2018.00011.3

Prasad, R. R. (1994). *Pastoral nomadism in arid zones of India: Socio-demographic & ecological aspects*. Discovery Publishing House.

Pushpa, C. O., & Akashraj, D. P. (2014). An investigation into land use dynamics in Karnataka–an ecological perspective. *Merit Research Journal of Agricultural Science and Soil Sciences, 2*, 047–052.

Qureshi, M. H. (2016). Land resource policies and agricultural patterns—Indian experiences. *Transactions, 38*(2), 147–162.

Rahmani, A. R., & Islam, M. Z. (1997). *Prioritisation of the Indian grasslands for conservation of biodiversity*. Bombay Natural History Society.

Ram, B., & Chauhan, J. S. (2019–20). Land use transformation in Thar Desert: A case study of barmer District, Rajasthan. *Journal of Global Resources, 6*(01), 96–101.

Ramnathan, S., & Rathore, M. S. (1994). Sustainability of Indira Gandhi canal and need for correct responses: A social scientist's perspective. In R. Hooja & P. S. Kavdia (Eds.), *Planning for sustainability in irrigation*. Rawat Publication.

Rathore, M. S., Ahuja, K., Singh, S. M., Rathore, J. S., & Kachawa, M. S. (2003). Biomass utilization and management, issues and strategies for the State of Rajasthan. Report prepared for The Energy and Resources Institute (TERI), New Delhi.

Rathore, M. S. (2003). Perspectives in biomass management in Rajasthan. Institute of Development Studies, Jaipur. Report.

Rathore, M. S. (2005). State level analysis of drought policies and impacts in Rajasthan, India. International Water Management Institute (IWMI), Colombo, Sri Lanka. 40p. (Working paper 93: Drought Series Paper No. 6).

Rathore, M. S. (2006). Droughts and the state failure: Unwilling to learn and unwilling to distribute. *Water Nepal, 12*, (1&2).

Rathore, M. S. (2007). Natural resource use-environmental implications. In V.S. Vyas, S. Acharya, S. Singh, & V. Sagar (Eds.), *Rajasthan—the quest for sustainable development*. Academic Foundation.

Rathore, M. S. (2009). Need for appropriate policy for prevention and management of invasive plants in Rajasthan—a case of *Prosopis juliflora*. Paper presented at the International Conference on Invasive Plants in the Tropics: Ecology, Management and Livelihoods, organised by ATREE, and SEC, Bangalore, from Jan. 4–6, 2009 at Bangalore, India.

Sharma, L. (2014). Implications of dynamics of land use shifts in Rajasthan. *Annals of Arid Zone, 53*, 177–183.

Sharma, U. K., & Pandey, V. K. (1992). Dynamics of land use in different states of India. *Agricultural Economics Research Review, 5*, 22–33. https://ageconsearch.umn.edu/record/265749/files/aerr-05-01-003.pdf

Singh, S., Kar, A., et al. (1994). Desertification problems in Western Rajasthan. *Annals of Arid Zone, 33*(3), 191–202.

Takle, P., Veerkar, P. D., Bhosale, S. S., & Malve, D. B. (2007). Dynamics of land use pattern in Maharashtra. *International Journal of Agricultural Sciences, 3*, 36–39. https://www.iaras.org/iaras/journals/ijas

Vasishta, A. K., Singh, R. P., & Mathur, V. C. (2003). Economic implications of land degradation on sustainability and food security in India. *Agropedology, 13*, 19–27.

Venkateswarlu, J., Anantha Ram, K., Purohit, M. L., Khan, M. S., Bohra, H. C., Singh, K. C., & Mathur, B. K. (1992). *Forage 2000 AD–the scenario for arid Rajasthan,* Central Arid Zone Research Institute.

Wani, M. H., Baba, S. H., & Yousuf, S. (2009). Land-use dynamics in Jammu and Kashmir. *Agricultural Economics Research Review, 22*, 145–154.

Chapter 7
Sustainable Management of Dry Lands of Africa: A Framework of the Holistic Participatory Approach

Boubié Vincent Bado and André Bationo

7.1 Introduction

The drylands of the semi-arid and dry sub-humid areas of Africa, covering 41 percent of the land surface, produce 44 percent of the crops, and contain over 2 billion people and half of the world's livestock. Two-thirds of the African continent is desert or drylands. About 73% of the African drylands are degraded and 51% are severely degraded. The population of Africa will double by 2050 (Dregne & Chou, 1992). The fragile land resources of Africa will face these high pressures from the growing population and livestock. However, rain-fed agriculture, livestock and agro-pastoralism systems remain the main sources of food, income and employment for the poor populations living in the drylands. Livestock plays a crucial role in generating income to invest in agriculture, purchase more diverse diets, paying for education, health and other household needs. In farming systems where the availability of fertilisers is limited and often unaffordable, manure from livestock is at the same time the main provider of nutrients to soils for staple crops. Food insecurity and poverty continue to be significant challenges in meeting the needs of a growing population.

The nutrient status of inherently low fertility soils has been further reduced by long-term cultivation with low or no nutrient input. These fragile and degraded soils of agricultural systems are unable to support the food needs of a growing population.

Being the main source of socioeconomic goods, services and livelihoods, the degradation of lands means high vulnerability to climate risk and change, food insecurity and poverty. Improving food security, the wellbeing and resilience of vulnerable people in the drylands will need more integrated technologies and management

B. V. Bado (✉)
Institut de l'Environment et de Recherches Agricoles (INERA), Bobo-Dioulasso, Burkina Faso
e-mail: Boubie.Vincent.Bado@gmail.com

A. Bationo
International Center for Research on Soil Fertility and Agricultural Development (IFDC), Accra, Ghana

M. S. Rathore et al. (eds.), *Desert Ecosystems*,
https://doi.org/10.1007/978-3-032-03237-9_7

options of crop-livestock farming systems including business opportunities in values chains. Nutrient-depleted soils, low soil organic carbon (SOC) and clay content, rapid SOC loss due to cropping, soil erosion, acidification, and nutrient mining pose significant challenges to agriculture in the West Africa Sahel (WAS) (Bado & Bationo, 2018a, 2018b; Bationo & Buerkert, 2001). Protecting and sustainably managing soil and land resources is critically important (Dewitte et al., 2013; Gisladottir & Stocking, 2005; Palm et al., 2007, 2010; Vlek et al., 2008). Long-term climate change projections (2050–2100) indicate higher temperatures, more uncertain climate and rainfall deficits (Periera, 2017). About 70% of African soils are degraded (Bationo & Mokwunye, 1991a, 1991b; Bationo et al., 2006). The high cost of fertilisers and the limited market for both fertilisers and agricultural products limit the utilisation of mineral fertilisers. Smallholder farmers are trapped in a vicious cycle of land degradation and poverty (Tittonell & Giller, 2013). With the decline of traditional fallow systems (10–15 years of fallow followed by 3–5 years of cropping) that once helped maintain soil fertility, innovative farming systems are needed to meet the growing demand for land.

A critical review of the main successful achievements and failures could help understanding the constraints and opportunities for sustainable management of dry lands. This includes crop-oriented or livestock-oriented approaches combining water harvest techniques, N_2-fixing legume crops and agroforestry systems. The aim is to investigate new strategies for sustainable land resources in dryland ecologies and resilience of smallholder farmers.

7.2 Management of Soil Fertility in Drylands

In general, the soils of WAS are originally poor in nutrients, low soil organic carbon (SOC) and clay contents. This is associated with a rapid loss of SOC with cropping activities (Bado & Bationo, 2018a, 2018b; Bationo & Buerkert, 2001), leading to soil erosion, acidification, and nutrient mining. Many concepts have been developed over the years, aiming to develop strategies or innovations for sustainable management of soil fertility. The paradigm of "external input" was first developed over the years 1960s and 1970s. This concept focuses on the utilisation of external inputs, mainly mineral fertilisers and improved germplasm, to improve agricultural production and productivity. The concept of external input has certainly boosted agricultural production in Asia and Latin America. It has certainly contributed to the "Green Revolution." On the contrary, the paradigm of external inputs did make a "Green Revolution" in Africa. The limited impact of the concept of external input was mainly attributed to the high fertiliser costs in Africa (Smaling, 1993). A new approach, named Low External Input Sustainable Agriculture (LEISA) was developed later in the 1990s with a focus on the importance of organic fertilisers resources. Considering the complementary role of organic and mineral fertilisers on nutrient use efficiency, the concept of Integrated Nutrient Management (INM) was developed. The INM was mainly focused on the combined applications of organic and mineral fertilisers. Taking account of the diversity of organic resources, sources of nutrients (biological nitrogen fixation), farming systems, the impacts of socioeconomic factors and policy on the sector of fertilisers, the INM concept shifted quickly to the concept of Integrated Soil Fertility Management (ISFM) (Vanlauwe, 2004).

7.2.1 Organic Fertilisers

The application of the LEISA and INM approaches pointed out the important role of organic fertilisers and organic amendment resources. Organic fertilisers or amendments are a broad term that comprises: farm yard manure (FYM), manure from different sources (cattle dung, excreta of other animals, crop residues (CRs), rural or urban wastes, composts and others. Organic fertilisers were traditionally the main sources of nutrients. They have been an important source to improve the chemical, physical, and biological properties of soils. This also includes the traditional crop-bare fallow system that contributes as mulching of residues to improve the recycling of nutrients, and soil protection against erosion.

7.2.1.1 Bare Fallow

In traditional cropping systems, lands are continuously cultivated for extended periods, with crop residues frequently removed and minimal or no nutrient inputs applied.

The practices of long-term fallow period (10–15 years) followed by a short cropping period (3–5 years) were the common technics to maintain soil fertility. Recycling the residues from a 10-year bare fallow, the system allowed a rebuilding of the level of SOC (Beets, 1990; Guillemin, 1956). The fallow system was sometimes combined with the burning of the 10–15 years of vegetation. By releasing most of the nutrients from burning, this practice increases the release of some nutrients through ash, allowing high crop yields during the first 3 years. This practice remains widespread, particularly in the humid forest zones of West and Central Africa. Research efforts have focused on evaluating the performance and long-term impacts of traditional fallow systems (Eckholm, 1976). The type and duration of fallow significantly influence soil fertility. In a long-term study, Bado (2002) and Bado et al. (2012) found that a 1-year bare fallow in rotation with sorghum improved SOC levels, increased nitrogen uptake by 120%, and boosted the grain yield of the subsequent crop by 57% compared to continuous sorghum monocropping. However, expanding fallow periods reduces the available land for cultivation. Given the rapidly growing population, rising food demand, and increasing pressure on land resources, traditional bare fallow systems are insufficient to maintain soil fertility and ensure sustainable land management.

7.2.1.2 Crop Residues

Crop residues (CRs) play a crucial role in maintaining soil fertility by protecting the soil from erosion while supplying organic carbon (OC) and nutrients. Bationo and Buerkert (2001) reported that applying 4000 kg/ha of CRs along with fertilisers in a cropping system-maintained SOC levels comparable to those of adjacent fallow plots, whereas plots without CR applications experienced rapid SOC losses. The agronomic efficiency of CRs can be enhanced through composting combined with agro-minerals like rock phosphate. For instance, composting CRs with local phosphate rock (PR) improves phosphorus solubilisation by organic acids, yielding higher-quality compost (Lompo, 1983). Both composting and direct CR applications increase phosphorus (P) availability (Sahrawat, 2004), enhance potassium (K) nutrition, improve water availability and use efficiency, reduce soil surface resistance to water infiltration by 65%, and lower soil temperature by over 4 °C (Buerkert & Stern, 1995). Additionally, organic amendments reduce soil P fixation, thereby increasing P availability and use efficiency (Buresh et al., 1997; Sahrawat, 2004). However, the impact of CR applications on SOC tends to be more pronounced in arid and semi-arid zones compared to humid agroecological zones (Bationo & Buerkert, 2001).

However, the availability of CRs for agricultural production is increasingly limited at the farm level due to competition for use as livestock feed, fuel, and building materials (Pieri, 1989), reducing their availability for field applications. In Niger and Nigeria, it is estimated that 50%–79% of stover is annually removed from fields for use as energy, building materials, and animal feed (Powell & Unger, 1998).

7.2.1.3 Farmyard Manure (FYM)

The FYM from different domestic animals is an important organic input in African Agro-ecosystems. But the availability and quality of manure varies with types of animals and feeds, collection, and storage methods. The yield potential production of crops, particularly of the improved germplasm cannot be achieved with the FYMs alone. Using manure is part of an internal flow of nutrients within the farm and does not add nutrients from outside the farm. Because of the weak integration of agriculture and livestock associated with high biomass demand for food and livestock feeding, CRs, and FYM alone cannot ensure sustainable management of soil fertility in the traditional farming systems.

Crop residues are usually leftovers from grazing by livestock utilisation by farmers for social needs (energy, ecosystem services), exposing the soils to water evaporation, soil erosion, increasing of soil temperature, reducing of formation and stability of aggregates, reducing of hydraulic conductivity, and air permeability (Bationo et al., 1993; Blanco-Canqui & Lal, 2009). The applications of organic residues (from fallow, CR, FYM) alone or associated with mineral fertilisers increase crop yields from up to 40% and 80%, respectively compared with the control without fertilisers (Bado et al., 2021; Whitbread et al., 2003). This is associated with an increase of water and nutrient use efficiencies (Suzuki et al., 2016; Yamoah et al., 2002).

7.2.2 *Mineral Fertilisers*

Nutrient deficiencies, especially nitrogen (N) and phosphorus (P), are major constraints in the soils of the WAS. Additionally, the positive effects of fertiliser applications are well-documented (Pichot et al., 1981; Pieri, 1989; Formoso, 1999; Bationo et al., 2003; Giko and Smithson, 2003; Bationo et al., 2006; Bado et al., 2010a, 2010b). P plays a critical role as the primary limiting nutrient in the weakly acid sandy soils of the WAS, as highlighted in numerous studies (Bado et al., 2010a, 2010b, 2021; Bationo et al., 2003; Sahrawat et al., 1997; Subbarao et al., 2000). For instance, a modest application of 13 kg/ha of P can boost pearl millet yields by up to 376%, while the combined use of nitrogen (N) and P fertilisers has been shown to increase yields by 600% (Bationo et al., 2006). P deficiency is identified as the primary constraint in the savannah zones (Bationo et al., 2006, 2012), whereas nitrogen deficiency tends to be more limiting than P in the humid forest zones (Sahrawat et al., 1997).

The use of mineral fertilisers is constrained primarily by their high cost, inefficient distribution systems, and inadequate policies and marketing strategies (Enyong et al., 1999; Zapata & Roy, 2004). However, the applications of mineral fertilisers alone lead to yield declines over time. This is frequently explained by the decreasing of SOC associated with long-term cropping without organic fertilisers, soil acidification, and decreasing of nutrient use efficiency (Pichot et al., 1981; Bationo & Mokwunye, 1991a, 1991b; Bado et al., 1997; Pieri, 1989; Bationo et al., 2003). Liming the soil is necessary to neutralise the acidity caused by the repeated application of mineral fertilisers (Bado et al., 1997; Bationo et al., 2012).

The combined application of mineral fertilisers and organic amendments is strongly recommended for the sustainable management of soil fertility. This approach helps delay the decline of soil organic carbon (SOC) (Bado et al., 1997; Bationo, 2008; Bationo & Mokwunye, 1991a, 1991b; Pichot et al., 1981) and is considered the most effective strategy for long-term soil fertility management. Recognising the crucial role of organic carbon in soil fertility, the LEISA paradigm was developed in the 1990s. Later, the Integrated Nutrient Management (INM) and Integrated Soil Fertility Management (ISFM) approaches emerged as integrated solutions for sustainable soil fertility management (Vanlauwe, 2004).

7.3 Nitrogen-Fixing Legume Crops as Source of Nutrients

Numerous studies have shown that cereals benefit from being grown in association with or in rotation with legumes (Bado et al., 2006a). In cereal-legume rotations, nitrogen-fixing legumes can increase subsequent cereal yields by 50 to 350%, thanks to the beneficial impact of legume residues, such as senescent leaves and underground biomass (Bado et al., 2006a; Bationo & Ntare, 2000). This improvement is attributed to enhanced soil mineral nitrogen and better fertiliser use efficiency for succeeding crops (Bado, 2002; Bado et al., 2006a; Bagayoko et al., 2000a, 2000b). For instance, Bado et al., (2006b, 2011) reported that cowpea, when grown in isolation, could increase soil mineral nitrogen by 52% and boost sorghum yields by 300%. In rotation with sorghum, cowpea can supply the equivalent of 42 kg N/ha to the following sorghum crop (Bado et al., 2012; Bado et al., 2013). Other benefits of nitrogen-fixing legumes include their role as cover crops, which help reduce soil erosion, prevent nutrient leaching, control weeds, improve water retention, and reduce water evaporation.

In addition to the positive effects of nitrogen-fixing crops on soil fertility, their inclusion in cropping systems offers numerous benefits, especially in the West African Sahel (WAS), where smallholder farmers rely on both livestock and farming for their livelihoods. Common feed sources for livestock in this region include open grazing and crop residues such as sorghum, millet, groundnut haulms, and cowpea hay (Ayantunde et al., 2007; Umutoni et al., 2021a, 2021b). Many nitrogen-fixing legume crops, such as cowpea, serve as both staple and cash crops, providing income and quality feed for livestock. This dual role is one of the reasons for the widespread use of millet-legume-based systems in the Sahel. The productivity of these farming systems can be enhanced through an integrated management approach that aligns with the concept of Integrated Soil Fertility Management (ISFM). This approach incorporates fertilisers, organic inputs, improved germplasm, and local knowledge on adapting these practices to specific conditions and farming systems (Vanlauwe et al., 2010a, 2010b; Waddington et al., 2007). Cereal-legume-based systems contribute significantly to food security and the resilience of smallholder farming systems (Whitbread et al., 2015).

7.4 Nutrients and Water Use Efficiency

Rainfall is often considered the primary limiting factor for agricultural production in the West African Sahel (WAS). However, the application of small doses of mineral fertilisers combined with organic residues, cover crops, and nitrogen-fixing legume crops has been shown to improve water use efficiency (WUE) and crop yields (Bado et al., 2021; Bationo & Mokwunye, 1991a). Several studies have indicated that nutrient limitations, rather than water, are the primary constraint to crop production (Twomlow et al., 2010). For example, Bado et al. (2021) demonstrated that a minimum of 350 mm of rainfall was sufficient to achieve the highest millet yields when NP fertilisers (15 kg N/ha and 13 kg P/ha) were applied. In contrast, traditional systems without mineral fertiliser applications failed to fully exploit the production potential, even with good rainfall (Fig. 7.1). Bationo et al. (2020) further highlighted that soil fertility, rather than rainfall, is the main limiting factor, noting that WUE for millet increased from 1.24 kg/mm without fertiliser to 4.14 kg/mm with mineral fertiliser application.

Environmental variations may arise from factors such as fertiliser applications, changes in soil chemical and physical properties, and fluctuations in rainfall (Ripoche

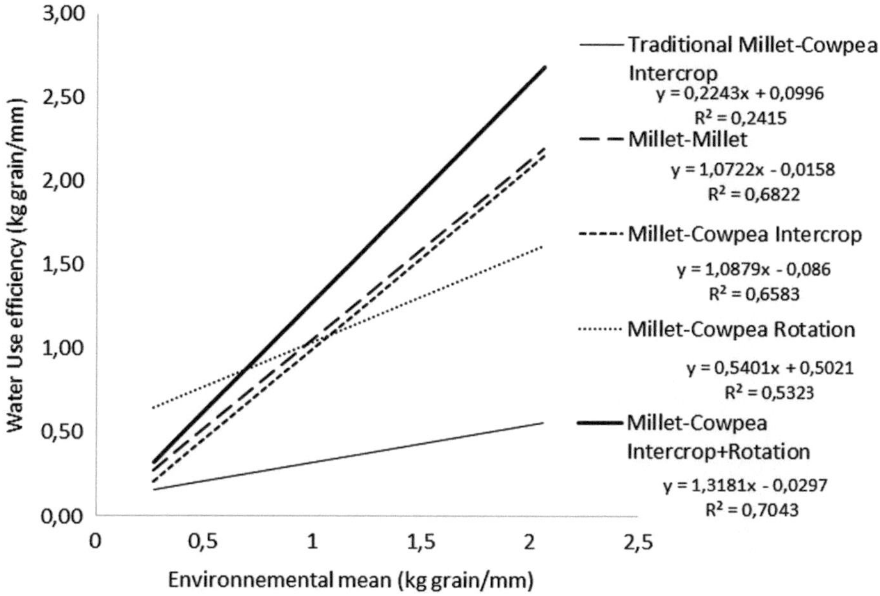

Fig. 7.1 Regression of millet grain yields (t/ha) on the environmental mean (t/ha) over 25 years (1995–2019) of cropping for the: **i** traditional millet/cowpea intercrop; **ii** millet-millet monocropping; **iii** millet/cowpea intercrop; **iv** millet-cowpea rotation and; **v** millet/cowpea intercrop and rotation with cowpea. Except the traditional M/C intercrop, P fertiliser was applied at 13 kg P/ha in the four improved cropping systems. *Source* Bado et al., (2021)

et al., 2015). These variations highlight the importance of water harvesting techniques in enhancing fertiliser use efficiency (Bationo et al., 2020). For instance, the widespread adoption of Zaï technology, which integrates water harvesting with micro-dosing of fertilisers (Ibrahim et al., 2015a), demonstrates the necessity of managing both water and nutrients in dryland areas of West Africa. An integrated approach to managing cereal-legume-based cropping systems with small doses of nitrogen and phosphorus can promote sustainable soil fertility and increase productivity in low-input cropping systems. This underscores the need for strategies that improve farmers' access to fertilisers and optimise the use of limited rainfall in dry ecosystems, particularly in the face of climate change.

7.5 Agroforestry Systems

The presence of trees through agroforestry systems can play a crucial role in the harsh environment of drylands. While annual crops are more vulnerable to rainfall variability, certain trees with high economic value, such as Ziziphus (*Ziziphus mauritiana*, commonly known as Pomme du Sahel in the WAS), are more resilient to dry spells (Ibrahim et al., 2015b; Bado et al., 2021). The presence of trees provides many advantages. Breman and Uithol (1984) noted a loss of up to 50% of rainfall through soil evaporation in the drylands of WAS. A reduction of run-off through the windbreak along with the increase in soil water holding capacity with the organic inputs from trees, protection of soil, reduction of soil erosion, and improvement of soil fertility trough build-up of soil organic carbon are some the possible beneficial impact of trees in the dry land ecologies (Bayala et al., 2006; Bado et al., 2021). For example, a fruit tree like Ziziphus tree increases soil organic carbon by 10 to 15% after 4 years of cropping, mitigates the impact of solar radiation and air temperature while improving both nutrient and water use efficiency (Tilander et al., 1999; Ndoli et al., 2017; Bado et al., 2021). Zhang et al. (2018) reported that maximum light area index (LAI) of millet was 17% higher in agroforestry than in sole millet system.

Moreover, fruit trees provide food and incomes (fruits) and wood as source of energy and for ecosystem services to smallholder households. An agroforestry system with fruit trees such as Ziziphus provides 2–3 times higher incomes (Bado et al., 2021) (Fig. 7.2). A significant increase of crop yields could be obtained through an integrated agroforestry system combining cereal crops, N_2-fixing legume crops and trees (Bado et al., 2021), associated with small quantities of mineral and organic fertilisers (Batioo and Mokwunye, 1991a, b; Bado et al., 1997; Bationo & Buerkert, 2001; Bationo et al., 2007a, 2007b; Ibrahim et al., 2015a; Bado & Bationo, 2018a, 2018b).

Agroforestry systems, due to their positive effects on biomass production, nutrient recycling, runoff reduction, windbreaks, and soil protection, can be considered sustainable land use options with significant potential for sequestering atmospheric carbon into the soil (Abbas et al., 2017).

7.6 Involving Local Communities

Farmers' decision-making is influenced not only by soil and climate but also by a wide range of factors across the biophysical, socioeconomic, and political domains. To address this complexity, the Integrated Natural Resource Management (INRM) approach was developed (Izac, 2000). Subsequently, the Integrated Soil Fertility Management (ISFM) approach was introduced, focusing on the effective management of soil resources while considering farmers' limited resources, the availability of technologies, and market opportunities (Bado et al., 2018; Vanlauwe, 2004) (Fig. 7.3).

7.6.1 The Challenge

A dilemma of sustainable management of the fragile resources of lands is the growing demand for lands, the quick growth of population and livestock, overgrazing of pasturelands with high pressure on lands and failure of smallholder farmers to

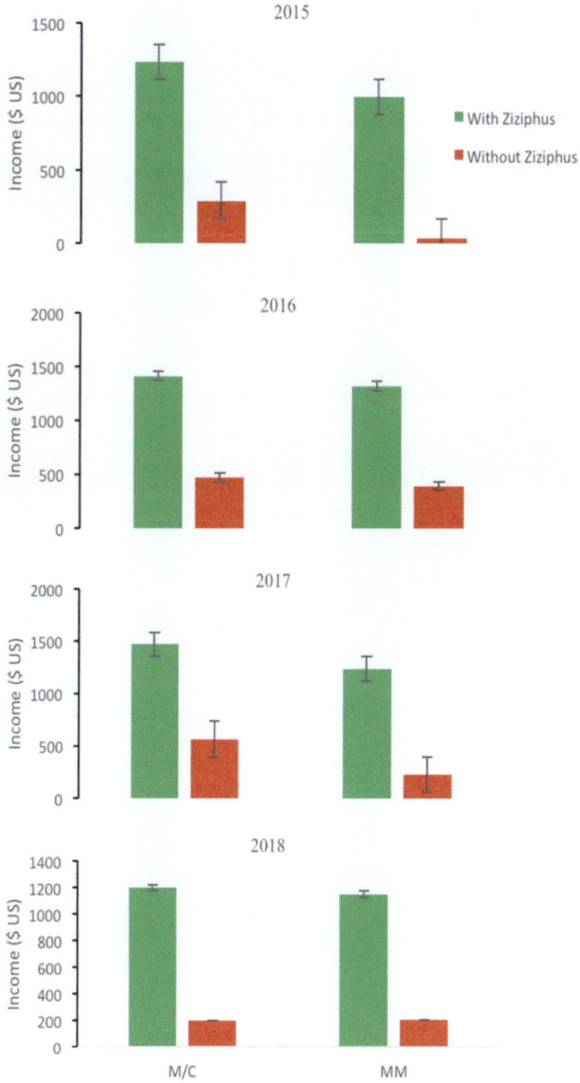

Fig. 7.2 Global revenues provided by cropping systems (mono cropping of millet, MM and millet/cowpea intercropping) as affected by the presence and absence of Ziziphus trees (80 plants ha^{-1}) over 4 years (2015–2018) in Sadoré, Niger. M/C: Millet/Cowpea Intercrop, MM: monoculture of millet. *The prices of grain, fodders and ziziphus on local market were used to calculate the global revenues. The references were: 1 US = 570 CFA; millet grain 0.32 \$ US kg^{-1}); millet Stover 0.03 \$ US kg^{-1}; cowpea grain, 0.53 \$ US kg^{-1}; cowpea fodder 0.16 \$ US kg^{-1}; fruits of ziziphus 1.75 \$US kg^{-1}.* *Source* Adapted from Bado et al. (2018)

afford fertilisers. This is associated with the declines of biomass, grass cover, soil erosion and nutrient mining, resulting in soil fertility loss, land degradation, and desertification (FAO, 1991).

Farm-oriented technologies alone cannot halt the decline in soil fertility, land degradation, or the decreasing productivity of ecosystems. Numerous studies have identified key constraints to livestock productivity, including shrinking rangelands, seasonal feed shortages, lack of livestock routes, and insufficient watering points (Ayantunde et al., 2007). In West Africa's Sahel, rangelands, which are open areas for unrestricted grazing, contribute to the degradation of vegetation and soils

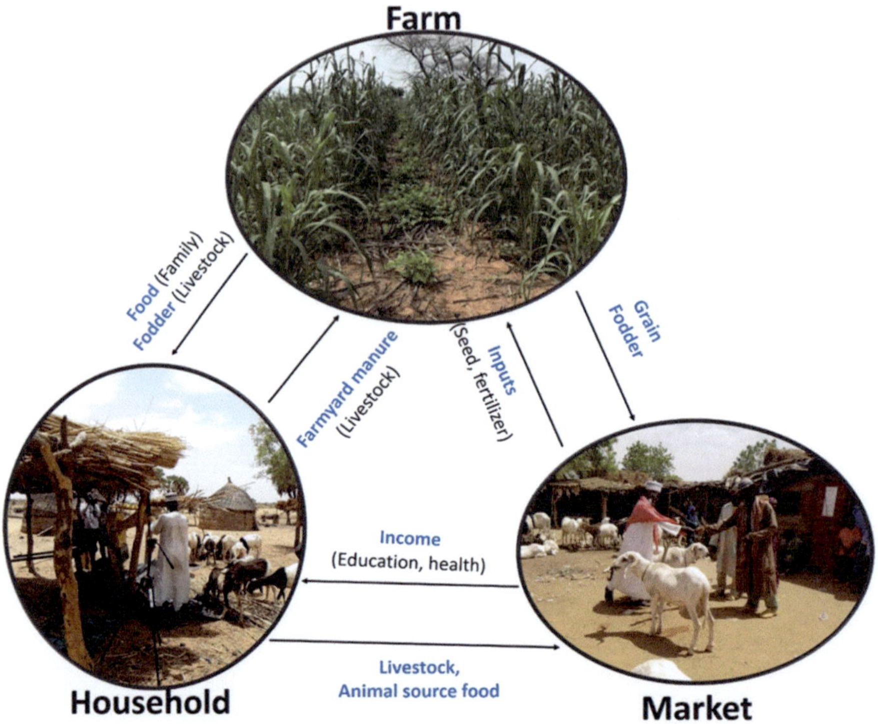

Fig. 7.3 Representation of resource flows among farm, market, and households in a typical crop-livestock farming system in West African Sahel. Adapted from Bado et al (2021)

(Milchunas & Lauenroth, 1993; Teague & Barnes, 2017), reducing productivity and compromising ecosystem resilience (Frank et al., 1998; Peterson et al., 1998). Herrero et al. (2014) reported that over 90 million smallholder farmers in West Africa are livestock keepers, with 80% engaged in mixed crop-livestock or agro-pastoral systems. Indicators of aridification and pasture degradation, such as biomass reductions and the spread of unpalatable species, are primarily driven by overgrazing (Holechek et al., 1999; Kiage, 2013; Mwendera et al., 1997; Penning De Vries & Djiteye, 1982; Savadogo et al., 2008). Contributing factors include poor soil quality, declining fertility due to cropping activities, overgrazing, erosion, and resource depletion, all exacerbated by the rising demand for food (Kiage, 2013). The increasing pressure on food production, limited resources for adopting new technologies, and the need for more land drive the degradation of land, deforestation, and desertification. Breaking this vicious cycle requires global, holistic, and integrated approaches. Few approaches have explored alternative new approaches for integration of management of both lands resources and eco-farming systems in the socio-economical context of smallholder farmers (Giller et al., 2011), availability and constraints for access to technologies, the social, institutional, and political environment. The challenge is to tackle this vicious cycle of land degradation, decrease of agricultural production, poverty and over exploitation of lands.

7.6.2 The Concept

Many studies remain focused either on crops or livestock separately. However, integrated crop and livestock systems, combined with strategic land resource management and active involvement of rural communities, enable more efficient use of natural resources and external inputs. A holistic approach is essential for the integrated management of land resources—encompassing cultivated lands, rangelands, and protected forests—on a global scale (Bridges & Catizzone, 1996, Yaalon, 1996; Bouma, Yaalon & Arnold, 2000; Adeel, 2003).

The land resources exploited by the entire local community (LC) should be perceived as the source of ecosystem services provider. There is a need to adapt an iterative approach by involving stakeholders and LC in our research approaches relate to these complex realities (Giller et al., 2011). Successful outcomes could be achieved on natural resource management at local and regional level with full participation of LC as primary actors (Blay et al., 2004). Whenever possible, local authorities and policymakers should be actively involved in the process of sustainably managing land resources, which include cultivated soils, pastoral lands, protected forests, and water resources. The proposed concept is based on three key principles: (i) land resource management is framed as an ecosystem service provider for the entire community; (ii) a decentralised and participatory approach is adopted to encourage the involvement of local communities, authorities, and other stakeholders; and (iii) the development of tools for the implementation and evaluation of the process.

7.6.3 The Conceptual Framework

The conceptual framework seeks to integrate scientific and local knowledge, along with the expertise of stakeholders, to identify appropriate technologies and practices and develop context-specific land management strategies or scenarios. Its goal is to connect researchers, development actors, policymakers, and local communities (LC) in a participatory and interactive process to create a global strategy for land resource management. To facilitate and coordinate contributions from key actors, tools such as stakeholder platforms (Teague & Barnes, 2017; Bampton, 2003) can be utilised. A lead institution (LI) is chosen through a participatory process among key actors to coordinate the stakeholder platform for land management (SP-LM). The process is led by local communities through committees at various levels to ensure sustainability. Thus, the stakeholder platform, LC committees, and the LI are key tools for implementation. The framework includes three main components: (A) the development/adaptation of land management options (LMOs); (B) testing of LMOs;

and (C) evaluation of LMOs, as shown in Fig. 7.4. The framework operates at the SP-LM level and scales down to community levels (region, commune, and village) in five steps.

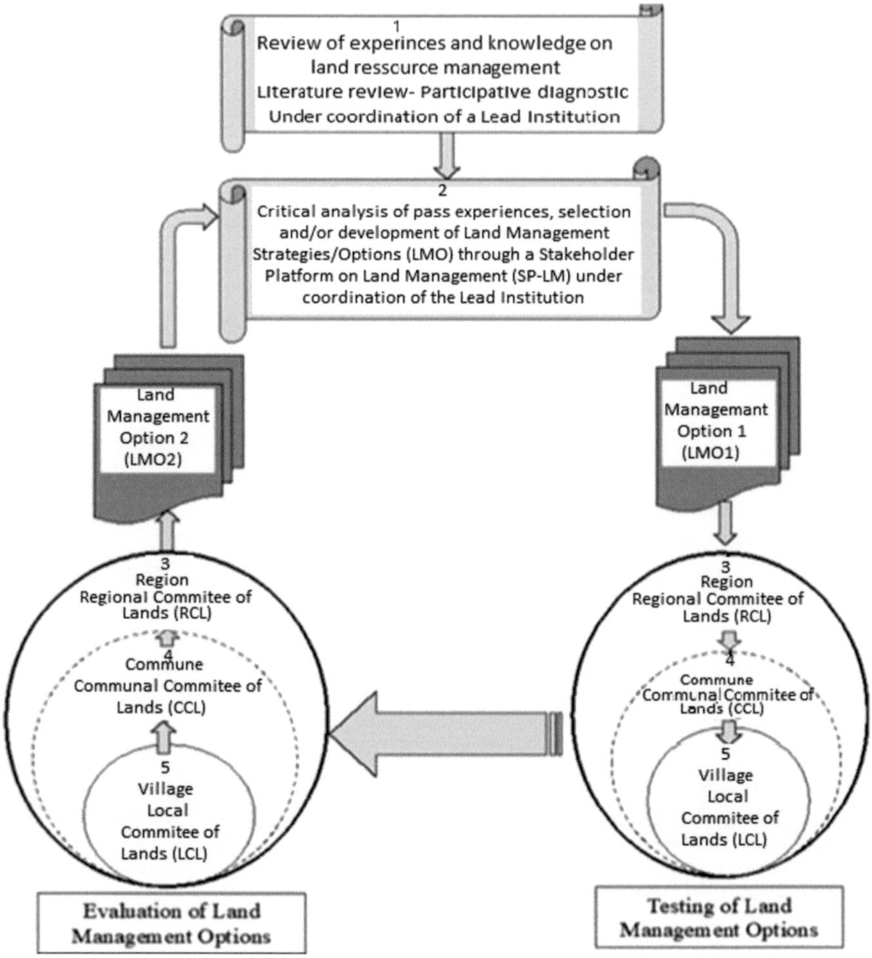

Fig. 7.4 A framework for participative diagnostic and development of integrated strategies, options, technologies and practices for management of land resources from village to commune, region and country levels with local communities. *The process starts with the review, identification of successful strategies, practices and technologies of land resources management at the country level (step 1); development of improved strategies and land management options (LMOs) (step 2); implementation (LMO1 for first testing) at regional, commune and village with local communities (step 3, 4, 5); evaluation of the performances from village to commune and region and refinement to generate new improved management options and strategies (step 3, 4, 3). Source* Bado and Bationo (2018a, 2018b)

7.6.3.1 State of the Art (Step 1)

A synthetic review of past experiences and achievements through literature reviews, coupled with interactions within actors, is the stated point of the process. A quick review, followed by a meeting of stakeholders, can be organised with the following terms of reference: (i) establishing the SP-LM, (ii) identifying a lead institution (LI) to coordinate the process, and (iii) defining terms of reference to document the experiences and achievements of stakeholders in land management. The LI will oversee the preparation of an initial report detailing the current status of land resources, management challenges, stakeholder engagement, policies, programs, successful experiences, failures, and innovations.

7.6.3.2 Identification/Development of LMOs (Step 2)

Under the coordination of the LI, the SP-LM reviews and validates the initial report on achievements and challenges in land resource management. An inventory of technologies, practices, successful experiences, and innovations is conducted to identify effective approaches for managing soils and rangelands at the farmer level and protected forests at the farmer and community levels. Promising practices, technologies, and management strategies are selected for initial implementation as the first phase of LMO1, with the potential for refinement or improvement in subsequent stages.

7.6.3.3 *Testing of LMOs (Steps 3–4–5)*

The implementation of selected options is structured at the regional level, aligned with decentralisation policies that typically progress from regions or districts to communes and villages. Each region and commune can develop local development policies, strategies, and plans for natural resource management. The implementation of Land Management Options (LMOs) occurs across three levels: regional, commune, and village. At the regional level (Step 3), the participatory implementation of LMOs will be presented to regional authorities through meetings and focus group discussions. Based on regional policies, specific communes will be identified in collaboration with regional authorities (Step 4). A similar participatory approach will be applied with communal authorities to select appropriate village sites (Step 5).The selected villages within a commune will establish Local Committees of Lands (LCL) at the village level. These LCLs will come together to form a Communal Committee of Lands (CCL) for the entire commune. Similarly, the CCLs across a region will collaborate to create a Regional Committee of Lands (RCL). At the national level, the Stakeholder Platform for Land Management (SP-LM) will serve as the overarching body. The implementation of Land Management Options (LMOs) and strategies will be coordinated across all levels—village, commune, region, and nation—by the respective LCL, CCL, RCL, and SP-LM.

7.6.3.4 *Evaluation of LMOs (Steps 5–4–3)*

The performance and weaknesses of LMOs will be evaluated progressively at the village (step 5), commune (step 4), and regional (step 3) levels. This evaluation will follow a participatory process involving land management committees (LCL, CCL, RCL, and SP-LM) across village, commune, region, and national tiers. Based on community feedback, the SP-LM will analyse and refine existing technologies, practices, and management innovations (step 1) to enhance the LMOs or develop new ones as the second generation of LMOs (LMO2). A workshop involving the SP-LM and other resource persons or institutions may be convened to further refine or design these improved LMOs. Consequently, LMO2 will represent either an improved version of LMO1 or entirely new LMOs and strategies, informed by the field experience of implementing LMO1. This iterative process will continue as a participatory and adaptive tool for inclusive and sustainable land resource management, engaging LC, communal, regional, and national authorities and stakeholders as needed.

Engaging local communities (LC) is crucial to the process, as it fosters empowerment through farmer participation (Dolinska & Aquino, 2016). By actively involving farming communities, farmers, extension workers, and researchers can collaboratively scale up agricultural practices, creating mutual benefits: researchers gain scientific insights, extension workers improve agricultural practices, and farmers achieve economic success while maintaining autonomy. A notable example of this approach is the rural research centre concept developed by ICRAF, which leverages participatory training and awareness-raising to promote the sustainable management of land resources (Takoutsing et al., 2014).

7.7 Conclusion

The primary drivers of food insecurity, malnutrition, and poverty in sub-Saharan Africa are the depletion of soil fertility, land degradation, desertification, and deforestation. By highlighting the progress made by different concepts over the last 30 years, we noted how local communities could play an important role for sustainable management of dry lands in a participative and holistic approach. We proposed a concept, a strategy and a framework for implementation. This conceptual framework is a dynamic process that allows for ongoing review and adjustments to strategies, tools, and methods to enhance the management process, considering the context, such as the number and profile of stakeholders, national policies, and the engagement of local communities (LC). It serves as a guideline for developing a comprehensive methodology for land resource management, recognising its role as an ecosystem services provider to the entire community.

References

Abalu, G. O. I. (1976). A note on crop mixture under indigenous conditions in northern Nigeria. *Journal of Development Studies, 12*, 11–20.

Abbas, F., Hammad´e, H. M., Fahad, F., Cerd, A., Rizwan, M., Farhad, W., Ehsan, S., & Bakhat, H. F. (2017). Agroforestry: A sustainable environmental practice for carbon sequestration under the climate change scenarios—A review. *Environmental Science and Pollution Research, 24*, 11177–11191. https://doi.org/10.1007/s11356-017-8687-0.

Adeel, Z. (2003). Approaches towards the integration of land management practices: UNU's experience in networking and capacity development. In *International Workshop on Combating Desertification: Rehabilitation of Degraded Drylands and Biosphere Reserves* (pp. 10–13). UNESCO, Aleppo, Syria.

Akponikpe, P. B., Michels, K., & Bielders, C. L. (2008). Integrated nutrient management of pearl millet in the Sahel combining cattle manure, crop residue and mineral fertilizer. *Experimental Agriculture, 44*, 453–472. https://doi.org/10.1017/S001447970800673X

Aune, J. B., & Bationo, A. (2008). Agricultural intensification in the Sahel—the ladder approach. *Agricultural System, 98*, 119–125. https://doi.org/10.1016/j.agsy.2008.05.002

Ayantunde, A. A., Delfosse, P., Fernandez-Rivera, S., Gerard, B., & Dan Gomma, A. (2007). Supplementation with groundnut haulms for sheep fattening in the West African Sahel. *Tropical Animal Health and Production, 39*, 207–216. https://doi.org/10.1007/s11250-007-9009-1

Baanante, C. A. (1998). Economic evaluation of the use of phosphate fertilizers as a capital investment. In A. E. Johnston & J. K. Syers (Eds.), *Nutrient management for sustainable agriculture in Asia* (pp. 109–120). CAB International.

Baanante, C. A., & Hellums D. T. (1998). An analysis of the potential demand for phosphate fertilizers: Sources of change and projections to 2025. In L. D. Currie & P. Loganathan (Eds.), *Long-term nutrient needs for New Zealand's primary industries: Global supply, production requirements and environmental constraints* (pp. 7–31). Occasional Report No. 11. Fertilizer and Lime Research Centre, Massey University, Palmerston North, New Zealand.

Bado, B. V., De Vries, M. E., Haefele, S. M., Marco, M. C. S., & Ndiaye, M. K. (2008). Critical limit of extractable phosphorous in a Gleysol for rice production in the Senegal River valley of West Africa. *Communications in Soil Science and Plant Analysis, 39*, 202–206.

Bado, B. V., Aw, A., & Ndiaye, M. (2010a). Long-term effect of continuous cropping of irrigated rice on soil and yield trends in the Sahel of West Africa. *Nutrient Cycling in Agroecosystems, 88*(1), 133–141. https://doi.org/10.1007/s10705-010-9355-7

Bado, B. V. (2002). Rôle des l_egumineuses sur la fertilit_e des sols ferrugineux tropicaux des zones guin_eenne et soudanienne du Burkina Faso (in French, with English abstract). PhD thesis, Universit_e Laval, Quebec.

Bado, B. V., & Bationo, A. (2018). Integrated management of soil fertility and land resources in Sub-Saharan Africa: Involving local communities. *Advances in Agronomy, 150*, 1–33. https://doi.org/10.1016/bs.agron.2018.02.001.

Bado, B. V., Bationo, A., & Cescas, M. P. (2006a). Assessment of cowpea and groundnut contributions to soil fertility and succeeding sorghum yields in the Guinean savannah zone of Burkina Faso (West Africa). *Biol. Fert. Soils, 43*(2), 171–176. https://doi.org/10.1007/s00374-006-0076-7

Bado, B. V., Bationo, A., & Cescas M. P. (2006). How do cowpea and groundnut improve soil fertility and succeeding sorghum yields in the Guinean Savannah Zone of Burkina Faso West Africa. In *Proceedings Series on Management Practices for Improving Sustainable.*

Bado, B. V., Bationo, A., & Cescas M. P. (2012). Rôles des légumineuses sur la fertilité des sols: opportunités pour une gestion intégrée de la fertilité des sols (p. 150). Éditions universitaires européennes, Saarbrucken, Germany.

Bado, B. V., Lompo, F., Sedogo, M. P., & Cescas, M. P. (2010b). Establishment of the critical limit of soil available phosphorous for maize production in low acidic ultisols of West Africa. *Communications in Soil Science and Plant Analysis, 41*(8), 968–976. https://doi.org/10.1080/00103621003646055

Bado, B. V., Sawadogo, A., Thio, B., Bationo, A., & Cescas M. P. (2006b). How do cowpea and groundnut improve soil fertility and succeeding sorghum yields in the Guinean Savannah Zone of Burkina Faso West Africa. *Proceedings Series*, 287–305. https://doi.org/10.1007/s00374-006-0076-7.

Bado, B. V., Sawadogo, A., Thio, B., Bationo, A., Traor´e, K., & Cescas, M. P. (2011). Nematode infestation and N-effect of legumes on soil and crop yields in legume-sorghum rotations. *Agricultural Science, 2*(2): 49–55. https://doi.org/10.4236/as.2011.22008.

Bado, B. V., Sedogo, M., Lompo, F., & Maman Laminou, S. M. (2018). Biological nitrogen fixation by local and improved genotypes of cowpea in Burkina Faso (West Africa): Total nitrogen accumulated can be used for quick estimation. *Adv. Agric., 2018*, 1–8. https://doi.org/10.1155/2018/9641923

Bado, B. V., Sedogo, M. P., Cescas, M. P., & Lompo F. (1997b). Effet a‘ long terme des fumures sur le sol et les rendements du mais au Burkina Faso. *Cahiers Agriculture, 6*(6), 571–575. http://revues.cirad.fr/index.php/cahiers-agricultures/article/view/30055.

Bado, B. V., Whitbread, A., & Manzo, M. L. S. (2021). Improving agricultural productivity using agroforestry systems: Performance of millet, cowpea, and Ziziphus-based cropping systems in West Africa Sahel. *Agriculture, Ecosystems & Environment, 305*, 107175. https://doi.org/10.1016/j.agee.2020.107175.

Bado, V. B., Bationo, A. (2018). Integrated management of soil fertility and land resources in Sub-Saharan Africa: Involving local communities. In *Advances in Agronomy* (Vol. 150, pp. 1–33). Academic Press. https://doi.org/10.1016/bs.agron.2018.02.001.

Bagayoko, M., Buerkert, A., Lung, G., Bationo, A., & Ro¨mheld, V. (2000). Cereal/legume rotation effects on cereal growth in Sudano- Sahelian West Africa: Soil mineral nitrogen, mycorrhizae and nematodes. *Plant Soil, 218*, 103–116. https://doi.org/10.1023/a:1014957605852.

Bagayoko, M., Buerkert, A., Lung, G., Bationo, A., & Ro€mheld V. (2000). Cereal/legume rotation effects on cereal growth in Sudano-Sahelian West Africa: soil mineral nitrogen, mycorrhizae and nematodes. *Plant Soil, 218*, 103–116.

Bampton, J. F. R. (2003). District forest coordination committee: An emerging multistakeholder platform for collaborative forest management in Nepal's Terai; Rural development forestry network, Overseas Development Institute (ODI), London, UK. *Journal of Forest and Livelihood, 2*(2), 35–47.

Bationo, A. (2008). Organic amendments for sustainable agricultural production in Sudano-Sahelian West Africa. In A. Bationo (Ed.), *Integrated soil fertility management options for agricultural intensification in Sudano-Sahelian Zone of West Africa* (p. 204). Academy Science Publishers.

Bationo, A., & A. Buerkert (2001). Soil organic carbon management for sustainable land use in Sudano-Sahelian West Africa. *Nutrient cycling in Agroecosystems, 61*, 131–142. https://doi.org/10.1023/A:1013355822946.

Bationo, A., Christianson, B. C., & Klaij, M. C. (1993). The effect of crop residue and fertilizer use on pearl millet yields in Niger. *Fert. Res., 34*, 251–258. https://doi.org/10.1007/BF00750571

Bationo, A., Hartemink, A., Lungu, O., Naimi, M., Okoth, P., Smaling, E., & L. Thiombiano (2006). African soils: Their productivity and profitability of fertilizer use. In *Background paper prepared for the African fertilizer summit* (Vol. 2006, p. 25). Abuja, Nigeria.

Bationo, A., Job, K., Bernard, V., Boaz, W., & Joseph, K. (2007a). Soil organic carbon dynamics, functions and management in west African agro-ecosystems. *Agricultural Systems, 94*, 13–25.

Bationo, A., Kihara, J., Vanlauwe, B., Waswa, B., & Kimetu, J. (2007b). Soil organic carbon dynamics, functions and management in West African agro-ecosystems. *Journal of Agriculture,*

Food Systems, and Community Development, 94, 13–25. https://doi.org/10.1016/j.agsy.2005.08.011

Bationo, A., & Mokwunye, A. U. (1991a). Role of manures and crop residues in alleviating soil fertility constraints to crop production: With special reference to the Sahelian and Sudanian zones of West Africa. *Fertilizer Research, 29*, 117–125. https://doi.org/10.1007/BF01048993

Bationo, A., & Mokwunye, A. U. (1991b). Alleviating soil fertility constraints to increased crop production in West Africa: The experience of the Sahel. *Fertilizer Research, 29*, 95–115. https://doi.org/10.1007/BF01048992

Bationo, A., Mokwunye, U., Vlek, P. L. G., Koala, S., & Shapiro, B. I. (2003). Soil fertility management for sustainable land use in the West African Sudano-Sahelian Zone. In M. P. Gichuru, A. Bationo, M. A. Bekunda, H. C. Goma, P. L. Mafongonya, D. N. Mugendi, H. M. Murwira, S. M. Nandwa, P. Nyathi, & M. J. Swift (Eds.), *Soil fertility management in Africa: A regional perspective* (pp. 253–292). Academy Science Publishers.

Bationo, A., & Ntare, B. R. (2000). Rotation and nitrogen fertilizer effects on pearl millet, cow-pea and groundnut yield and soil chemical properties in a sandy soil the semi-arid tropics, West Africa. *Journal of Agricultural Science, 134*, 277–284. https://doi.org/10.1017/S0021859699007650

Bationo, A., Singh, U., Dossa, E., Wendt, J., Agyin-Birikorang, S., Lompo, F., & Bindraban, P. (2020). Improving soil fertility through fertilizer management in Sub-Saharan Africa. In R. Lal (Ed.), *Soil and fertilizers: managing the environmental footprint* (pp. 67–102). CRC Press.

Bationo, A., & Vlek, P. L. G. (1998). The role of nitrogen fertilizers applied to food crops in the Sudano-Sahelian Zone of West Africa. In G. Renard, A. Neef, K. Becker, & M. von Oppen (Eds.), *Soil fertility management in West African land use systems* (pp. 41–51). Margraf Verlag.

Bationo, A., Waswa, B., Adamou, A., Bado, V., Bonzi, M., Iwuafor, E., Kibunja, C., Kihara, J., Mucheru, M., Mugendi, D., Mugwe, J., Mwale, C., Okeyo, J., Olle, A., Roing, K., & Sedego, M. (2012). Overview of long-term experiments in Africa. In A. Bationo, B. Waswa, J. Kihara, I. Adolwa, B. Vanlauwe & K. Saidou (Eds.) *Lessons learned from long-term soil fertility management experiments in Africa* (pp. 1–26). Springer, Dordrecht, The Netherlands. https://doi.org/10.1007/978-94-007-2938-4_1.

Bationo, A. and B.S. Waswa (2011). New challenges and opportunities for integrated soil fertility management in Africa. In Bationo et al. (Eds.) *Innovations as key to the green revolution in Africa: Exploring the scientific facts* (pp. 3–18). Springer, Dordrecht, The Netherlands.

Bayala, J., Balesdent, J., Marol, C., Zapata, Z., Teklehaimanot, Z., & Ouedraogo, S. J. (2006). Relative contribution of trees and crops to soil carbon content in a parkland system in Burkina Faso using variations in natural 13C abundance. *Nutr. Cycl. Agroecosys., 76*, 193–201. https://doi.org/10.1007/s10705-005-1547-1

Bayala, J., Sanou, J.H., Bazié, R., Coe, R., Kalinganire, A., & Sinclair, F. L. (2020). Regenerated trees in farmers' fields increase soil carbon across the Sahel. *Agroforestry System, 94*, 401–415. https://link.springer.com/journal/10457.

Becker, M., & Asch, F. (2005). Iron toxicity in rice conditions and management concepts. *Journal of Plant Nutrition and Soil Science, 168*, 558–573.

Becker, M., & Johnson, J. E. (2001). Improved water control and crop management effects on lowland rice productivity in West Africa. *Nutrient Cycling in Agroecosystems, 59*, 119–127.

Beets, W. (1990). *Raising and sustaining productivity of small-holder farming systems in the tropics.* AgBe Publishing.

Bekunda, M., Sanginga, N., & Woomer, P. L. (2010). Restoring soil fertility in sub-Sahara Africa. *Advances in Agronomy, 108*, 183–236.

Bielders, C. L., & Gerad, B. (2015). Millet response to microdose fertilization in south–western Niger: Effect of antecedent fertility management and environmental factors. *Field Crops Research, 171*, 165–175. https://doi.org/10.1016/j.fcr.2014.10.008

Blanco-Canqui, H., & Lal, R. (2009). Crop residue removal impacts on soil productivity and environmental quality. *Critical Reviews in Plant Sciences, 28*, 139–163. https://doi.org/10.1080/07352680902776507

Blay, D. E., Bonkoudndou, E., Chanshama, S. A. O., & Chikamai B. (2004). Rehabilitation of degraded lands in sub-Saharan Africa: Lessons learned from selected case studies. In *Forestry research network for Sub-Saharan Africa*. Special Programme for Developing Countries.

Bostick, W. M., Bado, V. B., Bationo, A., Soler, C. T., Hoogenboom, G., & Jones J. W. (2007). Soil carbon dynamics and crop residue yields of cropping systems in the Northern Guinea Savanna of Burkina Faso. *Soil and Tillage Research, 93*, 138–151. https://doi.org/10.1016/j.still.2006.03.020.

Bouma, J. (2001). The new role of soil science in a network society. *Soil Science, 166*, 874–879.

Bouma, J., Broll, G., Crane, T. A., Dewitte, O., Gardi, C., Schulte, R. P. O., & Towers, W. (2012). Soil information in support of policy making and awareness raising. *Curr. Opin., 4*(5), 552–558.

Brady, N. (1990). *The nature and properties of soils* (10th ed.). MacMillan Publishers.

Branca, G., Lipper, L., McCarthy, N., & Jolejo M. C. (2013). Food security, climate change, and sustainable land management. A review. *Agronomy for Sustainable Development, 33*, 635-650. https://doi.org/10.1007/s13593-013-0133-1.

Bray, R. H., & Kurtz, T. (1945). Determination of total, organic and available forms of phosphorus in soils. *Soil Science, 59*, 39–46. https://doi.org/10.1097/00010694-194501000-00006

Bridges, E. M., & Catizzone, M. (1996). Soil science in a holistic framework—discussion of an improved integrated approach. *Geoderma, 71*, 275–287.

Buerkert, A., Bationo, A., & Komlan Dossa, K. (2000). Mechanisms of residue mulch-induced cereal growth increases in West Africa. *Soil Science Society of America Journal, 64*, 346–358.

Buerkert, A. K., & Stern, R. D. (1995). Crop residue and P application affect the spatial variability of non-destructively measured millet growth in the Sahel. *Experimental Agriculture, 31*, 429–449.

Buresh, R. J., Smithson, P., & Hellums D. T. (1997). Building soil phosphorus capital in Africa. In R. J. Buresh, P. A. Sanchez & F. Calhoun (Eds.), *Replenishing soil fertility in Africa* (Vol. 51, pp. 111–149). SSSA Special Publication, Madison.

Cassman, K. G., De Datta, S. K., Olk, D. C., Alcantara, J., Samson, M., Descalsota, J., & Dizon M. (1995). Yield decline and the nitrogen economy of long-term experiments on continuous, irrigated rice systems in the tropics. In R. Lal & B. A. Stewart (Eds.), *Soil management: Experimental basis for sustainability and environmental quality. Advances in soil science* (pp. 181–222). CRC Press, Boca Raton, FL.

Chalk, P. M. (1998). Dynamics of biologically fixed N in legume–cereal rotations: A review. *Australian Journal of Agricultural Research, 49*, 303–316. https://doi.org/10.1071/A97013

Chapagain, T., Pudasaini, R., Ghimire, B., Gurung, K., Choi, K., Rai, L., Magar, S. B. K. B., & Raizada, M. N. (2018). Intercropping of maize, millet, mustard, wheat and ginger increased land productivity and potential economic returns for smallholder terrace farmers. *Field Crop Research, 227*, 91–101.

Chien, S. H. (1995). Chemical, mineralogical, and solubility characteristics of phosphate rock for direct application. In K. Dahanayke, S. J. Van Kauwenbergh, D. T. Hellums (Eds.), *Direct application of phosphate rock and appropriate technology fertilizers in Asia: What hinders acceptance and growth* (pp. 49–58). Institute of Fundamental Studies, Muscle Shoals, USA.

Chien, S. H., & Friesen D. K. (1992). Phosphate rock for direct application. In F. J. Sikora (Ed.), *Future Directions for Agricultural Phosphorus Research* (pp. 47–52). Bulletin Y-224. Muscle Shoals, USA, Valley Authority.

Chien, S. H., & Hammond, L. L. (1978). A comparison of various laboratory methods for predicting the agronomic potential of phosphate rock for direct application. *Soil Science Society of America Journal, 42*, 1758–1760.

Chien, S. H., Hammond, L. L., & Leon, L. A. (1987). Long-term reactions of phosphate rocks with an oxisol in Colombia. *Soil Science, 144*, 257–265.

Chien, S. H., & Menon, R. G. (1995). Agronomic evaluation of modified phosphate rock products: IFDC's experience. *Fertilizer Research, 41*, 197–209.

Chien, S. H., & Meno, R. G. (1995). Factors affecting the agronomic effectiveness of phos-phate rock for direct application. *Fertilizer Research, 41*, 227–234.

Chien, S. H., Sale, P. W. G., & Hammond L. L. (1990). Comparison of effectiveness of various phosphate fertilizer products. In *Proceedings of International Symposium on Phosphorus Requirements for Sustainable Agriculture in Asia and Oceania* (pp. 143–156). IRRI, Manila.

Dewitte, O., Arwyn, J., Otto, S., Henrik, B. M., Michel, B., Almami, D., Jozef, D., Tahar, G., Stephen, H., Robert, J., Method, K., Pieter, L. R., Erika, M., Luca, M., Thiombiano, L., Van Ranst, E., Martin, Y., & Robert, Z. (2013). Harmonisation of the soil map of Africa at the continental scale. *Geoderma, 211–212*, 138–153.

Diatta, S., & Sahrawat L. K. (2005). Iron toxicity of rice in West Africa: Screening tolerant varieties and the role of N, P, K, and Zn. In A. Audebert, L. T. Narteh, P. Kiepe, D. Millar, B. Beks (Eds.), *Iron toxicity in rice-based Systems in West Africa* (p. 75). Africa Rice Center (WARDA), Cotonou, Benin.

Diffenbaugh, N. S., & Giorg, F. (2012). Climate change hotspots in CMIP5 global climate model ensemble. *Climate Change, 114*, 813–822. https://doi.org/10.1007/s10584-012-0570-x

Dinesh, D., Campbell, B. M., Bonilla-Findji, O., & Richards M. (2017). 10 best bet innovations for adaptation in agriculture: A supplement to the UNFCCC NAP technical guidelines. CCAFS Working Paper no. 215. Wageningen, the Netherlands.

Dingkuhn, M., & Sow A. (1997). Potential yields of irrigated rice in the Sahel. In K. M. Miezan, M. C. S. Wopereis, M. Dingkuhn, J. Deckers, T. F. Randolph (Eds.), *Irrigated rice in the Sahel: Prospects for sustainable development*. West Africa Rice Development Association, Bouake.

Dobermann, A., George, T., & Thevs, N. (2002). Phosphorus fertilizer effects on soil phosphorus pools in acid upland soils. *Soil Science Society of America Journal, 66*, 652–660.

Dobermann, A., & White, P. F. (1999). Strategies for nutrient management in irrigated and rain-fed lowland rice systems. *Nutrient Cycling in Agroecosystems, 53*, 1–18.

Dolinska, A., & Aquino, P. (2016). Farmers as agents in innovation systems. Empowering framers for innovation though communities of practice. *Agricultural Systems, 142*, 122–130.

Dregne, H. E. (1990). Erosion and soil productivity in Africa. *Journal of Soil and Water Conservation, 45*, 431–436.

Dregne, H. E., & Chou N. T. (1992). Global desertification dimensions and costs. In H. E. Dregne, (Ed.), Degradation and restoration of arid lands. Texas Technical University, Lubbock, TX.

Dunstan, S. C. S., Matlon, P. J., & Lo€ffler H. (2004). African agricultural production and productivity in perspective. Background paper No. 1 was commissioned by the Inter-Academy Council (IAC) Study Panel on Science and Technology Strategies for Improving Agricultural Productivity and Food Security in Africa, Realizing the Promise of African Agriculture. ISBN: 90-6984-418-4.

Eckholm, E. P. (1976). Losing ground: Myth and reality in the humid tropics. *Environmental Sustainability, 4*, 552–558. Norton, New York.

Enyong, L. A., Debrah, S. K., & Bationo, A. (1999). Farmers' perceptions and attitudes towards introduced soil-fertility enhancing technologies in western Africa. *Nutrient Cycling in Agroecosystems, 53*(2), 177–187.

Fairhurst, T., Witt, C., Buresh, R., & Dobermann, A. (2007). *Rice: A practical guide to nutrient management. Nutrient management, nutrient deficiencies, mineral toxicities, tools and information* (2nd ed.). International Rice Research Institute.

Finlay, K. W., & Wilkinson, G. N. (1963). The analysis of adaptation in a plant breeding programme. *Australian Journal of Agricultural Research, 14*, 742–754.

Fixen, P. E., & Grove J. H. (1990). Testing soil for phosphorus. In R. L. Westerman, (Ed.), *Soil testing and plant analysis* (pp. 141–180). Soil Science Society of America, Madison, WI. https://doi.org/10.2136/sssabookser3, 3 ed.c7.

Food and Agriculture Organization of the United Nations (FAO) (1991). Guidelines: Land evaluation for extensive grazing. Soils Bulletin (Vol. 58, p. 158). Food and Agriculture Organization (FAO), Rome, Italy.

Food and Agriculture Organization of the United Nations (FAO) (1999). Fertilizer yearbook 1998. FAO Statistics Series No. 150, Rome.

Food and Agriculture Organization of the United Nations (FAO) (2000). Guide diagnostic participatif des contraintes et des potentialit_es pour la gestion des sols et des _el_ements nutritifs des plantes. FAO, Land and Water Development Division, Rome.

Formoso, M. L. L. (1999). Workshop on tropical soils. Brazilian Academy of Sciences, Rio de Janeiro, Brazil (p. 192).

Frank, D. A., McNaughton, S. J., & Tracy, B. F. (1998). The ecology of the earth's grazing ecosystems. *BioScience, 48*, 513–521.

Gachene, C. K. K., Karanja, N. K., & Mureithi, J. G. (2000). Effect of soil erosion on soil productivity and subsequent restoration of productivity using green manure cover crops: Data summary of a long-term field experiment. *Legume Res. Network Project Newsletter, 4*(4), 8–10.

Garrity, D. P. (1997). Soil carbon dynamics in different cropping systems in principal ecoregions of Asia. In R. Lal, J. M. Kimble, R. F. Follett, B. A. Stewart (Eds.), *Management of carbon sequestration in soil* (pp. 35–57). CRC Press, Boca Raton, FL.

GENSTAT 5 Committee, (1993). GENSTAT 5 Reference Manual. Release 3. Clarendon Press, Oxford, UK.

Gikonyo, E. W., & Smithson P. C. (2003). Effects of farmyard manure, potassium and their combinations on maize yields in the high and medium rainfall areas of Kenya. In A. Bationo (Ed.), Managing nutrient cycles to sustain soil fertility in Sub-Sahara Africa. AfNet-CIAT, 2004, Springer.

Giller, K. E., Rowe, E. C., De Ridder, N., & Van Keulen H. (2006). Resource use dynamics and interactions in the tropics: Scaling up in space and time. *Agricultural systems, 88*, 8–27. https://doi.org/10.1016/j.agsy.2005.06.016.

Giller, K. E., Tittonell, P., Rufino, M. C., van Wijk, M. T., Zingore, S., Mapfumo, P., Adjei- Nsiah, S., Herrero, M., Chikowo, R., Corbeels, M., Rowe, E. C., Baijukya, F., Mwijage, H. A., Smith, J., Yeboah, E., van der Burg, W. J., Sanogo, O. M., Misiko, M., de Ridder, N., Karanja, S., Kaizzi, C., K'ungu, J., Mwale, M., Nwaga, D., Pacini, C., & Vanlauwe, B. (2011). Communicating complexity: Integrated assessment of trade-offs concerning soil fertility management within African farming systems to support innovation and development. *Agricultural systems, 104*, 191–203.

Gisladottir, G., & Stocking, M. (2005). Land degradation control and its global environmental benefits. *Land Degradation and Development, 16*, 99–112.

Gomez, K. A., & Gomez, A. A. (1983). *Statistical procedure for agricultural research* (2nd ed.). John Wiley and Sons.

Gonzalez, P., Tucker, C. J., & Sy, H. (2012). Tree density and species decline in the African Sahel attributable to climate. *Journal of Arid Environments, 78*, 55–64. https://doi.org/10.1016/j.jaridenv.2011.11.001

Guillemin, R. (1956). Evolution de l'agriculture autochtone dans les savannes de l'Oubangui. Deuxie' me partie: L'agriculture Oubanguienne a ses origines. *L'agronomie Tropicale, 11*, 143–176.

Haefele, S. M., Johnson, D. E., Diallo, S., Wopereis, M. C. S., & Janin, I. (2000). Improved soil fertility and weed management is profitable for irrigated rice farmers in the Sahel. *Field Crops Research, 66*, 101–113.

Haefele, S. M., Saito, K., NDiaye, K. M., Mussgnug, F., Nelson, A., & Wopereis M. C. S. (2013). Increasing rice productivity through improved nutrient use in Africa. In M. C. S. Wopereis, D. E. Johnson, N. Ahmadi, E. Tollens, A. Jalloh (Eds.), *Realizing Africa rice promise* (pp. 250–264). CAB International, Wallingford, UK.

Haefele, S. M., Wopereis, M. C. S., Schloebohm, A. M., & Wiechmann, H. (2004). Long-term fertility experiments for irrigated rice in the west African Sahel: Effects on soil characteristics. *Field Crops Research, 83*, 61–77.

Haefele, S. M., Wopereis, M. C. S., & Wiechmann, H. (2002). Long-term fertility experiments for irrigated rice in the west African Sahel: Agronomic results. *Field Crops Research, 78*, 119–131.

Hardter, R., & Horst, W. J. (1990). Nitrogen and phosphorus use in maize sole cropping and maize/cowpea mixed cropping systems on an Alfisol in northern Guinea Savannah of Ghana. *Biology and Fertility of Soils, 10*, 267–275.
Hartemink, A. E., & McBratney, A. (2008). A soil science renaissance. *Geoderma, 148*, 123–129.
Henao, J., & Baanante C. A. (1999). Estimating rates of nutrient depletion in soils of agricultural lands in Africa, IFDC. Muscle Shoals, Alabama.
Hauggaard-Nielsen, H., Ambus, H., & Jensen, E. S. (2001). Temporal and spatial distribution of roots and competition for nitrogen in pea-barley intercrops; a field study employing 32P technique. *Plant and Soil, 236*, 63–74.
Helmke, P. A., & Sparks D. L. (1996). Lithium, sodium, potassium, rubidium and cesium. In J. M. Bigham, (Ed.), *Method of soil analysis. Part 3. Chemical methods* (pp. 551–574). Soil Science Society of America/American Society of Agronomy, Madison, WI.
Herrero, M., Havlik, P., McIntire, J., Palazzo, A., & Valin H. (2014). African livestock futures: Realizing the potential of livestock for food security, poverty reduction and the environment in Sub-Saharan Africa. Office of the Special Representative of the UN Secretary General for Food Security and Nutrition and the United Nations System Influenza Coordination (UNSIC), Geneva, Switzerland.
Holechek, J. L., Gomez, H., Molinar, F., & Galt, D. (1999). Grazing studies: What we've learned. *Rangelands, 21*, 12–16.
Ibrahim, A., Abaidoo, R. C., Fatondji, D., & Opoku A. (2015a). Hill placement of manure and fertilizer micro-dosing improves yield and water use efficiency in the Sahelian low input millet-based cropping system. *Field and Crop Research, 180*, 29–36. https://doi.org/10.1016/j.fcr.2015.04.022
Ibrahim, A., Abaidoo, R. C., Fatondji, D., & Opoku, A. (2015b). Integrated use of fertilizer micro dosing and *Acacia tumida* mulching increases millet yield and water use efficiency in Sahelian semi-arid environment. *Nutrient Cycling in Agroecosystems, 103*, 375–388. https://doi.org/10.1007/s10705-015-9752-z
Ibrahim, A., Pasternak, D., Guimbo, I. D., Saidou, A.S., & Amadou, M. (2015c). Rain-fed plantation of the domesticated *Ziziphus mauritiana* in the Sahel: Effects of varieties and rootstocks on yields and fruit quality. *Journal of Horticultural Research, 23*(1), 33–38. https://doi.org/10.2478/johr-2015-0005.
ICRISAT (1985–1988). ICRISAT Sahelian Center Annual Report, 1984. International Crops Research Institute for the Semi-Arid Tropics, Patancheru, India.
Izac, A-M. N. (2000). What paradigm for linking poverty alleviation to natural resources management? In *Proceedings of an International Workshop on Integrated Natural Resource Management in the CGIAR: Approaches and Lessons, August 21–25, 2000, Penang, Malaysia.*
Kermah, M., Franke, A. C., Adjei-Nsiah, S., Ahiabor, B. D. K., Abaidoo, R. C., & Giller, K. E. (2017). Maize-grain legume intercropping for enhanced resource use efficiency and crop productivity in the Guinea savanna of northern Ghana. *Field Crops Research, 213*, 38–50. https://doi.org/10.1016/j.fcr.2017.07.008
Kiage, L. M. (2013). Perspectives on the assumed causes of land degradation in the rangelands of Sub-Saharan Africa. *Progress in Physical Geography, 5*, 664–684.
Liu, E., Changrong, Y., Xurong, M., Wenqing, H., Bing, S. H., Ding, L., Liu, Q., Liu, S., & Fan, T. (2010). Long-term effect of chemical fertilizer, straw, and manure on soil chemical and biological properties in northwest China. *Geoderma, 158*, 173–180. https://doi.org/10.1016/j.geoderma.2010.04.029
Lompo, F. (1983). Probl_ematique de la matie're organique dans la zone du plateau Mossi. Etude de la disponibilit_e des re_esidus culturaux et de leur mode de transformation. In Memoire de fin d'etudes. ISP/Universit_e d'Ouagadougou, Burkina Faso, p. 108.
Lorenz, K., & Lal, R. (2014). Soil organic carbon sequestration in agroforestry systems: A review. *Agronomy for Sustainable Development, 34*(2), 443–454. https://doi.org/10.1007/s13593-014-0212-y

Mando, A., Ouattara, B., Somado, A. E., Wopereis, M. C. S., Stroosnijder, L., & Breman, H. (2005). Long-term effects of fallow, tillage and manure application on soil organic matter and nitrogen fractions and on sorghum yield under Sudano-Sahelian conditions. *Soil Use and Management, 21*(1), 25–31.

McLean, E. O. (1983). Soil pH and lime requirement. In A. L. Page, R. H. Miller, D. R. Keeney (Eds.), Methods of soil analysis. Part 2. Chemical and microbiological properties (2nd ed., pp. 199–223). American Society of Agronomy and Soil Science Society of America, Madison, WI. https://doi.org/10.2134/agronmonogr9.2.2ed.c12.

McNair Bostick, W., Bado, B. V., Bationo, A., Soler, C. T., Hoogenboom, G., & Jones J. W. (2007). Soil carbon dynamics and crop residue yields of cropping systems in the Northern Guinea Savanna of Burkina Faso. *Soil and Tillage Research, 93*, 138–151. https://doi.org/10.1016/j.still.2006.03.020.

Milchunas, D. G., & Lauenroth, W. K. (1993). Quantitative effects of grazing on vegetation and soils over a global range of environments. *Ecological Monographs, 63*, 327–366.

Mirasol, F. P., Eufrocino, V. L., Gines, H. C., & Buresh, J. R. (2008). Soil carbon and nitrogen changes in long-term continuous lowland rice cropping. *Soil Science Society of America Journal, 72*, 798–807.

Monjardino, M., McBeath, T. M., Brennan, L., & Llewellyn R. (2013). Are farmers in low- rainfall cropping regions under-fertilising with nitrogen? A risk analysis. *Agricultural System, 116*, 37–51. https://doi.org/10.1016/j.agsy.2012.12.007.

Mucheru-Muna, M., Pypers, P., Mugendi, D., Kung'u, J., Mugwe, J., Merckx, R., & Vanlauwe B. (2010). A staggered maize–legume intercrop arrangement robustly increases crop yields and economic returns in the highlands of Central Kenya. *Field Crops Research, 115*(2), 132–139.

Mwendera, E. J., Saleem, M. A. M., & Woldu, Z. (1997). Vegetation response to cattle grazing in the Ethiopian highlands. *Agriculture, Ecosystems & Environment, 64*, 43–51.

Nair, P. K. R., Kumar, B. M., & Nair, V. D. (2009). Agroforestry as a strategy for carbon sequestration. *Journal of Plant Nutrition and Soil Science, 172*(1), 10–23. https://doi.org/10.1002/jpln.200800030

Ndoli, A., Baudron, F., Schut, A. G. T., Mukuralinda, A., & Giller, K. E. (2017). Disentangling the positive and negative effects of trees on maize performance in smallholdings of Northern Rwanda. *Field Crops Research, 213*, 1–11. https://doi.org/10.1016/j.fcr.2017.07.020

Nelson, D. W., & Sommers, L. E. (1980). Total nitrogen analysis of soil and plant tissues. *Journal of the Association of Official Analytical Chemists, 63*, 770–778.

Nelson, D. W., & Sommers L. E. (1983). Total carbon, organic carbon and organic matter. In A. L. Page, R. H. Miller, D. R. Keeney (Eds.), *Methods of soil analysis. Part 2. Chemical and microbiological properties* (2nd ed., pp. 538–578). American Society of Agronomy and Soil Science Society of America, Madison, WI.

Nelson, W. C. D., Hoffmann, M. P., Vadez, V., Ro¨tter, R. P., Koch, M., & Whitbrea, A. M. (2021). Can intercropping be an adaptation to drought? A model based analysis for pearl millet–cowpea. *Journal of Agronomy and Crop Science, 00*, 1–18. https://doi.org/10.1111/jac.12552.

Nknoya, E., Kato, E., Crespo, S., Place, F., Pender, J., Mwanjololo, M., Okhimamhe, A., Ndjeunga, J., Traore, S., van Rheenen, T., & Ferguson, J. (2013). Climate change adaptation and sustainable land management in Sub-Saharan Africa. *Agro Environment, 12*(6), 1206.

Olk, D. C., Cassman, K. G., Randall, E. W., Kinchesh, P., Sanger, L. J., & Anderson, J. M. (1996). Changes in chemical properties of organic matter with intensified rice cropping in tropical lowland soil. *European Journal of Soil Science, 47*, 293–303.

Palm, C., Sanchez, P., Ahamed, S., & Awiti A. (2007). Soils: A contemporary perspective. *Annual Review of Environment and Resources, 32*, 99–129.

Palm, C. A., Smukler, S. M., Sullivan, C. C., Mutuo, P. K., Nyadzi, G. I., & Walsh, M. G. (2010). Identifying potential synergies and trade-offs for meeting food security and climate change objectives in sub-Saharan Africa. *Proceedings of the National Academy of Sciences of the United States of America, 107*, 19661–19666.

Penning De Vries, F. W. T., & Djiteye M. A. (1982). La Productivit_e des P^aturages Sah_eliens. Une Etude des Sols, des V_eg_etations et de L'exploitation de Cette Ressource Naturelle. Agricultural Research Reports, 918, Pudoc, Wageningen.

Peoples, M. B., & Crasswell, E. T. (1992). Biological nitrogen fixation: Investment, expectation and actual contribution to agriculture. *Plant and Soil, 141*, 13–39.

Peoples, M. B., & Herridge, D. F. (1990). Nitrogen fixation by legumes in tropical agriculture. *Advances in Agronomy, 44*, 155–223.

Pereira, L. (2017). Climate change impacts on agriculture across Africa. In *Oxford Research Encyclopedia of Environmental Science*. https://doi.org/10.1093/acrefore/9780199389414.013.292.

Peterson, G., Allen, G. R., & Holling, C. S. (1998). Ecological resilience, biodiversity and scale. *Ecosystems, 1*, 6–18.

Pichot, J. P., Sedogo, M. P., Poulain, J. F., & Arrivets, J. (1981). Evolution de la fertilité d'un sol ferrugineux tropical sous l'influence de fumures minérales et organiques. *Agronomie Tropicale, 36*, 122–133.

Pieri, C. (1989). Fertilit´e des terres de savane. Bilan de trente ans de recherche et de d´eveloppement agricoles au Sud du Sahara. Minist'ere De La Coop´eration Et Du D´eveloppment. CIRAD, Paris, France.

Powell, J. M., & Unger, P. W. (1998). Alternatives to crop residues for sustaining agricultural productivity and natural resource conservation. *Journal of Sustainable Agriculture, 11*(2–3), 59–84.

Powlson, D. S., & Olk D. C. (2000). Long-term soil organic matter dynamics. In G. J. D. Kirk, D. C. Olk (Eds.), *Carbon and nitrogen dynamics in flooded soils. Paper Presented at the Workshop on Carbon and Nitrogen Dynamics in Flooded Soils*. IRRI, Los Ban˜os, 19–22 April 1999.

Ripoche, A., Crétenet, M., Corbeels, M., Affholder, F., Naudin, K., Sissoko, F., & Tittonell, P. (2015). Cotton as an entry point for soil fertility maintenance and food crop productivity in savannah agroecosystems–Evidence from a long-term experiment in southern Mali. *Field Crops Research, 177*, 37–48.

Rusinamhodzi, L., Corbeels, M., Nyamangara, J., & Giller, G. E. (2012). Maize–grain legume intercropping is an attractive option for ecological intensification that reduces climatic risk for smallholder farmers in central Mozambique. *Field Crops Research, 136*, 12–22.

Sahrawat, K. L. (2004). Organic matter accumulation in submerged soils. *Advances in Agronomy, 81*, 169–201.

Sahrawat, K. L., Jones, M. P., & Diatta, S. (1997). Extractable phosphorous and rice yield in an Ultisol of the humid forest zone in West Africa. *Communications in Soil Science and Plant Analysis, 28*, 711–716. https://doi.org/10.1080/00103629709369823

Sahrawat, K. L., Jones, M. P., Diatta, S., & Sika, M. (2003). Long-term phosphorus fertilizer effects on phosphorus uptake, efficiency, and recovery by dryland rice on an Ultisol. *Communications in Soil Science and Plant Analysis, 14*, 999–1011.

Sanchez, P. A., Shepherd, K. D., Soule, M. J., Place, F. M., Buresh, R. J., Izac, A.-M. N., Mokwunye, A. U., Kwesiga, F. R., Ndiritu, C. G., Woomer P. L. (1997). Soil fertility replenishment in Africa: An investment in natural resource capital. In R. J. Buresh, P. A. Sanchez, F. Calhoun (Eds.), *Replenishing soil fertility in Africa*. SSSA Special Publication (No. 51, pp. 1–46). Soil Science Society of America, Madison, WIisconsin.

Savadogo, P., Tiveau, D., Sawadogo, L., & Tigabu M. (2008). Herbaceous species responses to long term effects of prescribe fire, grazing and selective cutting tree in the savanna woodland of West Africa. *Perspectives in Plant Ecology, Evolution and Systematics, 10*, 179–195.

Scherr, S. J. (1999). Past and present effects of soil degradation. In S. J. Scherr (Ed.), *Soil degradation—A threat to developing-country food security by 2020*. International Food Policy Research Institute 2020 Discussion Paper 27 (pp. 13–30), Washington, DC.

Schroth, G. (1998). A review of belowground interactions in agroforestry, focussing on mechanisms and management options. *Agroforestry System, 43*, 5–34. https://doi.org/10.1007/978-94-017-0679-7_1.

Singh, R. P., Saharan, N., & Ong, C. K. (1989). Above and below ground interactions in alley-cropping in semi-arid India. *Agroforestry Systems, 9*, 259–274. https://doi.org/10.1007/BF00141088

Sivakumar, M. V. K. (1986). Climate of niamey. Progress Report No. 1. ICRISAT Sahelian Center, Niamey, Niger.

Smaling, E. M. A. (1993). *An agro-ecological framework of integrated nutrient management with special reference to Kenya*. Wageningen Agricultural University.

Smaling, E. M. A. (1995). The balance may look fine when there is nothing you can mine: nutrient stocks and flows in West African soils. In H. Gerner, A. U. Mokwunye (Eds.), *Proceedings of a Seminar on the Use of Local Mineral Resources for Sustainable Agriculture in West Africa*. IFDC, Africa.

Steiner, K. G. (1991). Overcoming soil fertility constraints to crop production in West Africa: Impact of traditional and improved cropping systems on soil fertility. In A. U. Mokwunye (Ed.), *Alleviating soil fertility constraints to increased crop production in West Africa* (pp. 69–91). Kluwer Academic Publishers, Dordrecht, the Netherlands. https://doi.org/10.1007/978-94-011-3224-4_7.

Stoorvogel, J. J., Smaling, E. M. A., Janssen B. H. (1993). Calculating soil nutrient balances in Africa at different scales. I. Supra-National Scale. *Fertilizer Research, 35*, 227–235.

Subbarao, G. V., Renard, C., Payne, W. A., & Bationo, A. (2000). Long-term effects of tillage, phosphorus fertilization and crop rotation on pearl millet-cowpea productivity in the West-African Sahel. *Experimental Agriculture, 36*, 243–264. https://doi.org/10.1017/S0014479700002106

Suzuki, K., Matsunaga, R., Hayashi, K., Matsumoto, N., Tobita, S., Bationo, A., & Okada, K. (2016). Long-term effects of fertilizer and organic matter application on millet in Niger. *Agronomy Journal, 108*(2), 873–883.

Tabo, R., Bation, A., Hassane, O., Amadou, Fosu, M., Sawadogo-Kabore, S., Fatondji, D., Korodjouma, O., Abdou, A., & Koala, S. (2008). Fertilizer microdosing for the prosperity of resource poor farmers: A success story. In *Proceedings of the Workshop on Increasing the Productivity and Sustainability of Rainfed Cropping Systems of Poor, Smallholder Farmers*. Tamale, Ghana

Takoutsing, B., Tchoundjeu, Z., Degrande, A., Asaah, E., & Tsobeng, A. (2014). Scaling-up sustainable land management practices through the concept of the rural resource centre: Reconciling farmers' interests with research agendas. *The Journal of Agricultural Education and Extension, 20*(5), 463–483.

Teague, R., & Barnes, M. (2017). Grazing management that regenerates ecosystem function and grazingland livelihoods. *African Journal of Range & Forage Science, 34*, 77–86.

Thiombiano, L. (2000). Etude des Facteurs _edaphiques et P_edopaysagiques Dans le D_eveloppement de la Desertification en Zone Sahelienne du Burkina Faso. The‘se de Doctorat d'Etat (Vol. I, p. 200). plus Annexes, Universit_e de Cocody, Abidjan, Co^te d'Ivoire.

Tilander, Y., & Ong, C. K. (1999). Conservation of and competition for water and nutrients in semi-arid agroforestry. *Annals of Arid Zone, 38*(3–4), 309–334.

Tittonell, P., & Giller, E. K. (2013). When yield gaps are poverty traps: The paradigm of ecological intensification in African smallholder agriculture. *Field Crops Research, 143*, 76–90. https://doi.org/10.1016/j.fcr.2012.10.007

Twomlow, S., Rohrbach, D., Dimes, J., Rusike, J., Mupangwa, W., Ncube, B., Hove, L., Moyo, M., Mashingaidze, N., & Mahposa, P. (2010). Micro-dosing as a pathway to Africa's green revolution: Evidence from broad-scale on-farm trials. *Nutrient Cycling in Agroecosystems, 88*, 3–15. https://doi.org/10.1007/s10705-008-9200-4

Udawatta, R. P., Gantzer, C. J., Anderson, S. H., & Garrett, H. E. (2008). Agroforestry and grass buffer effects on pore characteristics measured by high-resolution X-ray computed tomography. *Soil Science Society of America Journal, 72*, 295–304. https://doi.org/10.2136/sssaj2007.0057

Umutoni, C., Bado, V., Whitbread, A., & Ayantunde, A. (2021a). Assessment of stovers of dual-purpose pearl millet varieties as feed for goats in the West African Sahel. *International Journal of Livestock Research, 11*(12), 37–43.

Umutoni, C., Bado, V. B., Whitbread, A. M., Ayantunde, A., & Gangashetty P. (2021). Evaluation of chemical composition and in vitro digestibility of stovers of different pearl millet varieties and their effect on performance of sheep in the West African Sahel. Acta Agriculturae Scandinavica, Section A—Animal Science (SAGA). https://doi.org/10.1080/09064702.2021.1919193.

UNDP/GEF (2004). Reclaiming the land sustaining livelihoods: lessons for the future. United Nations Development Fund/Global Environmental Facility.

Várallyay, G. (2007). Potential impacts of climate change on agro-ecosystems. *Agriculturae Conspectus Scientificus, 72*(1), 1–8.

Vanlauwe, B. (2004). Integrated soil fertility management research at TSBF: the framework, the principles, and their application. In A. Bationo, (Ed.), *Managing nutrient cycles to sustain soil fertility in Sub-Saharan Africa.* Academy Science Publishers, Nairobi, Kenya.

Vanlauwe, B., Bationo, A., Chianu, J., Giller, K. E., Merckx, R., Mokwunye, U., Ohiokpehai, O., Pypers, P., Tabo, R., Shepherd, K. D., Smaling, E. M. A., Woomer, P. L., & Sanginga, N. (2010a). Integrated soil fertility management. Operational definition and consequences for implementation and dissemination. *Outlook Agriculture, 39*, 17–24. https://doi.org/10.5367/000000010791169998

Vanlauwe, B., Bationo, A., Giller, K. E., Merckx, R., Mokwunye, U., Ohiokpehai, O., Pypers, P., Tabo, R., Shepherd, K. D., Smaling, E. M. A., Woomer, P. L., & Sanginga, N. (2010b). Integrated soil fertility management. Operational definition and consequences for implementation and dissemination. *Outlook Agriculture, 39*, 17–24.

Vlek, P., Le, Q. B., Tamene L. (2008). Land decline in land-rich Africa—A creeping disaster in the making (p. 55). CGIAR Science Council Secretariat, Roma, Italy.

Waddington, S. R., Mekuria, M., Siziba, S., & Karigwindi, J. (2007). Long-term yield sustainability and financial returns from grain legume–maize intercrops on a sandy soil in subhumid north central Zimbabwe. *Experimental Agriculture, 43*(4), 489–503.

Walkley, A., & Black, I. A. (1934). An examination of the Detjareff method for determining soil organic matter, and a proposed modification of the chromatic acid titration method. *Soil Science, 37*, 29–38. https://doi.org/10.1097/00010694-193401000-00003

Wallace, J. S. (1996). The water balance of mixed tree-crop systems. In C. K. Ong, P. A. Huxley (Eds.), *Tree-crop interactions. A physiological approach* (pp. 189–233). CAB International, Wallingford, UK.

Wang, G. H., Dobermann, A., Witt, C., Sun, Q. Z., & Fu, R. X. (2001). Performance of site-specific nutrient management for irrigated rice in Southeast China. *Agronomy Journal, 93*, 869–878.

West, L. T., Wilding, L. P., Landeck, J. K., & Calhoun, F. G. (1984). *Soil Survey of the ICRISAT Sahelian Center*. Tropsoils, Texas A&M University, College Station, TX.

Whitbread, A., Blair, G., Konboon, Y., Lefroy, R., & Naklang, K. (2003). Managing crop residues of rice and wheat cropping systems in Thailand and Australia. *Agriculture, Ecosystems & Environment, 100*, 251–263. https://doi.org/10.1016/S0167-8809(03)00189-0

Whitbread, A. M., Davoren, C., Gupta, V. V. S. R., Llewellyn, R., & Roget, D. (2015). Long-term cropping system studies support intensive and responsive cropping systems in the low rainfall Australian Mallee. *Crop Pasture Science, 66*, 553–565. https://doi.org/10.1071/CP14136.

Whitbread, A. M., Lefroy, R. D. B., & Blair, G. J. (1998). A survey of the impact of cropping on soil physical and chemical properties in north-western New South Wales. *Australian Journal of Soil Research, 36*, 669–681. https://doi.org/10.1071/S97031

Witt, C., Cassman, K. G., Olk, D. C., Biker, U., Liboon, S. P., Samson, M. I., & Ottow, J. C. G. (2000). Crop rotation and residue management effects on carbon sequestration, nitrogen cycling and productivity of irrigated rice systems. *Plant and Soil, 225*, 263–278.

Witt, C., Dobermann, A., Abdulrachman, S., Gines, H. C., Wang, G. H., Nagarajan, R., Woomer, P. L., Martin, A., Albretch, A., Resck, D. V. S., & Scharpenseel H. W. (1994). The importance and management of soil organic matter in the tropics. In P. L. Woomer, M. Swift (Eds.), *The biological management of tropical soil fertility* (pp. 47–80). Wiley-Sayce, Chichester, UK.

Wopereis-Pura, M. M., Watanabe, H., Moreira, J., & Wopereis, M. C. S. (2002). Effect of late nitrogen application on rice yield, grain quality and profitability in the Senegal River valley. *European Journal of Agronomy, 17*, 191–198.

Wopereis, M. C. S., Donovan, C., Nebie, B., Guindo, D., & Ndiaye M. K. (1999). Soil fertility management in irrigated rice systems in the Sahel and savanna regions of West Africa. Part I. Agronomic analysis. *Field Crops Research, 61*, 125–145.

Yaalon, D. H. (1996). Soil science in transition—soil awareness and soil care research strategies. *Soil Science, 161*, 3–8.

Yaalon, D. H., & Arnold, R. W. (2000). Attitudes toward soils and their societal relevance: Then and now. *Soil Science, 165*, 5–12.

Yamoah, C. F., Bationo, A., Shapiro, B., & Koala, S. (2002). Trend and stability analyses of millet yields treated with fertilizer and crop residues in the Sahel. *Field Crops Research, 75*, 53–62. https://doi.org/10.1016/S0378-4290(02)00008-4

Yanggen, D., Kelly, V., Reardon, T., Naseem A. (1998). Incentives for fertilizer use in sub-Saharan Africa: A review of empirical evidence on fertilizer response and profitability. MSU International Development Working Paper No. 70. http://www.aec.msu.edu/agecon/fs2/index.htm.

Zapata, F., Axmann, H., Braun H. (1986). Agronomic evaluation of rock phosphate materials by means of radioisotope techniques. In ISSS & BG (Eds.) Trans. 13th Int. Cong. Soil Sci. 3, Hamburg, Germany, pp. 1012–1013.

Zapata, F., & Roy R. N. (2004). Use of phosphate rocks for sustainable agriculture. In *FAO fertilizer and plant nutrition Bulletin 13*. Rome.

Zhang, D., Dub, G., Sun, Z., Bai, W., Wang, Q., Feng, L., Zheng, J., Zhang, Z., Liu, Y., Yang, S., Yang, N., Feng, C., Cai, Q., Evers, J. B., van der Werf, W., & Zhang, L. (2018). Agroforestry enables high efficiency of light capture, photosynthesis and dry matter production in a semi-arid climate. *European Journal of Agronomy, 94*, 1–11. https://doi.org/10.1016/j.eja.2018.01.001

Zhang, M., & He, Z. (2004). Long term changes in organic carbon and nutrients of an Ultisol under rice cropping in south-East China. *Geoderma, 118*, 167–179.

Zhang, W., Wang, B. J., Gan, Y. W., Duan, Z. P., Hao, X. D., Xu, W. L., Lv, X., & Li, L. H. (2017). Competitive interaction in a jujube tree/wheat agroforestry system in northwest China's Xinjiang Province. *Agroforestry Systems, 91*, 881–893. https://doi.org/10.1007/s10457-016-9962-7

Chapter 8
Challenges of Drought and Water Resources Management for Resilient Ecosystem

Rajendra Prasad Pandey

8.1 Introduction

The water bodies, landscape and soil nurture habitat and aquatic species due to the presence of water on the earth. Changes in various components of hydrologic cycle due to a variety of reasons are increasingly becoming a threat to the ecosystem. Therefore, to assess the impact of hydrology on the ecosystem, ecohydrology has emerged as one of the new scientific streams of interdisciplinary science between hydrology and ecology. Ecohydrology is related to the elements of hydrological processes which take place within water bodies, such as rivers and lakes, or on land, in forests, deserts, and other terrestrial ecosystems. Droughts have always been a major concern of the society and ecosystem. Drought-affected regions on the earth are primarily responsible for below-average-prosperity or poverty in regions. Despite many noteworthy achievements of science and technology in the twentieth century, extreme hydrological events, particularly droughts and floods, continue to hit society and natural ecosystems and weaken developmental efforts by breaking continuity (Pandey et al., 2016). The frequent occurrence of droughts causes adversaries in sustainable development. Droughts cause loss to agricultural crops, and vegetation/forest and induce the threats of economic uncertainties and water deficits. Global climate change is one of the key factors affecting the hydrological cycle (IPCC, 2007). A rise in temperature affects the atmospheric moisture, precipitation and circulation pattern of the atmosphere, e.g. changes in the rate of evaporation may intensify the hydrological cycle leading to more frequent hydrologic extremes (i.e. drought and floods) (Pandey et al., 2016). Global warming is one of the primary causes of the increased severity and frequency of drought events in recent decades (Dai, 2013; Trenberth et al., 2014; Liu et al., 2016). Further, the deviation and change in climate increases rainfall variability and leads to the occurrence of more frequent droughts.

R. P. Pandey (✉)
Department of Hydrology, Indian Institute of Technology (IIT) Roorkee, Roorkee, India
e-mail: rppanndey@gmail.com

M. S. Rathore et al. (eds.), *Desert Ecosystems*,
https://doi.org/10.1007/978-3-032-03237-9_8

Regions with greater inter-annual variability of precipitation are more susceptible to the occurrence of frequent and severe droughts (Ponce et al., 2000; Wilhite, 2000; IPCC, 2007).

Drought frequency, severity and persistence in successive years vary across the climatic regions (Dracup et al., 1980; Pandey & Ramasastri, 2001; Ponce et al., 2000). Drought is defined as a relative deficit in a given area compared to its average or usual water availability, either in the form of rainfall, river flow, surface/groundwater storage or due to a combination of these for a certain period of time (Pandey et al., 2016). This reveals that droughts are temporary climatic phenomena defined by the magnitude of relative rainfall deficit (Pandey et al., 2016). Often, the terms 'drought' and 'water scarcity' are used synonymously which is not correct. Water scarcity refers to long-term unsustainable use of water resources, associated with excessive exploitation of water resources. Thus, water scarcity is an anthropological phenomenon. Droughts differ from other natural hazards such as cyclones, floods and tsunamis. Unlike other disasters, they have a slow process of onset and are intricate to determine the beginning and termination of the event. The impacts of droughts are multi-sectoral, non-structural and spread over larger geographical areas and therefore, their quantification is more complex.

The distribution of rainfall in the monsoon season, the time of onset and withdrawal of monsoon rains and dry-spells during the monsoon period are the primary factors that cause droughts in India. Since the monsoon rainfall is highly erratic, highly variable and unevenly distributed across the climatic regions in India, the drought conditions may prevail in almost any year in one or the other parts of the country. Usually, rain-fed agricultural activities go on and off due to recurrent droughts in arid and semi-arid regions of India. Recent drought events in India have illustrated the vulnerability of all parts of arid, semi-arid and dry-sub-humid climatic regions of the country. For example, a persistent drought condition during 2001–2003 resulted in serious losses to crops and fodder and caused mass migration for population and livestock in Rajasthan and Gujarat state. Experiences of past drought management efforts at the state and national levels indicate that the efforts have been largely inadequate in the form of financial assistance or relief to the affected farming community and farm workers through temporary employment opportunities including suspension of loan recovery, etc. The drought affected areas need proactive preparedness and mitigation efforts to cope with drought.

The countries like United States of America (USA) and Australia have evolved acceptable tools and drought preparedness planning to reduce the risks of future events (Wilhite et al., 2000). Consequently, about 30 states in the USA have prepared their drought plans plus other states are at various stages of plan development. However, India does not have a comprehensive drought mitigation policy yet, even though one or the other part of India faces drought hardship almost every year. Yet, our country's present approach is to respond to drought using reactive ad hoc strategies and a crisis management approach.

Ministry of Agriculture and Farmers Welfare (MoAFW), Govt. of India, is the nodal agency for drought management in the country. The MoAFW has recently implemented prescribed guidelines for monitoring, assessment and declaration of drought at the district or sub-district level by the state government (i.e. "Manual for Drought Management-2016"). The National government aids and supports (Ministry of Agriculture & Farmers Welfare, 2016) through relief grants. Still, the prevailing drought management guidelines are more towards temporary relief and crisis management rather than long-term planning. Climatic region-specific drought mitigation plans are required to be prepared to deal with droughts in India (Pandey et al., 2016).

The purpose of this paper is to review India's present approach and future challenges for region-specific drought monitoring, assessment and mitigation planning to manage and cope with drought effectively.

8.2 Incidence of Droughts in India

The droughts recur in India due to erratic distribution and/or less than average rainfall during south-west monsoon season (June–Sept.). Also, the states of Tamil Nadu and Karnataka and parts of Andhra Pradesh receive rainfall in the succeeding north-east monsoon season (Oct.–Nov.). There are any one or more of the following causes of drought occurrence in India: (i) late arrival (or onset) of monsoon, (ii) less than average amount rainfall in space and time, (iii) occurrence of prolonged dry-spells during monsoon period, and (iv) early withdrawal of monsoon. Droughts cause relatively greater hardships to rural populations that are largely dependent upon rainfed agriculture and related occupations for livelihood. Mass migration from rural areas of India has emerged as one of the major challenges in drought-affected regions in the country.

The drought characteristics, viz., frequency, severity and persistence in successive years, have been analysed for the period from 1901 to 2013 (i.e. 113 years) across the climatic regions. The analysis revealed that drought characteristics of one climatic region differ from that of the other. The governing climatic factors of regional drought characteristics are mean annual precipitation, evapotranspiration, length of wet season and annual temperature variation. Relationships between the average return period of drought and the ratio of mean annual potential evapotranspiration (Ep) and mean annual precipitation (Pa) have been derived using 113 years' monthly rainfall data.

Data from 536 meteorological stations of the India Meteorological Department (IMD) all over India are used for the analysis. The relationship presented in Fig. 8.1 clearly indicates that the average drought return period decreases with an increase in wetness. Figure 8.2 shows the relationship between the Ep/Pa Ratio and drought return period. The arid and semi-arid regions in India face droughts with an average frequency of once in 3 to once in 4 years, covering the states of Rajasthan, Gujarat, Haryana, Punjab and parts of UP plus interiors of Karnataka and Tamil Nadu (Fig. 8.2). Further, the relationship of drought return period and normal annual temperature variation derived from monthly temperature data for 268 stations and presented in Fig. 8.3. It indicates that the frequency of meteorological drought has the

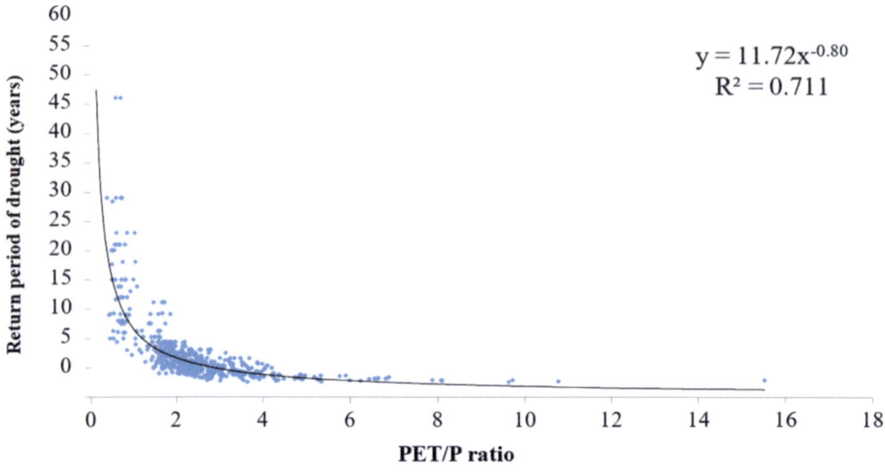

Fig. 8.1 Relationship of Ep/Pa ratio and drought return period

notable relationship with the normal annual temperature variation (T_d). The average drought return period decreases with an increase in the T_d (Pandey et al., 2016) as can be seen in Fig. 8.3.

From Fig. 8.3, it can be seen that the average return period decreases with an increase in the range of normal annual temperature variation (T_d). Since frequency is reciprocal of the return period, therefore, a decrease in the return period means an increase in the frequency of Meteorological drought events.

The magnitude of deficit, i.e. severity, (Fig. 8.4) varies in different climatic regions too. Socio-economic impacts are much higher in arid and semi-arid regions compared to other climatic regions of the country due to a greater degree of severity. The relationship of maximum rainfall departures with the range of average annual temperature variation is shown in Fig. 8.5. It reflects that the severity of meteorological drought events has notable relationships with the normal annual temperature variation (T_d) (Pandey et al., 2016).

In the arid regions followed by semi-arid and dry-sub-humid regions, the occurrence of persistent drought events for two or more consecutive years is much common. On the other hand, persistent drought events in sub-humid (with mean annual rainfall of 1200–1600 mm) and humid regions (with annual rainfall >1600 mm) are much rare.

From the studies, it emerged that the drought attributes differ from one climatic region to another and therefore, a drought mitigation plan devised for one climatic region may not be effective for another. For instance, the arid, semi-arid and dry-sub-humid areas need exhaustive mitigation plans to conserve water and increase groundwater recharge to minimise the adverse impacts of recurring droughts. Therefore, it is necessary to prepare area-specific drought mitigation plans for different climatic regions.

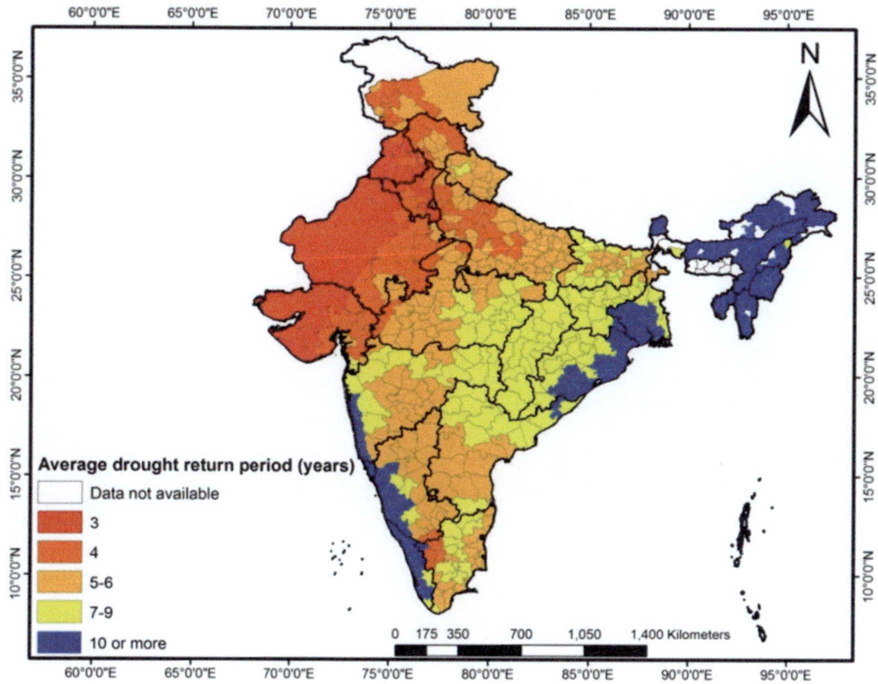

Fig. 8.2 Map showing average drought return period in India (*Map is not to the scale*)

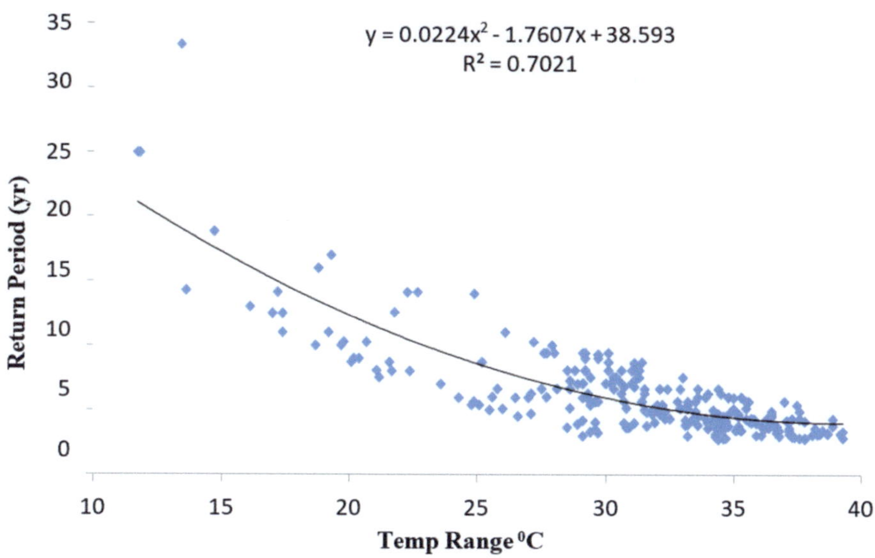

Fig. 8.3 Relationship of average drought return period with range of average annual temperature variation

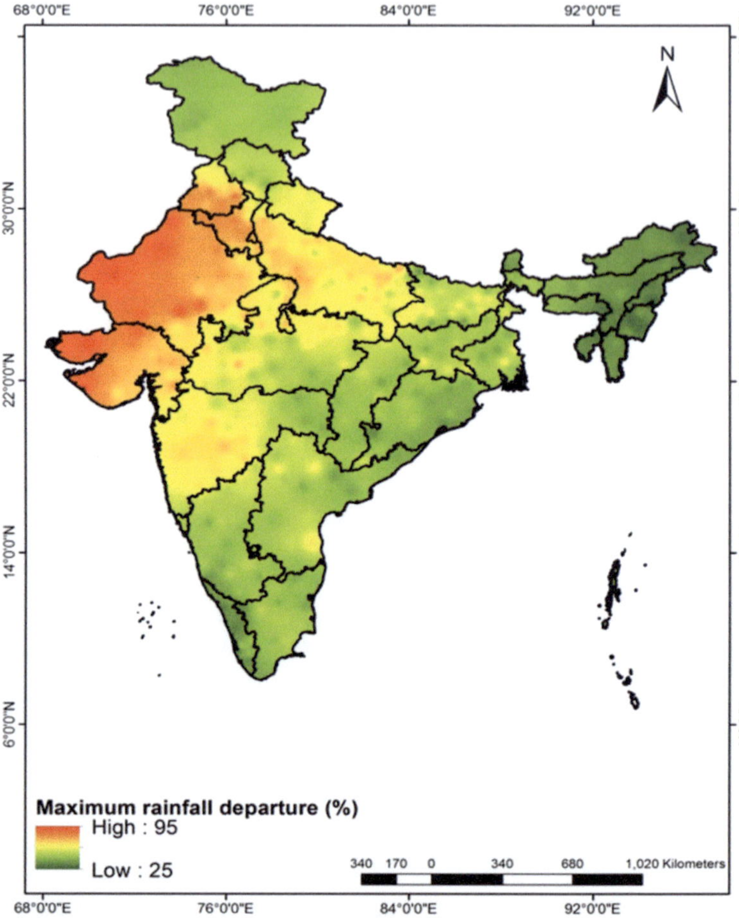

Fig. 8.4 Map showing distribution of observed drought severity (Maximum rainfall deficiency) in India. *(Map is not to the scale)*

8.3 Prescribed Procedure for Drought Declaration in India

In the Government of India's "Manual for Drought Management (2016)", the declaration of drought depends on multiple indicators. The deficit in monthly/fortnightly rainfall is the causative indicator and mandatory to trigger the initiation for assessment of other impact indicators such as crop area sown, vegetation index [Normalized Difference Vegetation Index (NDVI) or Vegetation Condition Index (VCI)], stream flow, groundwater level, and lake and reservoir storage as shown in Table 8.1 (Ministry of Agriculture & Farmers Welfare, 2016). However, during the drought

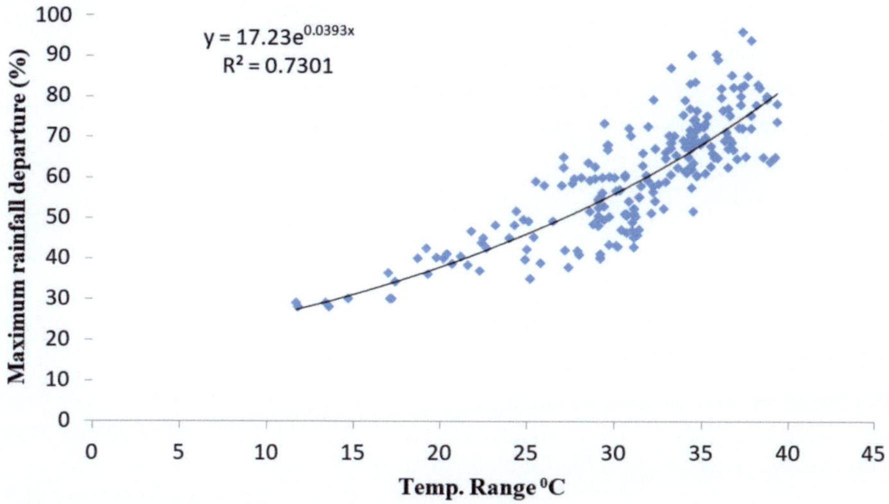

Fig. 8.5 Relationship of maximum rainfall departures with the range of average annual temperature variation

declaration, the state government is free to consider other associated impact indicators, such as migration of cattle, human migration, fodder availability, drinking water, animal health and employment opportunities in the agriculture sector.

The Drought Monitoring Cells (DMCs) in the States are responsible for regular monitoring of data on various key variables and indicators as above and apprise the State Executive Committee (SEC)/State Disaster Management Department on the spread and severity of drought. Steps for drought declaration prescribed in the manual are as follows:

Step 1: Matrix for application of mandatory Indicators, viz., RF deviation/SPI/Dry Spell is given in Table 8.2 to set the Trigger-1.

Table 8.1 List of variables and indicators for drought monitoring and assessment

S. no	Key variable	Indicator/index(ices)	Sources of data
1	Rainfall (RF)	Rainfall Deviation/SPI/Dry Spell	IMD, state govt
2	Crop area sown	Deviation from normal	State govt. (Department of agriculture)
3	Satellite-based crop condition	NDVI, NDWI deviation from normal VCI form of NDVI/NDWI	MNCFC, NRSC, ISRO and state remote sensing centres
4	Stream flow	SFDI	CWC/India-WRIS
5	Groundwater level	Groundwater levels	GWDI CGWB
6	Reservoir storage	Reservoir level/storage	CWC, irrigation department, water resources department

Table 8.2 Mandatory indicators' matrix for Trigger-1 (i.e. rainfall deficit/dry spells) (Ministry of Agriculture & Farmers Welfare, 2016)

RF Dev/SPI	Dry spell	Drought trigger
Deficit or scanty RF/SPI <−1	Yes	Yes
Deficit or scanty RF/SPI <−1	No	Yes if rainfall is scanty or SP <−1.5, else No
Normal RF/SPI >−1	Yes	Yes
Normal RF/SPI >−1	No	No

Step 2: In the event that the first drought trigger is set off in Step 1, the Impact Indicators are required to be examined as given in Table 8.3.

Explanation: The intensity of the drought will be contingent upon the values of at least three out of four Impact Indicators, viz., Agriculture, Remote Sensing, Soil Moisture and Hydrology in the following manner:

- Severe drought: if two out of 4 impact indicators are in Severe category.
- Moderate drought: if three of 4 impact indicators are 'Moderate'.
- Normal: for all other cases.
- Trigger-2 will be set off in the event of a finding of 'severe' or 'moderate' drought.
- If the irrigation percentage of the administrative region (District/Taluk/Block/Mandal), for which drought is being declared, is more than 75%, the State has an option to reduce the drought category by one rank (i.e. Severe to Moderate). However, in such a situation of reduction of drought intensity from 'Moderate' to 'Normal', the State Government will still be required to conduct field verification as prescribed in Step 3 below.

Step 3: In the event that Trigger-2 is set off, in order to make a final determination of drought the States have to conduct a sample survey for ground truthing and the finding of the field verification exercise will be the final basis for judging the intensity of drought as 'severe' or 'moderate'.

Early season drought declaration during August month may be carried out if the seasonal conditions signify drought-like situation. The other condition when there can be an early declaration of drought is deficit rainfall in June and July with prolonged dry spells leading to a significant reduction in crop-sown area. However,

Table 8.3 Impact indicators' matrix for Trigger-2

Mandatory indicators		Impact indicators				Category of drought
Rainfall indices		Agriculture	Remote sensing	Soil moisture	Hydrology	
Rainfall deviation (RFdev) or SPI	Dry spell	Crop area sown	VCI or NDVI deviations	PASM/MAI	SFI/RSI/SGW	
		Any two out of the four types of the impact Indicators can be used for drought declaration				

indicators shown in Table 3 continue to be mandatory for early season drought declaration as they are based on rainfall deviation or SPI and the dry spell.

Reduction in crop-sown area or failed sowing, MAI, groundwater or Reservoir water index are important impact indicators. When the canopy coverage is low, NDVI is less effective. In such a situation, Normalized Difference Wetness Index (NDWI), a surface wetness indicator, is preferred to NDVI.

Demand for relief assistance from the Central Government for the implementation of mitigation and response actions is possible only after the declaration of drought by the concerned state government.

8.4 Application of Indices for Drought Assessment

The major constraints in the timely identification of onset and assessment of drought severity level are because of the lack of universally accepted definitions and widely acceptable indices. Also, it is one of the most challenging jobs to develop suitable and acceptable hydro-meteorological drought indices to correctly describe the regional drought attributes. Various drought indices developed in other parts of the world have been introduced in the literature. These indices include the decile of precipitation for characterising droughts in Australia, the Effective Drought Index (EDI, Byun & Wilhite, 1999) and Palmer Drought Severity Index (PDSI) developed by Palmer in 1965. An algorithm for the generation of the time series of the Moisture Adequacy Index for agricultural droughts has been developed by the Food and Agricultural Organisation (FAO). However, none of these indices are fully accepted by the researchers due to the limitations associated with these indices. McKee et al. (1993) introduced Standardized Precipitation Index (SPI), which has relatively better acceptability in some regions.

There is a need for evolving suitable and appropriate regional indices which could be useful in analysis and the management of drought in the different climatic regions of the country.

8.4.1 Stream Flow Drought Index (SFDI)

Understanding the hydrological drought entails data on the amount of water flowing in streams and rivers in a region. As the flows in streams display high seasonal variability necessitating estimation of variable truncation levels for all twelve months in a year. The truncation level is an analytical interpretation of the expected availability of water flow in a river. The truncation level is defined as 75% dependable flow (discharge) at a given time and for a given site over a month-long time period. In order to derive flow duration curves for determination of monthly truncation level, average daily/monthly/monthly flow records are needed. Water flow in a river or stream below the truncation level is an indication of a drought-like scenario. The severity of drought events could be classified using a stream flow drought index (SDI) developed by Pandey et al. (2008). The SDI is defined as a function of (i) The ratio of deficit flow volume to corresponding volume at the truncation levels and (ii) The ratio of duration of deficit flow to the maximum expected duration of the independent stream flow drought event:

$$SDI = \frac{V_d}{V_{tl}} * \frac{d_e}{d_m} \tag{8.1}$$

where SDI = stream flow drought Index.

V_d = deficit flow volume.

V_{tl} = corresponding flow volume expected at TL flow.

d_e = duration of independent drought event and

d_m = maximum duration of an independent drought event (=365 days).

Stream flow drought can be estimated in several steps which are as follows:

1. Workout on 10 daily days of a month.
2. Give rank (r) in all items.
3. Probability of exceedance of stream flow is calculated using the Weibull plotting position as $P = r/(n + 1)$, where r = rank of flow value (for highest flow value $r = n$ and for lowest one $r = n$ and n = total number of flow records).
4. Plot graph between the discharge and probability of exceedance, and develop the flow duration curve.
5. Estimate truncation level at 75% exceedance by linear interpolation.
6. Arrange the truncation level according to the discharge.
7. Find deficit/excess flow with respect to truncation level.
8. Find out the types of drought and categorise them as given in Table 8.4.

WRD maintains stream flow records (average 10—daily or monthly discharge data) which can be used by the States for assessment of drought. In case of the river systems having perennial flow characteristics, SDI is straight away applicable and in the case of intermittent and ephemeral flow characteristics, a joint probability distribution approach can be used for the estimation of truncation level. However, the SDI suffers from limitation arising from the non-availability of historical data (30 years or more) (Table 8.5).

Table 8.4 Classification of streamflow drought index

SFDI	Classification of SFDI
<0.01	Weak
0.01–0.05	Mild
0.05–0.2	Moderate
0.2–0.5	Severe
>0.5	Extreme

Table 8.5 Drought category classification for stream flow drought

Groundwater Drought Index (GWDI)	Groundwater deficit class
>−0.15	Normal
−0.16 to −0.30	Mild
−0.31 to −0.45	Moderate
−0.46 to −0.60	Severe
<−0.60	Extreme

8.4.2 Groundwater Drought Index

The status of groundwater is useful for assessing groundwater availability for domestic and agricultural supply. For the computation of mean value of the monthly groundwater depletion rate, monthly groundwater (GW) table records are required for a minimum period of 10 years. When the rate of depletion of groundwater table in a given month/period is more than the corresponding mean value then it is an indication of water deficit. The computation procedure for groundwater drought index (GWDI) is as follows:

$$GWDI_{ij} = \frac{MGDWD_j - GWD_{ij}}{GWD_{i\max}} \tag{8.2}$$

where

$GWDI_{ij}$ = Groundwater drought index for ith month and jth year.

$MGWD_j$ = Mean depth to groundwater table below surface (in meter) in jth year.

GWD_{ij} = Depth to groundwater table in ith month and jth year (in meter).

$GWD_{i\max}$ = Maximum depth to groundwater table in ith month in available data set for n number of years (in meter).

$i = 1, 2, 3, 4\ldots12$ and $j = 1, 2, 3\ldots n$.

n = Total numbers of years for which monthly groundwater records are used.

8.4.3 Reservoir Storage Index

The Reservoir Storage Index is also one of the indices used for drought assessment. An effective foil against drought is the availability of water in reservoirs. The reservoir storage status derived from the percentage of storage deficit vis-à-vis long-term averages can provide an indication of drought:

$$RSI_{ij} = \frac{RS_{ij} - MRSI_j}{MRSI_j} * 100 \tag{8.3}$$

where

RSI_{ij} = Reservoir Storage Index for ith month and jth year (in percentage).
$MRSI_j$ = Mean Live storage of reservoir (in MCM).
RSI_{ij} = Live storage of reservoir in ith month and jth year (in MCM).
$i = 1, 2, 3, 4\ldots12$ and $j = 1, 2, 3\ldots$n.
n = Total numbers of years for which monthly live storage records are used.

8.5 Demarcation of Areas Vulnerable to Drought

The degree of susceptibility to a hazard either as the result of varying exposure to water deficit or because of variation in the ability to cope with its impact is an indicator of vulnerability to drought. The degree of vulnerability caused by the intensity of drought mainly depends on the availability of water in an area. It is now being realised that methodical water resource planning is essential including the following options to cope with drought: i) integrated information for strategic planning and management control and ii) reactive to proactive strategies for coping with drought (Table 8.6).

Risk of drought at a given place is a function of hazard and vulnerability where hazard refers to climatology, and the vulnerability includes physiographic, social, environmental and other factors like coping ability, awareness and regional economic activities (Hamouda et al., 2009). A similar approach was used in various parts of the world (Wilhelmi and Wilhite (2002) in Nebraska USA; Pandey et al., (2010) and Vinit et al., (2015) in India) for integrated assessment of vulnerability to drought. It is understood that geographical location, physical characteristics of watersheds, surface and groundwater availability, variability of rainfall, regional climatic factors and socio-economic factors have been responsible for water deficits and crop loss from time to time (Pandey et al., 2008, 2010). The National Institute of Hydrology (NIH) has developed a method for integrated assessment of vulnerability to drought using hydro-meteorological factors, topography, land use, soil type, special and temporal water demand, surface and groundwater availability, irrigation support and social-economic aspects expressing coping ability of the region (Pandey et al., 2010) as categorised in Table 8.7.

Table 8.6 Drought category classification for reservoir storage index

Percentage deficit in live storage volume of reservoir w.r.t. average storage of last 10 year	Category of deficit
Less than 20%	Normal deficit
20–30%	Mild deficit
30–40%	Moderate deficit
40–60%	Severe deficit
>60%	Extreme deficit

Table 8.7 Different categories of factors used to demarcate areas vulnerable to drought (Pandey et al., 2016)

Static factor of vulnerability (Physiographic factors which vary w.r.t. space)	Semi-static factors of vulnerability (which vary with the phase of development w.r.t. space and long-term temporal variability)	Variable factors of vulnerability (which vary w.r.t. space and time)
1. Topographic factors (General slope and drainage, etc.) 2. Soil (Soil water holding capacity) 3. Climatic components (Precipitation and ET)	1. Water resources development (irrigation supports) and in situ water conservation 2. Status of surface water storage availability 3. Status groundwater availability 4. Population density (Population concentration, industrial/commercial activities) 5. Land use 6. Regional cropping system/ crop water demand 7. Region-specificactivities (like cattle farming/wildlife preservation, etc.)	1. Rainfall (monthly/seasonal/ annual) 2. Stream flow 3. Storages in tanks/reservoirs (if any)

This method has got wider acceptance because of the use of spatial distribution of information/indicators on the regional topography, land use, soil, regional climate, water demand, water availability, irrigation support (if any), socio-economic conditions, coping ability, etc. Also these are integrated using a weighing scheme and Geographic Information System (GIS). The integrated values of weights of various indicators may be computed on grid scale in the spatial domain and maps may be prepared to represent the spatial distribution of integrated vulnerability to drought. The drought vulnerability Index (DVI) for each grid could be calculated for the rationalisation of the approach. The DVI has been defined as the ratio of the sum of the weights of factors to the sum of their maximum weight values. Integrated assessment of vulnerability to drought method may set a better direction in dealing with drought monitoring and mitigation. The application of this method has been made in the Tel sub-basin of the Mahanadi river system and Ken basin in central India to demarcate zones vulnerable to drought. As an example, the demarcation of areas vulnerable to drought in the Ken River basin is shown in Fig. 8.6. It is hoped that this method will bring objectivity in decision-making and effective dealing with drought mitigation activities (Pandey et al., 2016).

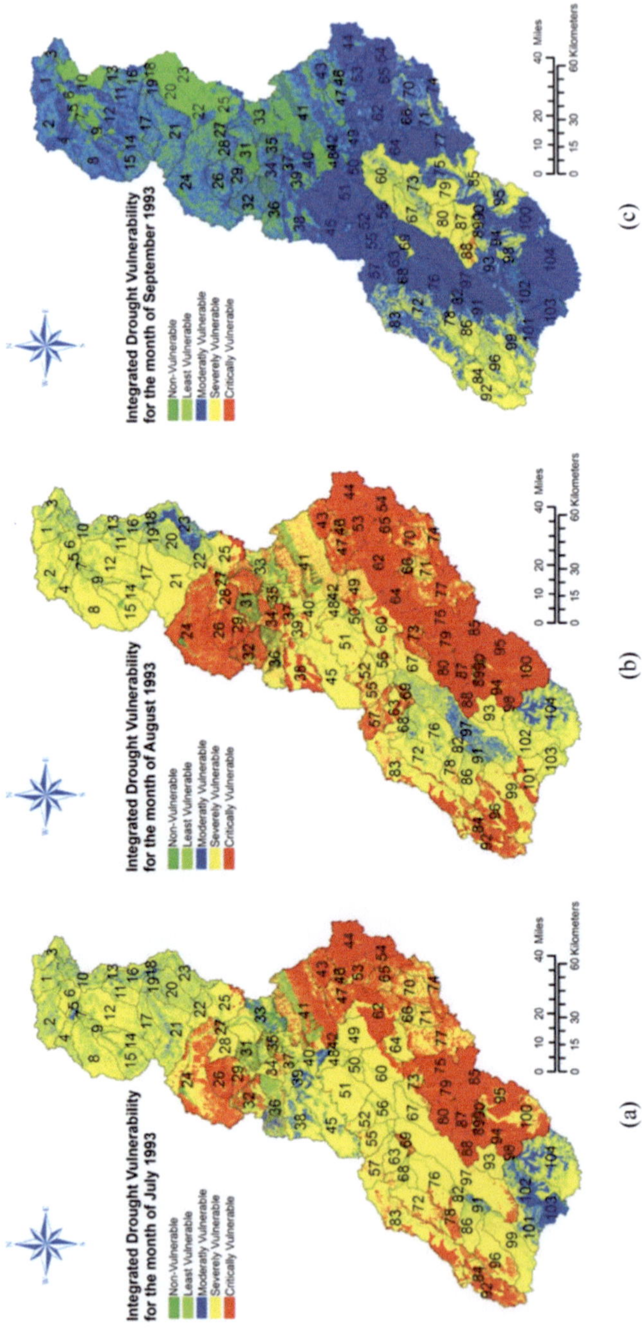

Fig. 8.6 A real example of demarcation of areas vulnerable to drought in Ken basin

8.6 Drought Risk Management and Mitigation

Region-specific planning and demarcation of zones/areas vulnerable to drought are important to prioritise the planning for drought risk reduction and implementation of mitigation actions.

Developing a drought management plan at the regional/national level can be a step to reduce or minimise the impact of drought. Based on the principal of disaster risk reduction, the implementation of the drought risk reduction approach needs to have the following components.

i. **Regular drought monitoring and early warning systems**:

- Monitor key indicators and indices of precipitation, temperature, soil moisture, vegetation cover condition, reservoir/tank water storage, stream flow and groundwater.
- Monitor the consequences of drought, especially the impacts on prime vulnerable sectors such as agriculture.
- Develop and evaluate reliable tools and method for seasonal forecasts and realistic impact assessment for impacted/affected sectors/regions.
- System for timely communication of reliable warning messages and activate decision-making system to timely respond to the event and implement mitigation measures.

ii. **Assess drought vulnerability and risk**:

- Identify and assess drought impacts on vulnerable economic sectors including food and agriculture (cropping and livestock), biodiversity and ecosystem, energy, tourism and health.
- Assess the physical, social, economic and environmental pressures on communities to identify who and what is at risk and why, before, during and shortly after drought.
- Assess conditions or situations that increase the resistance or susceptibility to drought and the coping capacity of communities affected by drought.
- Assess the extent of potential damage or loss in the event of a drought.

iii. **Implement measures to limit the impacts of drought and better respond to drought**:

- Identify and implement structural, or physical measures, and non-structural measures to limit the adverse impacts of drought, prioritised based on level of vulnerability.
- The response includes all efforts to meet the life preservation and basic subsistence needs of the communities and sectors most vulnerable and severely affected.
- Relevant sectors affected by drought, based on their vulnerabilities, particularly agriculture, water resources and the environment, including health, transport and tourism.

- Measures can be long-term, medium-term or short-term, depending on their implementation time.
- Biodiversity, land and ecosystem services play a vital role in reducing vulnerability and mitigating impacts of drought.

iv. **Water Management to deal with drought in different climatic regions**:

Based on water availability, delineation of different categories of areas vulnerable to drought would be helpful in the development of region-specific drought mitigation plans for different climatic regions as follows.

The Arid Areas have an average frequency of drought as once every 2 to 3 years, and the droughts are likely to persist for two or more consecutive years; and these areas need priority attention of the decision-makers. The mitigation options for these areas often are very limited. For such areas, mitigation options may be needed (Pandey et al., 2016):

- Intensive rain water harvesting/water conservation can moderate adverse impacts to some extent only.
- Recycling of wastewater particularly for supplemental irrigation.
- Promotion of Micro-irrigation systems (sprinkler/drop systems).
- Measures for evaporation reduction from tanks/reservoirs.
- Need to opt inter-basin water transfer as an effective alternative—to mitigate the impacts of drought and bring prosperity to the region.

The Semi-Arid Areas face drought with an average frequency of drought as once every 4–5 years and often encounter severe drought events (Pandey et al., 2016). The mitigation option for these areas may be as follows:

- Creation of water storage in tanks and reservoirs/in situ water retention through gully checks and stop dams
- Provisions of within-stream storage and water diversions
- Intensive water conservation/rain water harvesting
- Increase water use efficiency
- Provisions of extensive groundwater recharge
- Recycling of wastewater for non-drinking purposes/irrigation
- Inter-basin water transfer—remains a better mitigation option for bringing prosperity in semi-arid areas.

The Sub-Humid Areas face drought with an average frequency of once every 6–8 years and the occurrence of severe drought events about once in 10 or more years. The persistence drought events are also not very common (Pandey et al., 2016). The mitigation options for these areas may be as follows:

- Increase in water storage in tanks and reservoirs
- Water conservation/in situ water retention through gully checks and stop dams
- Intensive Groundwater Recharge may be a better option to enhance groundwater availability

- Within-stream water storages and water diversions may be used to enhance both surface and groundwater availability in such areas
- Rejuvenation of tanks/ponds/springs particularly in the case of hilly regions.

The Humid Areas face drought with an average frequency of once every 10 or more years, and the occurrence of severe drought is much rare. The persistence of drought events is very rare. There exist multiple options to manage drought like measures for rejuvenation of springs/ponds/lakes and in situ water conservation to enhance water availability in the lean period (Pandey et al., 2016).

8.7 Summary and Conclusions

This paper presents a systematic approach to coping with future challenges of drought and sustenance of the ecosystem. The approach is based on assessments of vulnerability and risk to drought and monitoring of hydro-meteorological data/indicators to assess drought risk, severity and impacts using precipitation, stream flows, groundwater, reservoir levels and soil moisture. The document also provides an inventory of the key indicators used to monitor and predict drought onset, duration, end, severity and impacts. Drought management in India needs proactive planning for different climatic regions. In most of the United States, there exists no system for regular monitoring of data/information and assessment of drought in near real time. Therefore, it is important to develop a comprehensive drought monitoring system for a regular collection of all relevant data, analyse, assessment and dissemination of information in a useable manner. The arid and semi-arid regions face more frequent, severe and persistent droughts of 2 or 3 successive years followed by dry-sub-humid areas. The occurrence of severe and persistent drought in humid areas is much less compared to other regions. The regions where droughts are likely to persist for two or more consecutive years need the priority attention of the decision-makers. The frequency, severity and persistence of successive droughts primarily depend on rainfall, potential evapotranspiration and annual normal temperature variation in the region. The integrated assessment of vulnerability to drought using physical, climatic and social factors provides proper input of the degree of water stress and possible impacts for decision-making. A precise demarcation of areas with a distinct level of vulnerability has prime importance in the selection of appropriate measures and prioritisation of drought mitigation actions. Therefore, it is essential to initiating actions for a demarcation of areas with different levels of vulnerability to drought. It is believed that this lead paper may pave the way forward for drought management in India.

There is a need to take up research studies in this direction on a long-term basis and make an attempt to devise regional thresholds/indices for near-real-time identification of the onset of drought, quantification of event severity and vulnerability to drought, and suitably utilise them to describe regional drought problems. A near-real-time monitoring system may lead to providing critical information for decision-making on a continuous basis to enable timely recognition of the occurrence of drought, its severity and management actions. In order to achieve the above goals of near-real-time drought monitoring and mitigation, specific studies are needed for the evaluation and improvement of drought vulnerability indices (DVI) and the development of techniques for the formulation of area-specific water management plans to deal with drought conditions in different climatic regions of the country.

References

Byun, H. R., & Wilhite, D. A. (1999). Objective quantification of drought severity and duration. *Journal of Climate, 12*, 2747–2756.

Dai, A. G. (2013). Increasing drought under global warming in observations and models. *Nature Climate Change, 3*(1), 52–58.

Dracup, J. A., Lee, K. S., & Paulson, E. G., Jr. (1980). On the definition of droughts. *Water Resources Research, 16*(2), 297–302.

Hamouda, M. A., Nour El-Din, M. M., & Moursy, F. I. (2009). Vulnerability assessment of water resources systems in Eastern Nile Basin. *Water Resources Management, 23*, 2697–2925.

IPCC (2007). Climate change–2007: Final report on climate change by Intergovernmental Panel. Published for the Intergovernmental Panel on Climate Change, Cambridge University Press.

Jain, V., Pandey, R.P., & Jain, M.K. (2015). Spatio-temporal assessment of vulnerability to drought. *Natural Hazards, 76*(1), 443–469. https://doi.org/10.1007/s11069-014-1502-z

Liu, N., Liu, Y., Bao, G., Bao, M., Wang, Y., Zhang, L., Yuxiang, G., Wurigen, B., & Tian, H. (2016). Drought reconstruction in eastern Hulun Buir steppe, China and its linkages to the sea surface temperatures in the Pacific Ocean. *Journal of Asian Earth Sciences, 115*, 298–307.

McKee, T.B., Doesken, N.J., & Kleist, J. (1993). The relationship of drought frequency and duration to time scales. In *8th Conference on Applied Climatology* (pp. 179–184). Anaheim, California.

Ministry of Agriculture and Farmers Welfare (2016). Manual for Drought Management, Dept. of Agriculture, Cooperation and Farmers Welfare, Govt. of India. New Delhi. pp 1–155.

Palmer, W.C. (1965). Meteorological drought. Research Paper No. 45, U.S. Department of Commerce Weather Bureau, Washington, D.C.

Pandey, R. P., & Ramasastri, K. S. (2001). Relationship between the common climatic parameters and average drought frequency. *Hydrological Processes, 15*(6), 1019–1032.

Pandey, R. P., Mishra, S. K., Singh, R., & Ramasastri, K. S. (2008). Streamflow drought severity analysis of Betwa River System (INDIA). *Water Resources Management Journal, 22*(8), 1127–1141.

Pandey, R. P., Pandey, A., Galkate, R., Hi-Ryong, B., & Mal, B. C. (2010). Integrating drought. *Water Resources Management, 24*(15), 4199–4217. https://doi.org/10.1007/s11269-010-9653-5

Pandey, R. P., Daradur, M., Jain, V. K., & Jain, M. K. (2016). Assessment of vulnerability to drought towards effective mitigation planning: The case of Ken River Basin in India. *Journal of Indian Water Resources Society, 36*(4), 36–48.

Ponce, V. M., Pandey, R. P., & Seza, E. (2000). Characterization of drought across climatic spectrum. *Journal of Hydrologic Engineering, ASCE, 5*(2), 222–224.

Sharma, K.D., & Ramasastri (2005). Drought Management (p. 397). Allied Publishers, New Delhi.

Trenberth, K. E., Dai, A. G., van der Schrier, G., Jones, P. D., Barichivich, J., Briffa, K. R., & Sheffield, J. (2014). Global warming and changes in drought. *Nature Climate Change, 4*(1), 17–22.

Wilhite, D. A., Knutson, H. C., & Smith, K. H. (2000). Planning for drought: Moving from crisis to risk management. *Journal of the American Water Resources Association, 36*(4), 697–710.

Wilhelmi, O. V., & Wilhite, D. A. (2002). Assessing vulnerability to agricultural drought: A Nebraska case study. *Natural Hazards, 25*, 37–58.

Chapter 9
Challenges and Opportunities for Livelihoods and Environmental Security in Desert Ecosystems of Rajasthan, India

V. P. Singh

9.1 Background

Rajasthan basically has two major zones (regions); the Desert (western side) and Aravali range and south eastern side zone. Each of the zones has 10 districts. The desert districts include, Ganganagar, Bikaner, Jaisalmer, Barmer, Jodhpur, Nagaur, Hanumangarh, Sikar, Pali, and Churu. The Aravali range and eastern side districts are, Alwar, Jaipur, Udaipur, Ajmer, Bhilwara, Chitaurgarh, Bharatpur, Kota, Jhalawar, and Banswara. The Western sidelands, which form part of the Thar Desert are dry and are mostly dominated by Acacias and scurvy vegetation on moving sand dunes. Aravali Range and south-eastern lands are generally better moist and more fertile. This region is home to the Kathiabar-Gir dry deciduous forests of tropical broad leaf species, including teak. The Indira Gandhi Naher Paryojana (IGNP) Canal also transacts this region. I had opportunity of visiting a couple of times the Desert National Park (DNP) for the protection of the Indian Great Bustard (*Ardeotis nigriceps*) and villages around it. The villages included, a.o, Damodar, Kanohi, Sam, Sudasari, Sukhdi, Gangaki basti, Khuri, Barna, Rithla, Vijaynagar (Wajeeron ki Dhani),and Dhauv, and on eastern side of Rajasthan, the villages, included, Amratiya, Kekaria, Chitrawas, Ugariawas, and Baraundi.

The case analysis of Rajasthan would highlight the point of severity imposed by the adverse conditions in the state. Rajasthan is India's largest state, occupying 10% of country's geographical area, and is a home for 5% country's population and 1% available water resources. More than half of the Rajasthan population lives in Thar Desert, the least developed area of the state; characterized by recurrent droughts, low

V. P. Singh (✉)
Former Director for South Asia of the International Centre for Research in Agroforestry (ICRAF), New Delhi, India
e-mail: V.P.Singh.cgiar@gmail.com

M. S. Rathore et al. (eds.), *Desert Ecosystems*,
https://doi.org/10.1007/978-3-032-03237-9_9

and erratic rainfall, high air and soil temperature, intense solar radiation, high wind velocity, and wide spread poverty. Severe droughts occur every three years, families face acute shortage of food, fodder, and water. Animals die, or are sold below market rates, children become ill, and parents are forced to separate from families because men migrate for employment elsewhere, leaving the women and children to fend for themselves.

Desert Development Program and Combating Desertification Program, which have been implemented over decades have somewhat achieved the sand dune fixation through only one tree species, *Acacia tortilis* on a limited scale, but have not been able to contribute to the livelihoods and other benefits to the public substantially. Paucity of multiple layers of vegetation is also now resulting in dune reactivation due to biotic and natural causes.

This paper suggests designing the required vegetation combinations for Livelihoods (food, nutrition, fuel, fodder, timber, and other purposes), and Environmental Security, and developing technologies for growing at the farm and landscape levels on lands owned by the small holders. The sequential restoration of vegetation and ecological services would be a multiplier benefit of this suggestion. The primary focus of this paper will be on the western region of Rajasthan, unless explicitly indicated otherwise.

9.2 Situation Analysis of Rajasthan

The fundamental difference between the two regions is that in Desert areas most of the livestock are reared by "open grazing" in community Gauchars, wherever they find Sewan grass (*Lasiurus scindicus*), a nutrition rich grass for fodder, and in the eastern region they are stallfed. Most of the pasture lands in the western side are deficient in phosphorus and other minerals. Many animals die due to consuming such a nutritionally poor fodder. Thus, there is the extensive problem of Pica; mineral deficiency in soil-fodder-animals nexus. The ready-made mineral mixture can also not be given to all animals as they are mostly not stallfed and remain in grazing lands for days.

Most of the livestock, especially the cows are low yielding, not more than two liters a day, and not every day. They are mixed breeds of poor/lower quality pedigree, poorly fed, and some of them affected by "botulism." In this area since the dead animals are left in open, they rot and develop *Clostridium botulinum* bacteria. The grazing livestock, due to severe protein and mineral deficiency (botulism), start licking or even eating the bones of the dead animals and get infected and die. Botulism is also common in productive animals because of their higher mineral requirement.

Cows and their young male and female as well as bullocks (oxen) are restricted from being brought out of Rajasthan State. Many agricultural and transport operations are done by mechanical means. This has rendered the young males and oxen unimportant. So, the communities have stopped tending them and let them lose all the time. These livestock roam and graze around anywhere at will, and by consuming poor and low amount of fodder are in poor health.

The livestock population also has a large number of stray bulls in the area, some villages have more than 40 stray bulls. They not only degenerate the breed, but also compete with the female livestock population for fodder, water, etc.

A new trend has been emerging with respect to livestock management. Camel has been declared as the state animal, so the camel raisers find it difficult to make income from the animal, and as a result, this animal is not properly attended to. Recently government has introduced Rupees ten thousand for its breeding. Hopefully, camel milk for health and cosmetic purposes, etc., may bring back the animal value.

Local communities keep large herds of livestock (average 100 units per house hold), mostly sheep, goat, and cows and their young calves (female calf). For example, in the Khuri village, which is situated next to DNP, has more than 10,000 animals. Such large herds outweigh the carrying capacity of the grazing lands, and exert extra pressure on nearby forest resources.

Overall, the animal load outweighs the bearing capacity of the open grazing lands due to low fodder availability in these lands, reduced grazing land area and high livestock population, especially in Jaisalmer. Therefore, keeping small number of improved breeds of livestock, stall feeding, improvement in pastures, producing fodder, including fodder trees on the farm, especially in gaushala lands and other areas along with social and legal aspects have emerged among of the best options to reduce the pressure on forest resources, including local biodiversity. This would apparently address many of the problems faced by the communities, government, and ecosystems in the area. Some of these options could be tried out on a pilot basis initially and scaled up afterward.

As can be seen from the preceding, that the livestock and field crops are the two main components of the sustainable agriculture, in fact livestock plays a greater role in western Rajasthan (Singh and Nayak 2013b). In Jaisalmer area, farmers grow Bajra/pearl millet (*Pennisetum glaucum*, also known as *Pennisitum typhoides*, and *Sitaria glaucum*), Mung (*Vigna radiata*), and Moth (*Vigna aconitifolia*). After the construction of IGNP, which transacts this region, farmers along the canal have been growing a number of other crops, mostly vegetables (personal observation).

After being aware of the high prices for "Guar gum," farmers started cultivating Guar (*Cyamopsis tetragonoloba*), in their allotted land and also in some portions of the restricted DNP lands as well as in Gauchar lands. The Guar productivity is reported to be on the decline after 3–4 cropping seasons across all the guar cultivated lands in Jaisalmer due to decline in soil fertility as the crop is generally grown year after year without replenishing the soil nutrients. The guar gum prices have also come down significantly since then. Grazing area is also reduced by the fencing of areas around wind mills set up for power generation. All this combined has reduced the grazing area significantly, and as a result communities have started exerting pressure on forest resources.

9.3 The Climate Change Adaptation

Adverse environmental conditions are wide spread throughout India, and those states which have low vegetation or forest cover suffer most from these conditions. Tree cover, including in combination with the field crops has been known to provide a significant insulation to the crop-livestock-human population through various mechanisms.

Although poverty is wide spread throughout India, the concentrations of poor are most visible in vulnerable ecologies, such as arid, semi-arid, sub-humid, and sub-temperate ecologies. These vulnerable ecologies are characterized by low resource base, fewer income generating opportunities, high degree of social conflicts and low level of education. These factors and inappropriate policies lead to food and income insecurity and make the community more vulnerable to all kinds of extreme events. The situation is further aggravated by the challenges of climate change and its impact is severely felt by the communities who are poor and have the least adaptive capacity.

Appropriate measures for adaptation to climate change can substantially reduce many of its adverse impacts, and thus contribute to food and livelihood security of the vulnerable rural population (Neufeldt et al., 2016). Such measures would need to provide integrated solutions that simultaneously address food security, livelihood improvements, and environmental sustainability (Sahoo et al. 2022). Adaptation to climate change has also to align itself to the national and sub-national plans for integrating it in various sector policy decisions to enhance the adaptive capacity of the most vulnerable sectors and groups.

Another consideration in this regard is that the adaptation to climate change has to be a community encompassing and community driven process (Singh 2014). It is to be undertaken by a constituency of local stakeholders—small holder producers, the women, out of school youth, input, credit and service providers, change agents, scientists, policymakers, donors (investors), etc.

Technologies and practices for farming sustainably under changing climates can gradually work (over three to five years) toward yielding higher resource productivity (land, water, biodiversity, labor, etc.) and stable income, and reduce the vulnerability level. Increased productivity and stable income are expected to provide food security, and by investing apart of the income in strengthening the resource base make the food security sustainable.

A number of positive developments that have been taking place recently favor taking this approach. First among them is the development and approval of the National Agroforestry Policy, and stronger focus in this area by the present government. The second is a clearer understanding of the climate change context, and the knowledge of specific adaptation measures which can be applied locally. The third is the realization that climate change and adaptation measures are not completely new phenomena, and there exists a lot of local traditional wisdom in this sphere. The fourth of them is the availability of knowledge, framework, and policy support.

The Government of India recognizes the critical importance of transforming agricultural practices to become more sustainable and a net contributor to conservation achievements. Improving agricultural practices is particularly important where productive landscapes are associated with places of high ecological value, such as the protected areas and/or key biodiversity areas.

Some Examples of Livelihood and Climate Change Adaptation-Related Tree-Based Systems Applicable to Western Rajasthan, India:

- **Fodder-based systems**: Both, the dry and green fodder are in short supply in India, more so in Rajasthan. Wheat straw is sold almost at the half price as wheat grain. Considering the large number of livestock, it is crucial to increase fodder resources and keep small number of improved breeds of livestock, and stall feed them.

Improvement in pastures, including fodder trees on the farm, especially in gaushala lands and other areas along with social and legal aspects have emerged among of the best options to reduce the pressure on forest resources and local biodiversity. This would apparently address many of the problems faced by the communities, government, and ecosystems in the area. Some of these options could be tried out on a pilot basis initially and scaled up afterward.

Some of the fodder trees, shrubs, and legume grasses that would do well in this area are Babool (*Acacia nilotica*), Shisam (*Dalbergia sisso),* Sirs (*Albizia lebbeck*), Eucalypts (*Eucalyptus globulus labill*), Neem (*Azadirachta indica*), Khejri (*Prosopis cineraria*), Ker (*Capparis decidua*), Kumbat (*Acacia Senegal Linn.*), Gunda (*Cordia dichotoma*), Urudu (*Ailanthus excellsa),* Aonla (*Phyllanthus emblica*), Indian fig (Gular, *Ficus racemosa*), *Faidherbia (Acacia) albida*; Moringa (drumstick, *Moringa Oleifera*), Calliandra (*Calliandra calothyrsus*), Leucaena (*Leucaena leucocephala*); and Staylo (*Staylosanthus hamata*), Buffel grass (*Cenchrus ciliaris*) and Sewan grass (*Lasiurus scindicus*), and many others (Nayak and Singh 2013; Singh and Nayak 2013b). Throughout in Rajasthan State and nearby surrounding areas, many farmers have been observed to grow *Prosopis cineraria* for livestock fodder (Singh 2014), and each tree is auctioned for lopping at as much as IR. 400/tree plus one day halting of the goat and sheep herd in farmers' field (Trees of Change, (2012). Planting of these species and other multi-purpose trees as shelterbelts around villages will be helpful in providing shade to the livestock close to the habitation. It will also provide relief to communities from sand storms and excessively high temperatures. Likewise, during winters, trees can protect them from strong winds and cold weather.

- **Fruit-based systems**: Fruit-based systems, particularly the Ber (*Ziziphus mauritiana)* and other species, such as the "jharberi" grown in Rajasthan are very popular in India and abroad. In addition, these species are also excellent source of fodder for small ruminants (Pushpakumara et al., (2007). Likewise, Jamun (*Syzygium cumini)* grown in Gujarat and pomegranate (*Punica granatum*) grown in Maharashtra are impacting in similar ways and scales, in the respective eco-regions throughout the country.

- **Multipurpose trees**: Multipurpose trees, like Moringa are the futuristic trees, enhancing the livelihood and income of the farmers. Moringa is visible throughout Rajasthan, especially in Jaipur and Pali districts. It is an excellent fodder source for the livestock, especially for goats and sheep. Every part (leaves, flowers and fruits) of Moringa is used in Indian cuisine.
- **Timber production and employment**: Three years ago, I visited several large-scale operational units in Jodhpur, making and exporting furniture using Babool (*Acacia nilotica*) wood. These units provide employment to a large scale skilled as well as unskilled labor force on regular basis. They get the sawn wood and dry it in huge chambers for six months before processing and making furniture. In 2022, I purchased some furniture made by one of the factories in Jodhpur from Babool wood and got it delivered in UP.

The timber based on—farm systems are meeting more than 60% of the country's timber demand, saving a substantial amount of foreign currency otherwise needed for the imports. In addition, they are generating significant amount of revenue for the state and are fully supporting the livelihoods of tens of thousands of skilled and unskilled labors engaged in various operations in this venture and allied services in the industry. As an example, Bhojvaid and Singh (2005) has indicated that in one of the markets in Yamuna Nagar town, Haryana State, 15,000 tons of poplar wood from small holder agroforestry systems is auctioned daily, valued at $ one million. In this market, the transport industry earns $ 100,000 daily and the State Gov. makes $ one million each month in revenues. Each ha land under agroforestry generates employment for 450 labors annually, 15,000 labor daily for logging, and 60,000 labor daily for plywood manufacturing.

Agroforestry can be practiced in many combinations and designs, viz crops in the main field and trees on field boundaries, or both in the main field together or in rotation, etc. (Singh, 2007). It can also be practiced in public lands such as canal embankments, e.g., IGNP canal, along roads, etc. Agroforestry practice on an average can provide a net supplementary income at least equal to the income from main field crop, depending on the tree species, density of tree planting, and market price (Bhojvaid and Singh 2005; Singh and Nayak 2013a). As has been said above, planting fast growing tree species for timber production and the integration of livestock in the system can further boost the supplementary income.

Some species can also be grown for their non-wood value. At times when the returns from other enterprises are down, a tree crop could provide a much-needed cash income. The timing of the harvest canals be very flexible (Singh et al., 2016). To make sure that there is a regular source of income from the plantation, the trees should be planted and harvested on a rotational basis, particularly if they are being grown for wood products.

Babool (*Acacia nilotica*), Shisam (*Dalbergia sisso),* Sirs (*Albizia lebbeck*), Eucalypts, Neem (*Azadirachta indica*), Khejri (*Prosopis cineraria*), Ker (*Capparis decidua*), Kumbat (*Acacia senegal Linn.*) Gunda (*Cordia dichotoma*), Aonla (*Phyllanthus emblica*), Urudu (*Ailanthus excelsa)* are some of the species that can be grown and processed in western Rajasthan to add to the farm income and provide employment.

- **Arresting soil erosion and stabilizing sand dunes**: A good example could be the halting of soil erosion and stabilizing of the sand dunes while continuing to provide livelihoods to the local population on a sustained basis. Eliminating soil loss enhances land productivity. Some reports show a drastic reduction in terms of soil loss due to erosion with trees on fields than without them, (Singh et al., 2018a).
- **Sequestering carbon**: All trees sequester carbon. The reports indicate that agroforestry systems sequester the same amount of carbon in below ground biomass as in well managed forests, and higher than in pastures, field crop lands and grasslands (Minang et al. 2015; Verchot and Singh 2009; Zomer et al. 2016). They also reduce emissions (Achten et al. 2013; Rao et al. 2015), and some of them also produce high quality saw logs, raw materials for plywood, particle board, paper, and pulp.
- **Reactivation of traditional watersheds and rehabilitation of degraded lands**: This may include using any water to rejuvenate "trees stock" which is reduced to look like root stock due to overgrazing, and making the land productive again. This process is known as Farmer Managed (Assisted) Natural Regeneration of trees (FMNR). Here, each "root stock" is irrigated twice a week, or as often as possible, and slowly the "rootstock" starts growing into the full-fledged tree, and underneath the tree shade farmers start growing millets, etc., using rain water. Ten of thousands of hectares have been rehabilitated using this practice in Niger. The actual amount of water that each tree uses depends on climatic conditions, their stage of growth and plant species (Zomer et al., (2007). Of course, if the rain or drainage water seeps through the soil to the water table, it will contribute to recharging the soil profile. Where the areas are badly degraded due to erosion, salinity, or other factors, trees, and other vegetation can be used in conjunction with other techniques to slowly return them to productive use. There are large scale examples from Madhya Pradesh, eastern Rajasthan, Haryana, Uttar Pradesh, and Odisha, where rehabilitation of degraded lands has taken place through agroforestry(Singh et al. 2018a & b).
- **Providing shade and shelter**: Windbreaks and shelters during high-speed winds and very hot summers can protect pasture and crops from heat stress, thereby improving productivity (Mohotti et al. 2020). I have personally observed four-degree lower temperature under shade than without it. Likewise, during the winter season, they can also protect the animals and plants from strong winds and cold weather and let them use their energy to grow rather than just staying alive.
- **Improving the general environmental conditions**: Trees are known to provide cleaner air, water, and natural biodiversity (both terrestrial and aquatic), and act as a noise barrier. Also, by planting local tree species, some of the natural characteristics of the landscape are retained. This helps to maintain the level of diversity of wildlife. One of the benefits that may result from doing this is the preservation of viable populations of insect-eating birds to protect crops and pastures. Natural

vegetation areas can be more effective if they are linked with the corridors of trees of suitable width to provide pathways for wildlife. The IGNP canal construction has been stated to have reduced the frequency of sand storms in nearby areas, including Delhi.
- **Increasing property's aesthetic values**: Trees improve the working environment, and generally make the property more pleasant to live on. Tree crops screen the air and serve as a noise barrier. Moreover, most people enjoy the natural beauty only a tree or a forest can provide. Trees can improve a property's appearance, making it more attractive to a potential buyer.

The field crops and trees can be of any species, their complementary combination is the science in it. The field crop and tree phenology, rooting pattern, canopy structure, etc., are to be studied (Singh et al. 2014). For example, any field crop, such as rice, wheat, millets, potato, oil seeds, forages and trees for timber, fruits, fodder, herbals and medicinal, multipurpose, have different phenology, rooting pattern, canopy structure, etc.

9.4 Summary

In order to deal with the challenges and opportunity for livelihoods and environmental security, we need to develop strategy at farm-house hold, landscape and community, regional and the state levels. However, they are not mutually exclusive and there maybe overlap in some of them.

- **At farm-household level**: Ecological rehabilitation; soil and mineral cycle, hydrology, biomass, and biodiversity. Mainstream the conservation into social and economic agenda. Identify ecology specific short-term and long-term livestock rearing and fodder resources development options, especially for replenishing soil fertility, nutrient recycling, and intercropping of fertilizer trees in integrated farming systems. Recycling and sequential use of resources in support of farm-based livelihoods, e.g., internal input dependent agriculture, reducing emissions, and sequestering carbon.
- **At landscape and community level**: Cover landscape: Common land and water resources are space for the poor and platform for their collective action. Focus on conserving land, water and forest resources, and optimal use of forest margins. Operate at community level to include common pool resources, poor, women, out of school youth and landless. Develop public–private partnerships and resolve community conflicts.
- **At regional level**: Strengthen resource conservation and livelihood linkages. Focus on ecosystem restoration separately for Thar and eastern regions including the Aravali range, especially in terms soil, water and biodiversity, e.g., water resources development and stabilization of sand dunes. Maintain land/soil cover throughout the year via natural regeneration of indigenous vegetation. Get research and development portfolios together.

- **At the state level**: Focus on arresting natural resources degradation and their over exploitation, ecological restoration at scale and conflict resolution. Reduce soot, smoke and GHG emissions, and build larger constituency through federating processes, bringing various stakeholders together.

9.5 Recommendations

- **Keeping fewer number and better breeds of livestock**: Raising fewer number of better/pure breed of livestock, e.g., Tharparkar cows, which are reported to be good milk yielder, have lower water requirement than other breeds and more suited to the kind of environment in western Rajasthan would provide input efficient opportunities for enhancing farmers' livelihoods and income and reducing pressure on forest resources.
- **Keeping milch animals in shelters**: Raising shelters to keep the cows at homestead could be an additional intervention to decrease open grazing to some extent. This would also allow easy collection of livestock manure, which can be used to maintain/raise the declining soil fertility for farming.
- **Enhancement of the stall feeding of milch animals**: Range land and Gauchar lands are reduced and are degraded. Sewan grass is destroyed by mechanization, and is in need of restoration. Pasture lands are deficit in phosphorus and other minerals, and there is the problem of Botulism and Pica. Therefore, communities are interested in stall feeding. This is possible through support for on-farm fodder production by growing Sewan and other grasses, especially those that require less water, and fodder trees; both to be grown by applying deficit minerals through fertilizers in the soil. Farmers are also interested in keeping the milch animals under shelters. However, this interest is subject to getting external support for on-farm fodder production and putting up of shelters for the milch animals at the homestead. As some NGOs, such as BAIF have done Sewan grass rehabilitation in Bikaner and in Jaisalmer, lessons from them can be utilized in this.
- **Establishing fodder banks**: This would involve fodder harvesting from areas having abundant fodder availability and storing the fodder in the village fodder banks for "akal (famine) time." The DNP management is willing to allow the community cut and carry the grasses from the fenced areas. Farmers of this village also identified the Sewan-belt toward the national boarder (in Dhanana area, which is 85 km from Jaisalmer and 45 km from Sam) which has lots of extra Sewan grass. This can be harvested and brought to the village. The village Panchayat at Khuri village showed willingness to form a village society/committee to manage this operation on a pilot basis. However, they would need support to create the infrastructure for the fodder bank. This program could be tried at Khuri village and may be replicated elsewhere.
- **Tree planting on farm and in public lands**: This should be encouraged and linked with the National Sub-Mission on Agroforestry of the Government of India. The trees that would do well in this area are: Babool (*Acacia nilotica*), Shisam

(*Dalbergia sisso*), Sirs (*Albizia lebbeck*), Eucalypts, Neem (*Azadirachta indica*), Khejri (*Prosopis cineraria*), Ker (*Capparis decidua*), Kumbat (*Acacia Senegal Linn.*), Gunda (*Cordia dichotoma*), and Aonla (*Phyllanthus emblica*). Planting of trees as shelterbelts around villages will also be helpful in providing shade to the livestock close to the habitation and also for protecting the communities from the sandstorms and excessively high temperatures.

- **Value addition, branding, and promoting sales of local produce**: There are two items mentioned to be special from the villages around Jaisalmer. These are the Ker-sangria, and desi ghee produced from Sewan grass-fed cows. The latter one is a unique product item of this area. This ghee is yellower than the egg yolk and preferred by the city dwellers and visitors. These two products need to be produced hygienically, aggregated, and linked to the right markets. Communities are willing to embark on this venture and need support for it.
- **Improving crop productivity**: Improving crop productivity through proper soil fertility management, including manure application and other agronomic practices could contribute to enhanced farmers income and reduce pressure on forest resources. This is where the Department of Agriculture can make a significant contribution through soil test-based recommendations for fertilizer application in cropped fields as well as in pasture lands. Such an intervention is also expected to reduce the Botulism and Pica problem.
- **Castration and/or taking out stray bulls**: Government Veterinary Doctors should be encouraged to castrate the bulls if the community members are willing to bring them to the Veterinary clinics. Government should pay a certain amount per bull to the farmer who brings the bull for castration. Additionally, stray bulls can be kept in certain specific fenced enclosures with sufficient provision of fodder and water. Furthermore, the bulls could be brought to designated Gaushalas and kept there. These measures would minimize the livestock breed degeneration and encourage farmers to raise pure breed bulls and adopt the artificial insemination procedures till these bulls come to maturity.
- **Promotion of local ecotourism**: Village level ecotourism with all the rules and regulations and training to operate it, e.g., the solid waste management, sanitation, and hygiene, may be looked into as a positive innovation for improving livelihoods. There are some examples of such an operation in few villages. However, they could be institutionalized through project support.
- **Restoring agriculture to the initially allotted land area only**: This may be difficult logistically, but can be tried out. DNP now has the GPS coordinates of each land, and after getting the community endorsement can try enforcing, initially at pilot scale.

9.6 Conclusions

Occurrence of extreme events is a reality. These are more frequent, severe, and extensive in coverage. Extreme events affect all across the lifescape and landscape systems. Fragile ecosystems and poor populations are more vulnerable to this nexus.

Caring and restoring natural resources (land, water and biodiversity) is one of the effective ways to lessen these effects and provide better opportunities for adaptation to climate change.

Integrating trees in food and pastoral systems with better and more resilient species through the application of agroforestry, ever greening agriculture, or conservation farming principles and practices opens the avenues for climate change adaptations. The 2000–2009 had been the driest decade in 100 years, 2004 had 13% less rainfall, yet from 2004 onward each year had a record production; meaning some innovations for climate change adaptation are taking place.

There are successful and good examples and case studies available right now, they can be tried, validated and scaled-up for the two major regions of the state—the Thar and eastern region including the Aravali range. Innovation is the name of the game for dealing with the climate change adaptation.

References

Achten, W. M. J., Trabucco, A., Maes, W. H., Verchot, L. V., Aerts, R., Mathijs, E., Vantomme, P., Singh, V. P., & Muys, B. (2013). Global greenhouse gas implications of land conversion to biofuel crop cultivation in arid and semi-arid lands—Lessons learned from Jatropha. *Journal of Arid Environments, 98*(2013), 135–145.

Bhojvaid, P. P., & Singh, V. P. (2005). Agroforestry provides opportunities for economic development and sustainable use of forest resources in Haryana, India. In *Proceedings of the 22nd IUFRO World Congress*. Brisbane, Australia.

Minang, P. A., Van Noordwijk, M., Freeman, O. E., Mbow, C., de Leeuw, J., & Catacutan D. (Eds.). (2015). Climate-Smart Landscapes: Multifunctionality in Practice (pp. 3–18). ISBN 978-92-9059-375-1.

Mohotti, A. J., Pushpakumara, G., & Singh, V. P. (2020). Shade in tea plantations: A new dimension with an agroforestry approach for a climate-smart agricultural landscape system. In R. P. De Silva, G. Pushpakumara, P. Prasada, J. Weerahewa (Eds.), *Agricultural research for sustainable food systems in Sri Lanka*. Springer, Singapore. https://doi.org/10.1007/978-981-15-3673-1_4.

Nayak, D., & Singh, V. P. (2013). Anthropological differential management of fodder trees and shrubs. In *Presented at the Science Forum-2013 ICRAF*. Nairobi, Kenya. Available online at http://intranet.worldagroforestry.org/events/sf2013.

Neufeldt, H., Christine, N., Hancock, J., Foster, K., Nayak, D., & Singh, V. P. (2016). Scaling up climate-smart agriculture: Lessons learned from South Asia and pathways for success. World Agroforestry Centre (ICRAF), Nairobi, Kenya, Working Paper No. 299, https://doi.org/10.5716/WP15720.PDF.

Pushpa kumara, D. K. N. G., Gunasena, H. P. M., & Singh, V. P. (eds). (2007). *Underutilized fruit trees in Sri Lanka* (Vol. 1). ICRAF- South Asia, University of Peradeniya, Sri Lanka Council for Agricultural Research Policy, Multi-Purpose Trees Network, Sri Lanka and Asian Centre for Underutilized Crops, Colombo, Sri Lanka Pub., Colombo, Sri Lanka. 494p

Rao, S. C., Ravichandra, K., Nagajyoti, C., Prasad Babu, B. M., Raju, B. M. K., Venkateswarlu, B., & Singh, V. P. (2015). Greenhouse gas fluxes from sorghum (Sorghum bicolour) and pigeon pea (*Cajanus cajan*) under rainfed conditions in semi–arid regions of India. *Journal of Agrometeorology, 17*(1), 17–22.

Sahoo, A. K., Baisakh, B., Pattnaik, P., Nayak, D., & Singh, V. P. (2022). Adaptation and mitigation of climate change through agroforestry. *International Journal of Tropical Agriculture, 39*(4), 393–401.

Singh, V. P. (2007). Agro-horti-silviculture in hill slopes for enhanced and sustained production and hill conservation (pp. 70–73). In Z. Islam, M. Hossain, T. Paris, B. Hardy & J. Gorsuch (eds) *Technologies for Improving Rural Livelihoods in Rainfed Systems in South Asia* (p. 124). IRRI, Los Baños, Laguna, Philippines.

Singh, V. P. (2014). Addressing the challenges of climate change in India (pp. 5–19). In A. Behera, et al. (Eds.), *Proceedings of the Odisha Environment Congress*. Bhubaneswar, Odisha, India.

Singh, V. P., & Nayak, D. (2013a). Integrating trees in the food producing systems/agricultural landscapes. *Technology Note* (No. 6). Published by ICRAF-South Asia, New Delhi.

Singh, V. P., & Nayak, D. (2013b). Enriching livestock wealth through enhanced fodder production and supply systems. *Technology Note* (No. 4). Published by ICRAF-South Asia, Delhi.

Singh, V. P., Sinha, R. B., Nayak, D., Neufeldt, H., Van Noordwijke, M., & Rizvi, J. (2016). The national agroforestry policy of India: Experiential learning in development and delivery phases. CRAF Working Paper No. 240. New Delhi, World Agroforestry Center (p. 26). https://doi.org/10.5716/WP16143.PDF.

Singh, V. P., Sinha, R. B., Singh, Y. P., Nayak, D., Singh, A. K., & Rizvi, J. (2018a). Impact assessment of anicuts and afforestation in chambal-ravines, Madhya Pradesh, India. In *Ravine Lands: Greening for Livelihoods and Environmental Security* (pp. 523–547). Springer Nature, Singapore.

Singh, V. P., Verchot, L., Martius, C., Nayak, D., & Rizvi, J. (2018b). Rehabilitation of degraded lands in semi-arid and sub-humid ecologies in India. In *Ravine Lands: Greening for Livelihoods and Environmental Security* (pp. 505–527). Springer Nature, Singapore.

Singh, V. P., Nayak, D., Prasad, J. V. N. S., Sahoo, A. K., Vasishth, A., Mahanta, D., & Tewari, N. D. (Eds.) (2014). Enabling small holders to improve their livelihood and benefit from carbon finance. In *NAIP Funded Research Project #103601 Final Report*. World Agrofroestry Centre (ICRAF), New Delhi, India.

Trees of Change: Taking the Heat Out of Farming (2012). A Booklet, Series No. 10. The World Agroforestry Centre, Nairobi, Kenya (p. 32).

Verchot, L. V., & Singh, V. P. (2009).Carbon sequestration opportunities with small holder communities: Forestry, agriculture and agro-forestry. In *Proceedings of the 4th World Congress on Conservation Agriculture* (pp. 351–361). New Delhi, India.

Zomer, R., Neufeldt, H., Xu, J., Ahrends, A., Bossio, D., Trabucco, A., van Noordwijk, M., & Wang, M. (2016). Global tree cover and biomass carbon on agricultural land: The contribution of agroforestry to global and national carbon budgets. *Science and Reports, 6*, 29987. https://doi.org/10.1038/srep29987

Zomer, R. J., Bossio, D. A., Trabucco, A., Yuanjie, L., Gupta, D. C., & Singh, V. P. (2007). Trees and water: Small holder agroforestry on irrigated lands in Northern India. A Research Report No. 122, of IWMI and ICRAF (p. 51).

Chapter 10
Mapping Ecological Concepts on to the Landscape of Rural Economy: What Makes for Resilience in Agropastoral Western Rajasthan?

Purnendu Kavoori

10.1 Introduction: Uncertainty and Dryland Adaptation

There is an increasing awareness that we are entering a long-term phase of "unpredictability" in our ecological and social environments. At the global level, we have an accumulating body of evidence that indicates that the earth's bio-ecosystem is likely to become increasingly volatile with consequences that are difficult to anticipate. This uncertainty is likely to be a part of the scenario for the century that lies ahead. It is important to note however that this insecurity is unequally distributed geographically, as well as economically. We may expect a considerable unevenness with respect to the way "uncertainty" is likely to be socially and ecologically distributed, which aspect often gets masked when we look at the problem primarily from the scale of the global alone.[1]

In this paper, our focus is on desert-type ecosystems. Arid and semi-arid ecosystems come in a wide range of landscapes, temperatures, altitudes and of course aridity. These landscapes, however, have one feature in common, namely the presence of a high degree of variability. Arid and semi-arid landscapes invariably have considerable fluctuation in temperature, rainfall, and local bio-physical gradients which make

[1] Acknowledgements. I am grateful for the support provided by the Foundation for Ecological Security, Anand for enabling field work required for this research. I am also thankful to Anjali Aggarwal, Nayntara Lakshman and Saurabh Levin for support during the research and data analysis. I am grateful to Chaitali Rathore for undertaking a survey of the river during its flooding phase. Special thanks to Shri Om Prakash Vaishnav who was a critical interlocutor and field guide over the course of fieldwork Last but not least my sincere thanks to Mr. Aman Singh for his companionship and contribution in the development of these ideas during field work and beyond.

P. Kavoori (✉)
Centre for Social Ecology, Jaipur, India
e-mail: purnendukavoori@censeindia.org

M. S. Rathore et al. (eds.), *Desert Ecosystems*,
https://doi.org/10.1007/978-3-032-03237-9_10

them unpredictable environments to live in. There are two consequences to this from the point of climate change. One that these habitats are likely to experience the impact of uncertainty in ways that will be different from other kinds of landscapes. Secondly, the nature of adaptation in these habitats has historically evolved to deal with uncertainty, and consequently, hence there is much that we can learn from the ways people have adapted to risk and the practices that enabled people living in these habitats to cope.

Western Rajasthan is one of the great desertic regions of the world with a long history of human occupation. Although it is the most densely populated desert region in the world, compared to other parts of India it is much less densely settled. As in many other desert environments, the nature of human adaptation is dominated by an agropastoral system of resource use. Agriculture in recent years has become dependent on deep ground water, historically however the region was mostly characterized by rain fed cultivation. A considerable component of the productive economy was in the past and continues to be today, its livestock, based on open range grazing. Pastoralism in varied degrees formed part of almost all the strategies of adaptation. There exists a range of mobility among herders extending from short migrations to long distance movements over months. Camels, sheep, goats, and cattle were the primary species herded, and all of them invariably incorporated some degree of movement and were otherwise maintained on open range grazing.

To a considerable extent, thus the landscape of western Rajasthan has been shaped by the way it has been used and modified by people keeping livestock. While agriculture forms historically a very important part of the rural land use system, it is the less influential and subordinate role. Even though the period after Independence saw a change or decline in its importance, it is important to understand why pastoralism and in general livestock had such a dominant role in shaping the landscape and the institutions of the region.

10.2 Pastoralism and Its Ecological Logic

In this section, we briefly discuss key characteristics that make pastoralism of such importance in dry land environments and why it remains critical to think strategically about coping with uncertainty and risk.

10.2.1 Pastoralism, Scale, and Rationality

Although pastoralism and pastoralists are found across the globe, it is difficult to generalize on pastoralism at the global level, because even while pastoralist strategies are in many ways comparable across different parts of the globe, they are embedded in ecologies that are specific. At the same time, however, no pastoralist is an island, and it is difficult to understand the rationality of pastoralism at the elementary level of a flock or a household in isolation from the larger systems they are part of. This is also the subjective lived experience of pastoralists, frequently on the move, pastoralists traverse considerable distances and multiple landscapes. Conventional reductionist methods that seek to understand their logic in terms of the economics of the individual flock or household decisions regarding movement do not get the point.

Both in terms of their ecology and the rationality of production pastoral livelihoods make sense only if we grasp them at a sufficiently inclusive scale. Pastoral systems can thus be quite extensive, traversing economies, ecologies, and societies and while forming part of them retaining their own identities and autonomy. They constitute an example of systems of adaptation that cohere, operate, and reproduce not at the very local scale or that of the global, but at what we may call the "middle range" of a bio-region. It is perhaps precisely because they operate at this scale that they provide us with a point of departure for grasping the consequence of climate uncertainty as these unfold in their concreteness and engage strategically with the challenge.

10.2.2 Pastoralism and Uncertainty

Equilibrium is a state where the relationships between elements of a system remain constant. Laboratory experiments are the ideal type of example where all conditions are kept constant so that outcomes are replicable. Scientific or industrial farming are examples from real-world contexts where predictability is realized by controlling the conditions under which production is undertaken. We know today, however, that these systems of control have a pattern of escalation and breakdown which may become increasingly unpredictable and virulent.

Pastoral systems may thus be contrasted to systems that control the conditions under which production is undertaken and are fundamentally a strategy for coping with variability. Unlike in a laboratory, rather than control the variables, pastoralism seeks to achieve the same outcomes by making all elements of the system move together at a similar rate, viz., it seeks to achieve an equilibrium that is dynamic rather than static. Pastoralists thus cope with perturbation by adapting to it rather than controlling it. Pastoralists do not seek to control variability, rather they accept disturbances as a gradient that is beyond control. What they have evolved are strategies that enable them to traverse the disturbance or to reduce the possibility of breakdown.

10.2.3 Pastoralism and Complexity

James Scott's work "Seeing Like a State" introduced a counter-intuitive perspective on what constitutes the processes of complexity in the context of development. Inverting the standard notion of equating development with complexity, he argued that industrial modernity drives not complexity but the opposite processes of "simplification." He observed that the rationality of industrialism is realized through standardizing the conditions of production and controlling them for deviation or variation, that also creates precisely the conditions necessary for systemic and structural breakdown. He contrasts this with "traditional" systems, techniques, and practices that work with existing gradients of landscape, rainfall, soil type, etc., in a way that a highly diversified production system comes into being and suggests that the latter type of structure is in fact the more complex and stable system with fewer negative externalities and greater capability of coping with unpredictable events (Scott, 1999).

Pastoral production systems are an excellent example of the kind of complexity that Scott is talking about, in that they maintain equilibrium through producing complexity at a certain scale. We have already noted the dynamic nature of equilibrium in pastoral systems, complexity is the counterpoint to their dynamic nature. Pastoralists adapt to diverse ecological gradients, thereby preserving its information content rather than flattening or overriding it. This means that pastoralists respond to landscapes in highly graduated ways that "work with" differences in ecology and social diversity. Pastoral systems are also complex in the interactions that sustain them. In South Asia, at any rate, all pastoral populations are embedded in broader agrarian communities and form part of them. It is our experience that the relationship between scale and complexity is also not a unilinear one and it is at the middle range of scale that complexity in structure and interconnectedness is perhaps at its greatest.

10.2.4 Pastoralism, Reproduction, and Autonomy

Through the years of colonialism, and even prior to that, pastoralists frequently acquired a "bad name" by antagonizing state powers. To an extent, this remains true even today. The standard explanation for this antagonism has been that pastoralists are difficult to "police" and need to be settled and "pacified." It may well be that historically pastoral populations were successful in eluding state authorities, even confronting them from time to time; but there is in our view a deeper reason behind the antagonism, going beyond the standard tropes of "taxing and policing." What the standard explanation does get right, however, is that pastoralism enables or makes it possible for a certain kind of "autonomy" and capacities to resist incorporation.

The fundamental basis of pastoral autonomy lies in our view of its capacity to sustain reproduction under the most adverse conditions. There are multiple levels at which "reproduction" may be understood. First undoubtedly is the capacity to biologically reproduce the capital, namely the herd or flock. The herd is both the source of value and the form of value. While pastoralists do seek to optimize the product of their animals, it is notable that they resist inputs that might increase productivity and returns. The reason for this is that the pastoralists are resistant to prioritizing production over reproduction. It is not that pastoralists are averse to making more money or innovation, what they do resist is the introduction of reproductive vulnerability into their system.

Similarly, while pastoralists are deeply enmeshed in the market economy viz in the exchange of commodities, they are equally embedded in relationships that are not realized through commodity exchange. Relations based on reciprocity, accommodation and embodied exchange form an important part of most pastoral communities, although these are no longer as strong today as in the past, and in many cases, reciprocities are now being monetized. There remains nonetheless a large domain of relations that are ecological as well as social, where rights, claims, obligations, and reciprocities continue to be expressed in forms that are non-commoditized. This is even true of the predominant forms in which pastoral labor appears, such that a complete separation between the worker and owner remains the exception rather than the rule. It is this non-commoditized realm that both sustains pastoral production and enables it to reproduce socially and ecologically. If indeed the totality of pastoral production were to be commoditized, in our view pastoralism would cease to be viable.

To summarize then, the critical feature of pastoralist capacities to survive uncertainty is that it is not just a system of production but that the fundamental strength of pastoralism lies in its resilience as a system of reproduction, and it is that resilience that lies at the core of its autonomy. If we have to strengthen pastoralism, we have to not only improve its productive capabilities but also conserve the essential elements that are necessary for its reproduction.

10.3 The Luni River: *A Pulse Modulated Ecosystem*

The purpose of this section is to provide a concrete example from a specific landscape in order to illustrate, at what we have termed the middle range level of a bio-region, the tension between the ecological and the developmental logics over the medium and long term.

Rajasthan is a large area of land, and it is difficult to talk about it "as a whole" without overgeneralizing. In the section below we confine our discussion to one substantial and defining feature of its landscape, namely that defined around the drainage of the river Luni. The Luni forms the sole drainage running across the Great Indian Desert. A rain-fed river falling in a region of low rainfall, the Luni has always been a seasonal stream, never carrying water for more than two to four months in a year. As is evident from its very name, much of the tract through which the Luni[2] runs are saline, a condition that increases as we dig deeper into the soil. It appears nonetheless that in times past the annual flow of the river was sufficient to lay the basis for the development of an interesting and innovative system of agriculture based on the exploitation of shallow open wells. Dependent on annual recharge by the river, all along the banks of the Luni were to be found unlined and masonry wells rarely more than twenty to thirty feet deep, which supported a flourishing agriculture growing wheat and gram. The water from these wells could only be used for three

[2] From the Sanskrit term *Lavana* meaning salt.

to four months, after which they became too saline for cultivation. Given an annual recharge by the river, this was nonetheless a stable and reliable system of exploitation.

Once we moved beyond this strip, there lay extensive tracts of land that cultivators had claimed from the desert through generations of labor. Surface topography comprised vast open tracts of sandy soil, scrub and grassland, hilly and rocky terrain. Wells in these tracts were a rarity and where they did exist, they would have been very deep. Cultivable soils were shallow and had to be carefully nurtured. The greater part of this land was a rain crop tract growing bajra millet (pearl millet) and dotted with fodder-growing trees such as the well-known Khejri *(Prosopis cineraria)*, large parts of which remained fallow for periods lasting two to three years. These fallow tracts combined with the presence of extensive uncultivated rangelands laid the basis for the development of a pastoral system of resource use involving the movement of animals through different parts of the habitat over the course of the year. Substantial cattle populations were thus supported on an open range system of grazing. Small ruminants, especially goats, survived on pasturage in ravine lands and stony tracts. Sheep herding was associated with nomadic and transhumant pastoral groups, as were also camels. The one animal, which was a rarity, was the buffalo.

This pattern of agropastoralism based on open range grazing, fallow pasture and rain crop farming coupled with a restricted amount of irrigated agriculture, laid the basis for a drainage friendly landscape. Open fields and pastures provide fewer barriers that impede water flows. Local rules and regulations protected the encroachment of drainage channels. Landscapes were deliberately shaped to encourage the flow of water over fields. Distinct parts of the habitat were defined by the characteristic manner in which water made its way across them. All these went into ensuring that the meager rainfall of the desert monsoon made its way into the river: agriculture and pastoralism interlocked in a cycle that began and ended with the life of the river.

At the turn of India's independence, the Luni was far from "dead." Not infrequently, villages retain memories and anecdotes of getting isolated during the rainy season when the movement became difficult and roads would get flooded. Conversations with community elders conducted thirty years ago suggest that the Luni River was very much an active stream during the fifties and even into the early sixties. In the decades following independence, however, the pattern of resource use, and with it, the lay of the land underwent an irretrievable transformation and how an ecosystem-based pattern of resource use and human adaptation was replaced by one characterized by technological overdrive and profit. Seen chronologically, these changes can be characterized as having taken place in three phases and demarcate changes in the organization and structure of *production* as well as in the *reproductive* autonomy and ecological integrity of the habitat.

10.3.1 Post-Independence—1950 to 1980

An intervention specific and limited to this period was the attempt at land reforms. Even in a watered-down form, in a semi-arid area dominated by pastoralism and open range grazing the introduction of "Ceilings on land" had complex consequences on the way resources were managed, accessed, and used. Large tracts of land owned under former Jagirdari grants were affected by the introduction of restrictions on land ownership. Many of these substantial tracts were used as pastures for horses, cows, and other animals of former ruling strata. With ceiling restrictions, these lands were frequently ceded to former client populations, who then changed the way these lands were used, leading to the beginning of reductions in the availability of open grazing. Incrementally and in the longer run these also contributed to changing the micro watershed regimes. This is also the period when physical interventions on the river were made by way of the construction of embankments, with the objective of flood control. Noteworthy at the time was the introduction and early spread in the landscape of an invasive species, *Prosopis juliflora.*

10.3.2 Impact of Development: 1980s to 2000

Whereas the previous phase was noted for the construction of bunds and embankments for control flooding, the focus during these subsequent years was on the construction of a number of small and medium sized dams built on the river channel. Even more widespread than these were the innumerable end cuts or check dams built across the landscape. These check dams were usually built on small water channels and may be found in open tracts, in pastures, in sacred forests, and even in agricultural fields. This extensive effort at capturing moisture across the landscape had impacted incrementally on the entire river system, by the mid-1990s. The reduction in flow of the river also meant the deterioration of a very important source of pasture, the riverbed itself. For when the river used to come annually, the water would bring new seeds and new growth. With the drying of the river, the bed became dominated by old growth and invasives.

The years from the mid-1980's to the turn of the century also witnessed the expansion of agricultural lands at the expense of fallow and pastoral lands. This was possible largely due to the introduction of tractors in significant numbers during these years. As a result, rocky outcrops, and hilly and stony tracts were gradually enclosed for cultivation. The incremental encroachment of pastoral lands, the reduction of fallow, the extension of cultivation into streambeds, ponds, etc., began to show their effects on the deteriorating condition of the region's cattle population. The latter part of this period was also marked by a series of droughts that witnessed the abandoning of cattle on a large scale.

This latter period also saw the extension of electrification and the spread of deep tube wells leading to a plummeting of water levels from 150 to 500 feet in a matter of a decade. Expanding cultivation also impacted the drainage of the landscape as foothills and open tracts were gradually blocked and surface flows ceased to reach the river bed. Finally, these years also witnessed the weakening of the many institutional practices and cultural norms regulating the flow of water and access to resources such as nadis, talabs, gauchars, and sacred groves.

10.3.3 Post-Liberalization: 2000 to 2022

The most dramatic and in a way culminating feature of the ground situation at the present time is the manner in which the river and the broader landscape have been impacted by mining. While mining, especially for river sand, is not new to these parts, what is new has been the scale and the rate at which extraction has taken off. Today there are only deep craters and gullies where there once was a vast sandy swathe of pastureland. Much of this extraction is illegal, but most is not. Additionally, the bed of the Luni is now also being used as a dumping ground for waste. In recent years, unused open wells and step-wells have become another space for throwing domestic garbage.

The drying of the riverbed created an opportunity for encroachment by communities, with cultivators gradually extending the boundaries of their fields into the river bed. Apart from the riverbed, encroachment is rampant also in other kinds of water bodies and their catchments. In certain cases, where the catchment has been entirely blocked, and the water bodies no longer carry any water, these ponds have been leveled over and converted into fields. The riverbed has also sometimes been taken over for a variety of interventions ranging from the building of stadiums and playgrounds for schools to dispensaries and other government constructions.

Most of the open-wells, the traditional sources of drinking water, are today saline and have been abandoned. The domestic supplies of drinking water today (including the public lines) come entirely from deep aquifers tapped into by tube wells or borings. Apart from drinking water, most of the water used in agriculture today also comes from deep groundwater aquifers; many of these wells are also now saline. Where salinity has not spread to the extent of making cultivation impossible, the crops which be grown are moving from more rewarding cash crops to less profitable ones.

All these developments have had an effect on the livestock ecology of the region which may be seen most dramatically in the near disappearance of the Nagori breed endemic to the region; cows in general are no longer looked at in favor. There has been a mushrooming of gaushalas in the area in response to the abandoning of cows and the general crisis of open range pastoralism. The severity of the crisis we see is connected directly to the drying of the river, a general disintegration of the landscape structure and loss of species diversity, and cultural and institutional collapse (See Appendix 1 for a summary of the above description in a tabular form).

10.4 A Breaking of the Pattern: The Flood of 2024

A dramatic and unexpected event such as a flood has a revelatory character to it, such that every aspect of the human nature relationship pertaining to that landscape becomes highlighted both for its positive and its negative consequences. In the quarter century that I have known the Luni, the kind of flooding we saw in this year was unprecedented. In terms of community recall this extent of flooding last occurred forty years ago. Alniyawas the village where I happen to live became accessible in a manner reminiscent of conditions half a century ago when for three months in a year the village would be isolated and people would need to stock up on necessities. There had been occasional years (e.g., in 2021) when the river would collect water in short patches at different points, but this was not comparable to what we witnessed in the year 2024. In the few paragraphs below, I am quoting verbatim from a short report prepared by a young researcher who travelled about seventy-five kilometers across the upper course of the river, to put together a description of the physical condition and the reactions of people to this experience of extensive flooding.

> The Luni flooded after almost 40 years. While this was the first sighting of the river flowing to its full might by many from the younger generation; it revived the memories of the old. During its dry days, the extensive sand mining was lamented by everyone, addressing it as the elephant in the room, a big cause for the deteriorating river health. This year, however, the 100 feet deep ditches caused by mining could not obstruct the flow of the raging river. The advent of the river after almost half a century has now changed the context of saving it.
>
> The event being a significant one, echoed with several voices and perspectives of the people around the Luni River. It not only filled the important dams but also raised the water table, reaching out to the problems faced due to scarcity of potable water and irrigation in the fields in the region. Its flooding also caused problems in communicating across different villages and casualty of lives and property.
>
> The river was welcomed by the villagers, who celebrated and performed rituals, by offering chunhari and taking the river water as a souvenir, as Gayatri ji from Ladwa called it their "own gangajal." The annual mela in Birantiya Khurd was held in the running river, with several visitors from Punjab and Haryana. The waste from the celebrations however made it difficult to recognize one of the parts of the river. A woman, in her late thirties, in village Birantiya was found planning to visit her natal village Giri to see the famous Nandiya dam, which is now 27 feet deep after 25 years, since her childhood. The famous Jaswant Sagar dam in Pichiyak flooded after 17 years.
>
> The farmers were joyful about the wells recharging to the level of sustaining agriculture for next 5–10 years. The visitors at Jaswant Sagar dam, coming from distant villages like Khejarla and Ransi said that the water table has recharged in their region, which was about 70–80 kms from the dam. While the privately owned wells were in a good state, the public wells did not recharge, as they were clogged with garbage. People were also seen fishing at certain places. The swimmers in Sukhri river, near Banjagudi could be seen spending time in the river. The heavy rains and flood, at the same time, cost the farmers their current crops of moong, guar and moth which require less water. However, looking at the prospects of growing rai, wheat, rijka and jau (barley), they are ready to forgo the current loss. Lekhraj, a 23-year-old daily wage worker and farmer in Alniyawas was however affected differently by the flooding river, as he did not get enough work to supplement the poor agriculture, this time.
>
> An old man from Kanecha told how the river destroyed the trees, reminding him of his younger days, when it flooded more often. According to him, the current generation underestimated the might of the river leading to several casualties at different places. Policemen were deployed at several spots with higher probability of accidents. There was an instance of three youths dying by drowning in Pichiyak during Ganesh Visarjan celebrations. In other villages like Kanwliya, Alniyawas and Lototi, trucks and motorcycles drowned.
>
> Young students going to school in bigger villages like Riyan had to either cross the river on foot or skip classes for almost a month. Many tried getting back to schools without their parents' knowledge, to continue their learning. In one instance, students from Lungiya were

> being carried in a JCB to the other side of the river. The damaged roads connecting the villages turned them into islands, disrupting the connectivity between them.
>
> Most importantly, the river asserted itself by reclaiming its course, as it flooded the structures built in its bed. Houses, cropped land, Gaushalas and even transformers submerged in the river. In places like Anandpur Kalu, the former dumping grounds have now come under the purview of the river. In Alniyawas, the villagers demanded restoration of the bridge connecting Alniyawas and Kod, but without pipes, to let the flow become stable, and to allow the groundwater to recharge.
>
> The river flooding after 40 years is a significant event. Heavy rains in Marwar, due perhaps to climate change, makes the pattern of such events unpredictable. However, for the current generation, which witnessed the river for the first time, the flooding has brought up an alternate scenario – How to improve the river ecosystem, irrespective of the erratic rainfall?
>
> (Chaitali Rathore, Intern, Centre for Social Ecology).

These notes, uncut and raw, express powerfully the visceral nature of the event that the floods of 2024 have been for the inhabitants of Luni. It is important, however, to look at this event in the long term rather than react to it as a "crisis to be controlled and mitigated." As we noted in an earlier part of the paper, flooding is a natural property of this river system and played an important role in shaping the way people evolved their adaptive strategies. Through the sixties, however, when the river did maintain an annual cycle of flooding, the government visualized the pattern as something to be controlled, managed and eventually subdued. It would be unsurprising if the same "diagnoses" again gain traction in the state's view of these events, although this is precisely not how we should see the flooding. As the reflections, observations and actions of people who interacted with the above indicate, flooding was recognized in the long term as an event with a great positive impact on the landscape, the resources, and the lives of people and received positively in the communities' cultural and spiritual life. In place after place there have been celebrations, poojas, music, dancing, and festiveness in the way people have responded to the event. No doubt there was a certain amount of damage and loss of lives. Most of this, as in damage to fields or cultivated lands, to buildings was a result of encroachment on the riverbed which had taken place over the previous decades.

There seems to be a recognition of the flooding as an event or experience that repealed the interconnectedness of communities and ecology at the larger landscape level. It does seem that people were more than aware of the favorable and formidable impact of the flood whether it was the filling up of reservoirs of small and medium dams in the tract, the recharging of open wells and the long-term impact on groundwater in locations at a considerable distance from the main course of the river. So, alongside the fresh deposit of sand, there would also have been a replenishing of its seeds that would grow in the riverbed and beyond once the river water subsided. There was insignificant damage to livestock reported and owners of herds and flocks clearly recognized that the flooding would benefit their animals and pastures in the year to follow.

At the same time, while not discounting this larger positive narrative around the coming of the river, it was also evident from various conversations however that younger people did not always perceive this nuanced view. For people below the age of forty, this was a first-in-lifetime experience and a critical one for they only now began to see the river as an active partner and contributor to the ecology of their landscape. The flood of 2024 was thus an important event in jogging cultural memory, a marker and reference point to carry through their lives and thus ensuring an intergenerational transfer of knowledge within the population.

10.5 Conclusions

As we have sketched above, the form in which life and adaptation in this landscape had evolved prioritized the dimensions of livelihood security and capacities for coping with extremes of variability. These capacities, evolving over extended periods of time, were an expression of the character of the people, and the values of their society expressed in the systems of resource utilization. The historical engagement with "development" was illustrative of how this "sustainable" socio-ecological regime was replaced by a regime far more vulnerable to perturbation.

The transition from open range animal husbandry with rain fed cultivation to one where groundwater becomes the critical variable was a move that reduces the capacity to cope with variability. Accompanying these changing strategies of adaptation and resource use was a moral shift from a risk-averse to a risk-taking culture and economy. We would say that the human and ecological resilience of the bio-region was far more vulnerable at the end of our period than at its beginning.

Secondly, there was a definite reduction in the information content at the landscape level indicating a general process of simplification. The overall decline in the diversity of water sources, the reduction in flow and the complexity of drainage networks were matched by the reduction in the diversity of livestock keeping and the atrophying of supportive institutional complexes and mark the broader undermining of the system's capacities for returning to equilibrium following a disturbance. Finally, we can see a definite loss in the systems' capacity for reproduction. In contrast to resilience, which is the system's capacity to return to normalcy after disturbance, reproduction is the capacity of a system to "replenish or restore" that which has been consumed in the processes of production (of resources as well as human capabilities). What we see instead are escalating dependencies on externalities beyond the reproductive capabilities of the system, thus reducing autonomy and ultimately sustainability.

The "structural crisis" of the Luni thus cannot really be grasped at the level of the village, much less an individual household. At the same time however, if we see the events narrated above simply as a process of "destructive development" at a national or even global level then we completely lose sight of the specificities in the way these processes unfolded. Frameworks for intervention or even strategic thinking on paths to recovery have to be articulated at a level in between these two, which may be called a middle-range scale of the bio-region. It is at this scale of the bio-ecological regions—in this case, that of the Luni—perhaps that we can best grasp the granularities of the process and at the same time cover a critical mass of both population and landscape.

To conclude, the accumulating evidence of climate change at the global level, and the probability of experiencing increasing uncertainty and random events at the bio-regional and even local scale has forced not just researchers but those directly engaged in the more pragmatic fields of governance and policy-making to rethink our conventional or mainstream strategies of development intervention. Researchers working in diverse landscapes where different types of ecosystems intersect such as forests with a long history of human activity or coastal landscapes with multiple user communities, recognize the critical role of movement and resource use flexibility in sustaining their complex intersectional character. The arid and semi-arid lands of western Rajasthan are also complex intersectional spaces. The Luni bioregion that we have described above is an example of how complex risk-minimizing strategies for coping with variability laid the basis for the resilience of both livelihoods as well as ecological elements. The floods of 2024 brought back into the picture the complex interconnections that persisted in a landscape that had been ostensibly "simplified" beyond repair. Semi-arid landscapes and livelihood strategies remain a vivid example of how non-equilibrium systems work and are ill-suited to a politics of governance based on principles of command and control.

Acknowledgements I am grateful for the support provided by the Foundation for Ecological Security, Anand for enabling field work required for this research. I am also thankful to Anjali Aggarwal, Nayntara Lakshman, and Saurabh Levin for support during the research and data analysis. I am grateful to Chaitali Rathore for undertaking a survey of the river during its flooding phase. Special thanks to Shri Om Prakash Vaishnav who was a critical interlocutor and field guide over the course of fieldwork. Last but not least my sincere thanks to Mr. Aman Singh for his companionship and contribution in the development of these ideas during field work and beyond.

From the Sanskrit term *Lavana* meaning salt.

Annexure 1—Changes in the Luni river landscape over the last six decades

Categories	Pre-1980	1980	1990	2000	2010	2020
River drying	• Last flood of the Luni in 1979 • The river came many times but never flooded	• Building of embankments for flood control • Building of anicuts on streams	• Reduced duration and flow of the river • Water stopped coming in large sections of the river	• River has largely ceased to flow • No water except for a few small ponds • Sand mining begins with Govt. giving licenses	• The river has completely ceased to flow except in parts of the upper reach • Youth aged 20–25 years have not seen the river flowing	• Riverbed marked by immense craters • Rainwater is no longer retained due to lack of sand
Water crisis	• 1–1.5 kms near the river, water was available at 10–15 m of depth • The water was sweet in wells for 8–9 months. The bigger taalabs retained water for a long duration (uptil summer)	• Open wells start to become saline • Diesel motorised tube wells are introduced	• Tube wells begin to spread across the landscape • Open wells get abandoned to tube wells • Widespread electrification drives deep tube wells	• Pumping and piping of water across long distances • The trading of water for irrigation and drinking purposes started	• Salinity in tube wells being to spread • Duration of water reduces • Open shallow wells are now abandoned	• Agricultural land becomes saline in parts • Tube wells become deep (700 ft) • The shallow well now used for dumping garbage
Changing agricultural practices	• Millets dominate and no hybrids. Bajra was dominant	• Millets largely remain native	• Hybrid millets begin to replace natives	• Large areas converted from rainfed to irrigated	• Onion cultivation reduced	• Due to increase in soil and water salinity,

(continued)

(continued)

Categories	Pre-1980	1980	1990	2000	2010	2020
	• Local varieties of wheat and winter crops dominate • Pesticide use very low • Fodder production extremely important	• Hybrid wheat is there • Hybrid oil seeds begin • Cotton and sugarcane are introduced	• High value crops (eg sugarcane, oilseed, pulses, wheat) are hybrid • Vegetables like onions, calbbage, chillies become important • Widespread fertilizer use	• 3 crops are grown annually in some places • High value commercial crops replace areas formerly growing millets • indigenous millets replaced by hybrids	• Sugarcane has disappeared • Pesticide use is widespread • Productivity begins to fall • Deep irrigation is widespread	Water salinity, agriculture production is falling
Livestock crisis	• Large camel herds • Nagori bullocks widespread and sold across India	• Sheep herds increase • Size of camel herds reduce • Open range grazing in river • Good prices for animals in fairs • Out Migration of sheep to haryana and malwa started for three months	• Meat market for sheep and goats overtake bull market • Bullock traction begins to decline • Hybrids introduced, buffaloes increase • Nagori is being replaced by milk breeds	• Bullock traction declines • Sheep migration increases • Buffaloes increase Cows are abandoned and run wild • Cattle number and begin to value decline	• Nagori bullocks are virtually extinct • No large herds of camels • Sheep + goat number increase	• Cattle and camels have disappeared • Cows are no longer in households, only Gaushala

(continued)

(continued)

Categories	Pre-1980	1980	1990	2000	2010	2020
Traditional institutional collapse + replacement by state institutions	• Institutions are strong Strict enforcement of rules and regulations laid out by community institutions	• Wakening of commons institutions • Dominant groups lose interest in commons	• Collapse + alienation of traditional institutions • Anghors used as latrines • wood cutting in Orans • Tractors take water from Nadis	• PRI (Panchayati Raj Institutions) start taking over traditional functions • JFM begins to be implemented	• Hybrid institutions start to come into importance • NREGA started and the interest of community in MNREGA increases due to the associated livelihood benefits • Gaushalas increase in number	• Bottom up institutions are replaced by top-down institutions
Encroachment	• Areas of commons + river body largely unencroached although utilized seasonally and managed by communities	• Encroachment of gauchars begins along with diversion of common land for govt activities	• Foot tracks and mud roads begin to get encroached • Anghors also get encroached by individuals • Encroachment starts to get regularized (Sect. 92)	• River encroachment begins for agriculture • Encroachment into orans and anghors for agriculture is now extensive	• Encroachment on river for sand mining gains momentum • Encroachment on stony and hilly tracts for stone mining is common	• Foot tracks have disappeared • Walling and fencing of agricultural land is common • Most parts of river have been encroached by cultivation

(continued)

(continued)

Categories	Pre-1980	1980	1990	2000	2010	2020
Invasives	• Great diversity of trees as welll as grasses • Vilayti babul is introduced by government and planted in forests under the control of the FD	• Intensive aerial seeding by the world bank	• Govt. forestation of Totalis (Israeli Babul), slowly replacing Vilayati babul • Spread into riverbed	• Charcoal manufacture + contracts for VB harvesting • Vilayati Babul becomes dominant, spreads in to commons	• Active govt. removal efforts through NREGA to remove Vilayati Babul	• Vilayati babul continues to remain the single largest species dominating the landscape, except in agricultural land
Mining	• Extraction for domestic purposes	• Extraction for domestic purposes	• The sale of sand extracted begins which increases the rate of extraction • Granite mining has not started yet	• Royalty for sand mining instituted and given to local contractors • Stone quarrying on small scale commenced	• Extensive sand mining, legal and illegal • Large companies begin to acquire mining contracts	• Large scale sand mining through govt contracts • Big companies are given contracts. SEZs established • Extensive employment to households for shipping of sand

References

Federici, S. (2019). *Re-enchanting the World*. PM Press.
Gadgil, M., & Guha, R. (1995). *Ecology and equity: the use and abuse of nature in contemporary India*. Penguin India.
Kavoori, P., & Singh, A. (2022). Environment, Development and Livelihoods in a "Pastoral' Landscape: The Luni River Ecosystem of the Thar Desert. Collaborative study undertaken by Centre for Social Ecology, Jaipur and Foundation for Ecological Security, Anand. Submitted to the Nudge Foundation.
Kavoori, P. (1999). *Pastoralism in expansion*. OUP
Kavoori, P. (2003). *Salt river, ecology and adaptation on the banks of the Luni*. Report Submitted to Ford Foundation, IDS.
Mies, M., & Bennholdt-Thomsen, V. (1999). *The subsistence perspective: beyond the globalised economy*. Zed Books.
Rao, V., & Appadurai, A. (2008). Scale and mobility in defining the commons. In P. Bardhan, & Isha Ray (Eds.), *The contested commons: conversations between economists and anthropologists*. OUP.
Scott, J. (1999). *Seeing like a state: How certain schemes to improve the human condition have failed*. Yale University Press.
Scott, J. (2017). *Against the grain: A deep history of the Earliest states*. Yale University Press.
Taylor, M. (2015). *The political ecology of climate change: Adaptation, livelihoods*. Agrarian Change and Conflicts of Development.
Wall, D. (2004). *The commons in history: Culture*. MMIT Press.
Weiss, T. (2013). *The ecological hoofprint, the global burden of Industrial Livestock* (p. 2013). ZED.

Chapter 11
Towards Achieving Sustainable Development in Agriculture with Rainfed Ecosystem

P. M. Chuahan and V. P. Joshi

11.1 Introduction

India is undergoing a significant transformation as it seeks recovery from the economic, political, and social disruptions caused by the pandemic. While most sectors faltered, agriculture proved to be the economic backbone, especially in rural areas. The investments in irrigation, despite their inherent limitations in rainfed and drought-prone regions, have been particularly beneficial. According to a high-level government committee, 'Our review thus far shows that the limits to further expansion of surface and groundwater irrigation through big dams and tube wells are being reached rapidly' (Government of India, 2006). The committee emphasised the need for a revised strategy that addresses the unique conditions of India's drylands and respects ecological boundaries.

Historically, while the Indian economy aspired for double-digit GDP growth, agriculture frequently cushioned the economy during downturns. Yet, the benefits of growth were uneven, favouring irrigated regions over rainfed ones, which suffered from neglect both historically and in policy terms. This neglect has exacerbated the vulnerability of resource-poor regions, increasing their socio-economic challenges.

11.2 Environmental Impact and Climate Change

Global warming represents one of the most severe threats to humanity. The Intergovernmental Panel on Climate Change (IPCC) reports that recent global temperature increases are likely due to intensified greenhouse gas emissions, including CO_2,

P. M. Chuahan (✉) · V. P. Joshi
Department of Renewable Energy Engineering, College of Agricultural Engineering and Technology, Junagadh Agricultural University, Junagadh, Gujarat, India
e-mail: pmchauhan@jau.in

M. S. Rathore et al. (eds.), *Desert Ecosystems*,
https://doi.org/10.1007/978-3-032-03237-9_11

CH_4, and N_2O (Intergovernmental Panel on Climate Change (IPCC), 2007). These emissions are primarily from burning fossil fuels, which is expected to rise with the global population projected to reach 9 billion by 2050 (Perea-Moreno et al., 2019). India aims to reduce its carbon emissions by 20–25% from 2005 levels, utilising cleaner technologies to lower CO_2 emissions.

11.3 Desert Reclamation and Agriculture

About one-quarter of Earth's land is desert, often marked by extreme climates and infertile lands. Historically, converting deserts into arable land has been a strategy to combat desertification and support population growth. For instance, the irrigated farmland in China's Shiyang River watershed has expanded significantly, leading to water scarcity and ecological issues. While desert farming offers a potential solution to food shortages, it requires substantial water resources, which poses a challenge for sustainability.

11.4 Water Scarcity and Agricultural Sustainability

Water remains a critical natural resource, essential for agriculture, food security, and environmental sustainability (Harken et al., 2019; Liu et al., 2019). With the global population and economic development increasing, optimising water use in agriculture becomes crucial. Areas dependent on irrigation face challenges in water scarcity (Fig. 11.1), necessitating improvements in water use efficiency to ensure sustainable development.

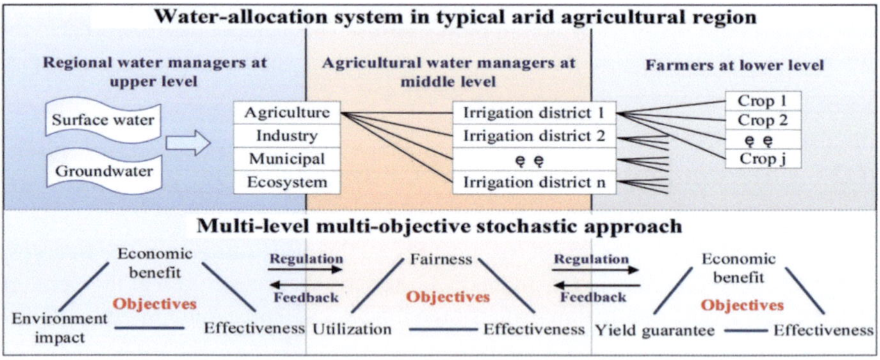

Fig. 11.1 Distribution of water in arid agriculture [adapted from Doe and Smith (2020)]

11.5 Overview of Rainfed Agriculture in India

Since achieving independence, the agricultural landscape in India has experienced significant fluctuations, largely influenced by external events such as wars and natural adversities like droughts. The early 1960s, post the Sino-Indian conflict, saw a financial strain that was exacerbated by consecutive drought years during 1965–67 and conflicts with Pakistan, which severely impacted agricultural productivity and overall food security.

The dependence of Indian agriculture on rainfall cannot be overstated, with the majority of the farming areas being categorically rainfed. This dependency necessitates distinct development policies tailored for diverse geographical regions like dry lands, arid, and semi-arid areas. Historically, the Royal Commission on Agriculture in 1928 highlighted the hardships faced by cultivators in rainfed areas and stressed the need for focused attention on these regions, a sentiment that has often been echoed in subsequent policy frameworks.

11.6 Socio-Economic Impact and Policy Challenges

Despite a decreasing contribution to the Gross Value Added (GVA) over the past five decades, agriculture remains a pivotal sector in reflecting and influencing India's overall economic health. The sector's performance is directly correlated with the nation's economic stability, particularly evident during periods of agricultural prosperity or distress. The Economic Survey by the Government of India frequently cites poor monsoon seasons as a primary reason for the subdued economic performance, reflecting the critical role of rainfed and drought-prone areas in national economic health.

Rainfed areas, which constitute over half of India's total agricultural land, are particularly susceptible to socio-economic challenges, including poverty and vulnerability to climate shocks. These areas are significantly impacted by natural water scarcity, which is increasingly viewed as a central issue in the rural crises of poverty, land degradation, and migration. Lyla Mehta's studies in Gujarat exemplify how drought-related scarcity has fostered a 'drought industry' where water scarcity is manipulated to serve the interests of the powerful, exacerbating inequity and hardship for the less privileged.

11.7 Agricultural Employment Trends and Economic Shifts

The structural shift in India's economy, with agriculture employing fewer people and an expanding service sector, mirrors a global trend where traditional farming is no longer the predominant employer due to its declining economic share. This shift has resulted in a massive movement of labour from rural to urban areas, often into precarious employment situations, as observed by several economists including. These migrations are largely driven by inconsistent income from agricultural practices that rely heavily on erratic rainfall patterns and inadequate irrigation facilities, highlighting the need for robust policy interventions in rainfed regions.

11.8 Policy Recommendations

The challenges faced by rainfed agriculture require comprehensive policy attention that goes beyond short-term relief measures to include long-term strategies aimed at enhancing water conservation, improving irrigation infrastructure, and supporting sustainable agricultural practices. The integration of modern technologies and the strengthening of research and development in agriculture could play pivotal roles in mitigating the risks associated with rainfed farming.

11.9 Economic Fluctuations and Agriculture's Role

While the services sector has shown notable growth, the variations in India's Gross Value Added (GVA) are still prominently swayed by the agricultural sector. The correlation between agricultural performance and national economic stability underscores the fundamental role of agriculture in shaping economic outcomes. These fluctuations suggest that any significant changes in agriculture, whether due to climatic conditions or policy shifts, invariably affect the broader economic landscape.

11.10 Agricultural GVA Versus National Economic Trends

The GVA for agriculture and the broader economy tends to hover around a 5% growth trajectory as indicated in Fig. 11.2. This observation raises questions about the robustness of the claim that India's GDP grew at approximately 9% between 2004 and 2019. According to Deshpande et al. (2004), while the agricultural sector maintained a growth rate of about 3.5% over the long term, the periodic fluctuations have posed serious challenges. Although the GDP figures post-2000 show a higher growth rate, GVA, which more accurately reflects economic health, did not exhibit the same levels of growth. This discrepancy highlights concerns regarding the stability and consistency of economic growth, particularly when considering the agricultural sector's vulnerability to external factors like weather conditions and global market changes.

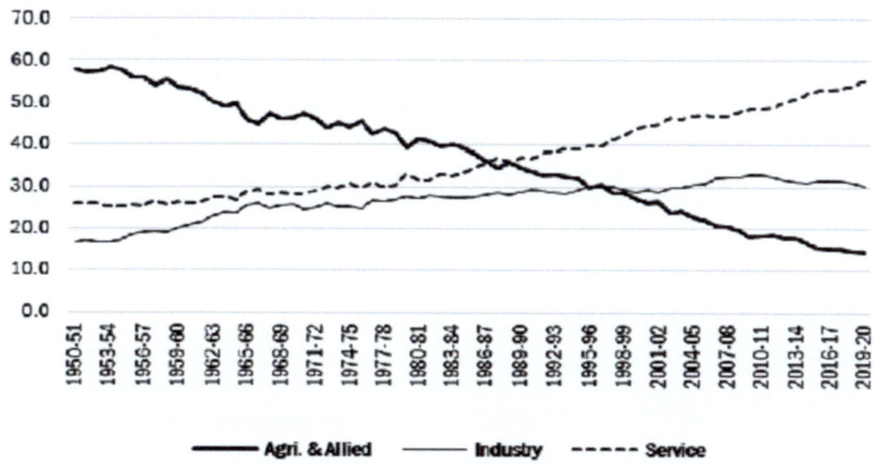

Fig. 11.2 Share of agricultural sector in aggregate GVA (%)

11.11 Historical Performance and Vulnerability to Climatic Variability

Analysing the period from 1951 to 2020 reveals 18 peaks and 19 troughs in economic performance, with agriculture's output deeply impacted by environmental factors, particularly in rainfed areas as indicated in Fig. 11.3. These regions, susceptible to droughts and inconsistent rainfall, often drag down the overall agricultural performance, reflected directly in the national GVA figures. The dependency on rainfed agriculture not only highlights the sector's vulnerability but also its significant influence on the economic stability of the country.

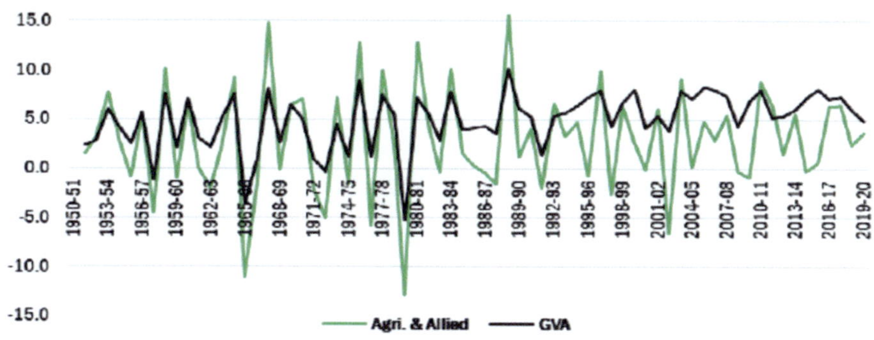

Fig. 11.3 Growth performance of agriculture sector and GVA of India (in %)

11.12 Implications for Policy

The persistence of such fluctuations necessitates targeted policy interventions aimed at mitigating the risks associated with agricultural dependency on rainfall. Enhancing irrigation infrastructure, investing in weather-resistant crop varieties, and improving water management practices could buffer the agricultural sector from some of the negative impacts of climatic variations. Additionally, diversifying the economic base in rural areas can reduce the direct impact of agricultural volatility on the overall economy.

The National Bureau of Soil Sciences (NBSS) has consistently conducted research in the field of regional climatic classifications (refer to MAP. 11.1). In 1981, Ramaswamy proposed that the threshold for gross irrigated areas should be capped at 30%, with the annual rainfall criterion reduced to 375 mm from an initial 1,125 mm. Although he categorised these levels, he did not justify the rationale behind these specific thresholds in his publication. In a similar vein, concurred with Bapna's recommendation to set the gross irrigated area threshold below 25%, with rainfall ranging between 500 mm and 1,500 mm. Adopted comparable methodologies to identify rainfed regions. Kelly and others in 1997 applied the agro-climatic regional classification to describe the pattern of rainfed agriculture in India, echoing previous findings. In a more recent effort, the IMD/NBSS produced a map in 1992 that detailed various climatic zones at the district level (see MAP. 11.1), advancing upon earlier work. This map illustrates that areas like the central Deccan Plateau along with Rajasthan and Gujarat are categorised under arid and semi-arid zones, aligning with the zones identified as vulnerable rainfed areas by ICRISAT. Additional details concerning the total area under rainfed agriculture and its classification are shown in Table 11.1.

The National Agro-climatic Regional Planning (NARP) project pinpointed 127 distinct agro-climatic zones distributed across 17 states and six union territories. Within these, the experts from NARP recognised 73 specific zones (refer to MAP. 11.2) as being particularly susceptible, either as traditionally known or newly classified vulnerable rainfed areas. These zones were selected based on factors such as rainfall levels, irrigation practices, soil characteristics, topographical details, and crop patterns as indicated in Fig. 11.4.

11.13 Dragon Fruit Cultivation in India

In India, the cultivation of Kamalam, commonly known as dragon fruit, remains relatively niche but is expanding across various regions including Gujarat, Karnataka, Kerala, Tamil Nadu, Maharashtra, Chhattisgarh, Odisha, West Bengal, Andhra Pradesh, the Andaman and Nicobar Islands, Mizoram, and Nagaland. According to the Ministry of Agriculture and Farmers Welfare, the land allocated for dragon fruit cultivation spans approximately 3,000 hectares. In the 2021–2022 period, under

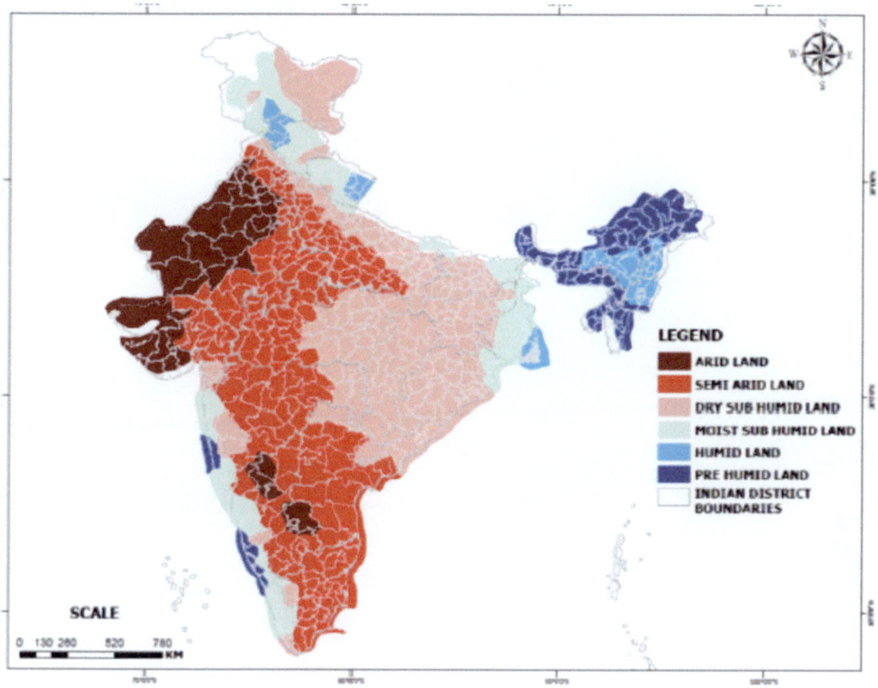

Map. 11.1 Map showing climatic region of India up to district boundaries

Table 11.1 Total cropped area falling under rainfed area

Sr. no	Average annual rainfall range	Typology	Share of area (%)
1	Less than 750 mm	Low rainfall rainfed	33
2	750 to 1125 mm	Medium rainfall and rainfed	35
3	1125 to 2000 mm	Assured rainfall	24
4	Above 2000 mm	Sub-humid and per humid	8

Source Government of India Report on Drought 2022, Ministry of Agriculture, New Delhi

the Mission for Integrated Development of Horticulture (MIDH), efforts to expand dragon fruit cultivation were initiated, with a goal to increase the cultivation area to 4,133 hectares across different states. The importance of dragon fruit is recognised in the government's 5-year plan under MIDH, which aims to boost the cultivation area from 4,000 hectares to 50,000 hectares within the next five years as part of the Atmanirbhar Bharat Campaign, promoting self-sufficiency by substituting imports as indicated in Fig. 11.5. Dragon fruit is increasingly valued not only for its adaptability to less fertile lands but also for its health and medicinal properties.

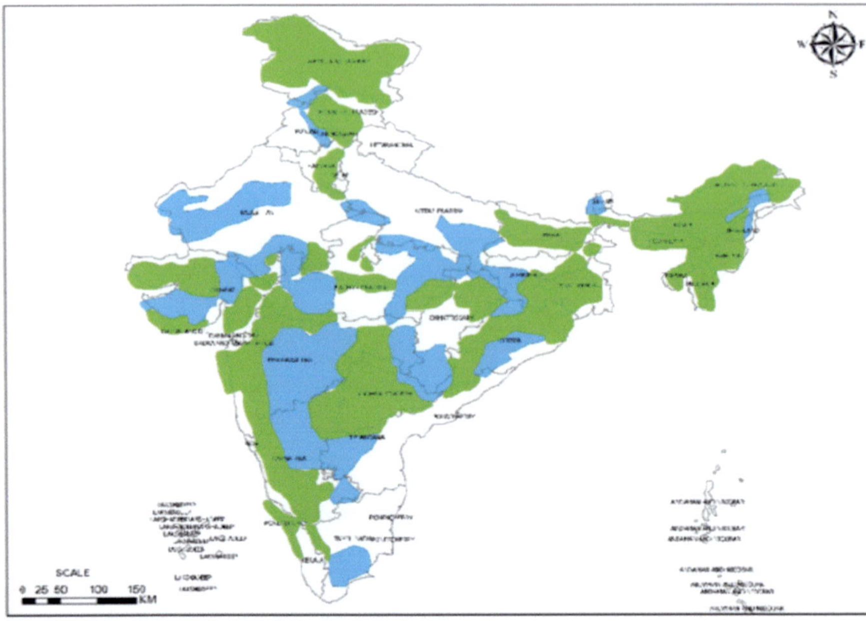

Map. 11.2 Rainfed areas identified by NRAA

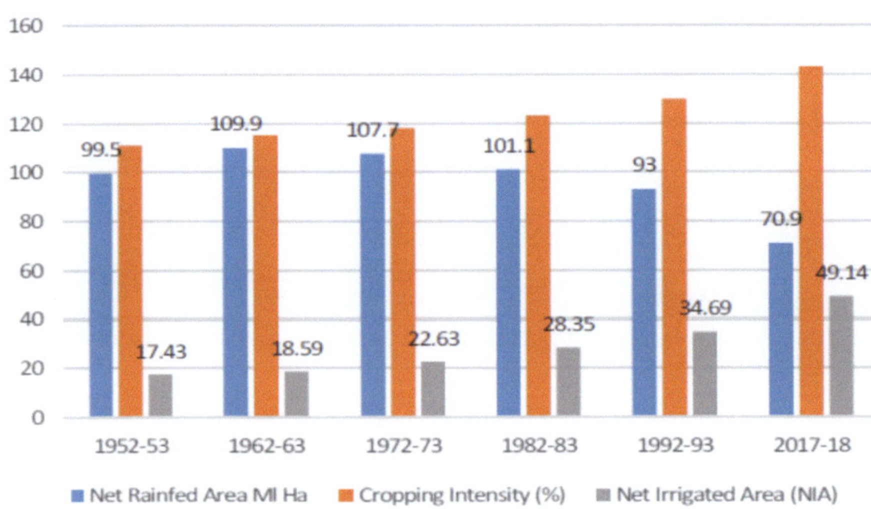

Fig. 11.4 Rainfed area (million hectares, cropping intensity (in %), and net irrigated area (%))

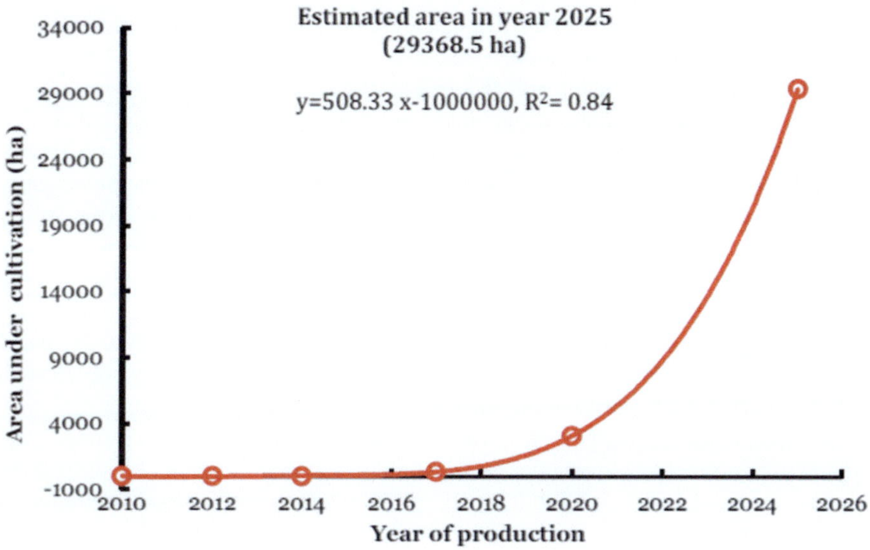

Fig. 11.5 Projected area under dragon fruit for next four years (2026)

11.14 Case Study: Dragon Fruit as a Lifeline for Farmers in Kachchh

Located in Gujarat, Kachchh is the second-largest district in India, accounting for 25% of the state's area. Positioned in a semi-arid zone, it experiences severe weather extremes, with summer temperatures soaring to 50 degrees Celsius and winter temperatures dropping to 3 degrees or below. Water resources are scant, constituting only 3% of Gujarat's total, which significantly impacts the agricultural practices and lifestyle of the region. The region's agriculture, coupled with animal husbandry, forms the backbone of local livelihoods, despite the erratic monsoon and limited rainfall averaging 300–350 mm annually.

Rivers in Kachchh are non-perennial, leading to frequent droughts and water scarcity. The district's proximity to the Arabian Sea exacerbates groundwater salinity, further complicating farming due to high levels of fluoride, chloride, iron, and nitrate, as reported by the Central Ground Water Board (CGWB). Following the earthquake, there has been a concerted effort to rejuvenate Kachchh's economy, particularly through horticulture. Many locals, including those who had previously moved to Maharashtra or overseas, are now returning and investing in horticultural ventures, bringing new technologies aimed at enhancing productivity with limited resources.

The district's harsh climatic conditions surprisingly benefit the farmers by reducing the prevalence of pests and diseases, making it conducive for growing a variety of horticultural crops such as mangoes, sapotas, citrus fruits, date palms, bananas, dragon fruit, guavas, pomegranates, papayas, and custard apples. Areas such as Anjar, Abdasa, Nakhatrana, Bhuj, and Mandvi have relatively better aquifers, supporting diverse crop cultivation.

Introduced in 2014, dragon fruit (Kamalam) has quickly become a favoured crop among Kachchh farmers due to its resilience in extreme temperatures. The cultivation of Kamalam requires careful water management, often achieved through micro-irrigation systems that provide just the right amount of moisture. As the plants grow to about five to six feet tall, they require physical support, initially provided by five-foot poles and concrete rings. Today, over 200 farmers across the Kutch area of Gujarat have embraced dragon fruit farming, covering an area of approximately 700–800 acres.

Regarding the marketability of dragon fruit in India, there is currently a stable and growing market due to the limited number of commercial producers coupled with high demand. The popularity of dragon fruit is expected to continue rising, driven by its appealing taste and recognised nutritional and medicinal benefits. In response, the Gujarat government has initiated a financial plan to support dragon fruit cultivators, offering financial assistance totalling Rs. 1,000 lakhs under a broad horticultural development scheme.

In terms of international trade, dragon fruit from Gujarat has recently made its debut in foreign markets, with initial shipments reaching London, the United Kingdom, and the Kingdom of Bahrain. This marks a significant step in diversifying the export bases for this exotic fruit, known locally as Kamalam.

Prime Minister Shri Narendra Modi highlighted the cultivation of dragon fruit in Gujarat's arid Kutch region during his 'Mann Ki Baat' radio broadcast. He praised Kutch's farmers for their role in advancing India's self-sufficiency in agricultural production. The export of this fruit to the UK and Bahrain is seen as a realisation of his vision for the country's agricultural independence.

Globally, dragon fruit is traded across more than 40 countries and territories. Beyond traditional markets like Vietnam, China, Thailand, Indonesia, Malaysia, the Netherlands, and Taiwan, dragon fruit has found its way into the demanding markets of the US, Australia, New Zealand, Chile, the European Union, Japan, Korea, and Singapore. Noted as the fifth most imported tropical fruit in Asia following lychee, longan, banana, and mango, the export dynamics of dragon fruit are detailed in Table 11.2, showing estimated quantities and values for major exporting countries.

Table 11.2 Estimated export quantity and value of dragon fruit of major countries

Sr. No	Major importers	Quantity (tonnes)	Value (Lakh US$)
1	Vietnam	5,11,000	3,940
2	Malaysia	4,256	20
3	Taiwan	111	3.1
4	Indonesia	92	2.6
5	India	12,000*	120*

Source Ahmed et al. (2019); Chen (2018); Muas et al. (2019)* Estimated value

According to recent estimates, India's dragon fruit production increased drastically to more than 12,000 tonnes over an area of 3,000–4,500 ha in 2023 according to estimates are based on the first–hand information collected by the ICAR–NIASM, Baramati.

11.15 Transforming Fortunes Through Date Palm Cultivation in Kutch

India holds the distinction of being the largest importer of date palms, accounting for 38% of the global market. In 2018, the country imported dates valued at approximately $248.88 million from 40 different countries, with Pakistan, Iraq, the United Arab Emirates, Iran, and Oman as leading suppliers. In the Kutch-Bhuj region of Gujarat, local date palm varieties have traditionally been grown from seeds. However, this method has not been widely successful due to the dioecious nature of the plants, which complicates natural seed-based propagation, and the unavailability of high-quality planting materials.

Data from the Gujarat state horticulture department indicates that Kutch alone encompasses about 1.89 lakh hectares of the 1.99 lakh hectares dedicated to date palm cultivation in the state. During 2020–21, this area contributed to 1.79 lakh metric tonnes out of the 1.85 lakh metric tonnes of fresh dates produced in Gujarat. Approximately, 60% of this cultivation area in Kutch is devoted to native varieties, with the remaining land cultivating the Bahri variety, originally from Israel. Remarkably, Kutch is home to around 20 lakhs date palm trees, with 17 lakhs being indigenous varieties, predominantly grown from seedlings.

The cultivation of date palms has not only markedly improved the socio-economic conditions of farmers in the region but has also played a critical role in addressing nutritional deficiencies. The increased local availability of dates has reduced India's reliance on imported dates, thereby conserving foreign exchange. Furthermore, date palm cultivation has influenced local agricultural practices, contributing to a change in cropping patterns and assisting in the fight against desertification. Farmers in the area have successfully introduced intercropping with crops like Green Gram, Moth Bean, and Sesame under the date palm orchards during the first four years of cultivation, enhancing biodiversity and land use efficiency in the region.

11.16 Conclusions and Strategic Directions for Agricultural Development

Over the past fifty years, the agricultural sector's contribution to India's Gross Value Added (GVA) has consistently been lower than that of the manufacturing and services sectors. However, fluctuations in agricultural performance are deeply intertwined with the overall economic health of the nation. When agriculture thrives, the entire economy benefits, whereas downturns in agriculture directly precipitate declines in broader economic performance. Despite its lower GVA contribution, agriculture remains critical, particularly in rainfed areas where diversification options such as animal husbandry, poultry, and piggery are limited.

The absence of varied industries in these rainfed regions means fewer employment opportunities for the rural youth, many of whom migrate to urban areas in search of better prospects. To counter this trend, the Mahatma Gandhi National Rural Employment Guarantee Scheme (MGNREGS) plays a pivotal role by providing employment for 100 days annually to rural labourers, though challenges remain during off periods. To enhance this support, the establishment of a Labour Supply Corporation (LSC) has been proposed. This entity would manage the registration and deployment of labourers, not only for government projects but also for private sector endeavours, thus facilitating fair wage distribution and skill development.

Furthermore, there is a critical need for a bottom-up approach in technology transfer for watershed projects, which currently operate under a predominantly top-down methodology. This shift would ensure more active and voluntary participation from stakeholders, fostering a sustainable agricultural ecosystem (Singh & Singh, 2010).

References

Ahmad, H. B., Mohd, M. H., & Nur, S. J. (2019). *Status and challenges of dragon fruit production in Malaysia* (pp. 1–8). FFTC Agricultural Policy Platform (FFTC-AP).

Chen, L. I. (2018). *Profile of the dragon fruit industry and its assistance measures in Taiwan.* pp. 1–8.

Deshpande, R. S., & Rao, V. M. (2004). *Coarse cereals in a drought–prone region, monograph no. 2.* Bangalore: Institute for Social and Economic Change.

Doe, J., & Smith, J. (2020). Towards sustainable water management in an arid agricultural region: A multi- level multi-objective stochastic approach. *Agricultural Systems, 191*, Article 103142. https://doi.org/10.1016/j.agsy.2020.10

Government of India. (2006). *Review of irrigation potential and strategies for India.* Ministry of Water Resources.

Intergovernmental Panel on Climate Change (IPCC). (2007). Climate change 2007: The physical science basis. In *Contribution of Working Group I to the Fourth Assessment Report of the IPCC.* https://doi.org/10.1017/CBO9780511546013

Muas, I., Mansyah, E., Hendri, & Yuliat, S. (2019). *Dragon fruit production and marketing in Indonesia: Standard quality in the global and regional levels* (pp. 1–10). FFTC Agricultural Policy Platform (FFTC-AP).

Perea-Moreno, A.-J., Samerón-Manzano, E., & Perea-Moreno, M. A. (2019). Biomass as renewable energy: Worldwide research trends. *Sustainability, 11*(3), 863. https://doi.org/10.3390/su1103 0863

Singh, S. P., & Singh, D. (2010). Biodiesel production through the use of different sources and characterization of oils and their esters as the substitute of diesel: A review. *Renewable and Sustainable Energy Reviews, 14*(1), 200–216.

Web Sources

http://www.afdc.energy.gov/fuels/natural_gas_basics.html
http://www.merriam-webster.com/dictionary/alternative%20fuel
http://www.ucsusa.org

Part II
Cold Desert Ecosystem

Chapter 12
Himalayan Cold Desert Environment, People and Change Process

Tej Partap

12.1 Chapter Overview

The Himalayan cold desert ecozone is a vast highland mountain area with having cold dry and sunny climate and few people. The region has an area of 79,388 sq kms and includes the whole of the Union Territory of Ladakh and two districts in Himachal Pradesh, namely Lahul and Spiti and Kinnour (Fig. 12.1). In history, harsh climate and hardy highland people mattered to the outside world only as an international trade route, China-Tibet Ladakh—Samarkand—Rome. Even today, most parts of the cold arid region are uninhabitable and human settlements are largely along river systems. The sparsely spread mountain communities have been making a living on the indigenously evolved subsistent farming cultures but strains of socioeconomic and ecological unsustainability of these agroecosystems are forcing a multidimensional change process.

The mountain communities that live here have been making a living on nomadic pastoral and agropastoral farming systems, supplemented by tourism, services and other non-farming livelihood systems. The ecozone is witnessing an interplay of the three driving forces that have become the cause factor of their uncertain future. Further, the geopolitical factor has made the ecozone an area of national strategic significance since 2020. This in combination with socioeconomic aspirations of local communities, aggravated by the economic growth of the country and harnessing of dormant niches of cold arid regions and the unintended impacts of climate change on the overall environment of this zone, are collectively contributing to creating conditions of an uncertain future for the land and its people.

Observations and analysis presented in this paper are based on the published accounts and more often on the first-hand knowledge and information gathered by the author, over a period of three decades (1990–2022). Efforts have been made to

T. Partap (✉)
Sri Sri University, Cuttack, Odisha, India
e-mail: tpartap52@gmail.com

M. S. Rathore et al. (eds.), *Desert Ecosystems*,
https://doi.org/10.1007/978-3-032-03237-9_12

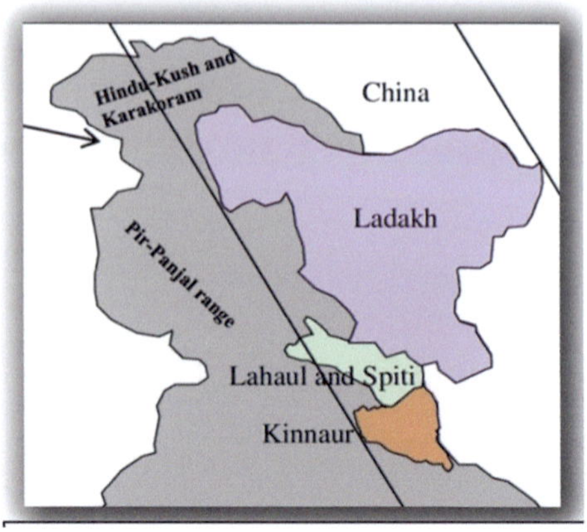

Fig. 12.1 Himalayan cold desert zone comprising areas of Ladakh, to districts of Himachal Pradesh

present a situation analysis and a perspective of the dynamics of the change process of the land, environment and people of this zone.

12.2 Introducing the Himalayan Cold Desert

The Himalayan cold desert, covering an area of 79,388 sq kms, is a vast stunning bare landscape that receives abundant sunlight and snowfall but little rainfall. The clear sky, lofty mountains, clear streams, deep gorges and little vegetation are the unique features of the Himalayan cold desert, a biome of its own kind. A large part of it, i.e. 59,146 sq kms falls in the Union Territory of Ladakh and the remaining 20,242 sq kms is in Himachal Pradesh. It is located along the northern borders of the Western Himalayan region. It includes the Union Territory of Ladakh with its two districts Leh and Kargil, and the two districts of Himachal Pradesh, Lahul-Spiti and Kinnaur-. It is the most elevated (2,900 m to 5,900 m msl), and coldest region of India. It has a vast stretch of harsh, inaccessible and arid temperate areas where due to sub-zero temperatures (winter temp. drops to −35 °C max) several natural springs and river systems freeze during winters. Its cold and dry climate is because of two factors. One, its location on the leeward side of the Himalayas, which makes it a rain-shadow zone, inaccessible to the annual south eastern monsoon winds, receiving only low levels of precipitation that create desert-like conditions. Second, its very high elevation (ranging from 3000 to 5000 m ASL) adds to the coldness of its environment. There is wide seasonal variation in the climatic conditions, ranging from short and dry summers with harsh sunlight, where the maximum temperature reaches up to 36 °C during the day. Winter nights are long, windy and freezing, experiencing a lot of blizzards, snowstorms and avalanches.

The cold climatic conditions make way for a short growing season during which only a few plants/crops can grow and complete their life cycle. Glacier-fed streams and springs and lakes are major sources of water available to plants, animals and people. The most famous lakes in this zone are **Pangong lake** at an elevation of 4,225 m *locally called Pangong Tso*) in Chan-gthang, **TsoMoriri** lake located south east of Pangong lake, and Chander-tal Lake in Lahul Spiti, Himachal Pradesh. Among these three lakes, Pangong lake is the largest **endorheic lake, 134 km long, 5 km wide** and about **100 m deep**. The Himalayan cold arid zone is home to several rare and endangered varieties of flora and fauna, so much so that many national parks and wildlife reserves have been established for the protection of the indigenous biodiversity of this ecozone.

The Himalayan cold desert, as an ecological zone, represents an extremely fragile ecosystem that shows a complex relationship between the climatic and geomorphological processes and exhibits very little but highly endemic diversity. Many rare and special varieties of flora and fauna are found in the wildlife reserves established in this zone. It looks like, that the physiographic peculiarities and ensuing harsh climatic conditions have made native mountain communities evolve unique farming cultures suited to its highlands, sloping lands and valleys.

In Himachal Pradesh, the district of Lahul Spiti and the District of Kinnour fall under the cold desert zone. Lahul is geographically separated from Spiti by high mountains and the two areas are connected through KUNZUM PASS.

12.3 Demography and Socio-Economy

The main cold desert area—Ladakh, was historically, a trade route from Punjab to Kashmir, and beyond to Baltistan (Skardo), Kashgar, Yarqand, Khotan (Eastern Central Asia or Xinjiang), Gartok, Lhasa (in Tibet) with Leh acting as an important trade centre. It not only contributed to the economy of the local people but also helped maintain strong linkages with the outside world. The native people of Ladakh are believed to have migrated from Tibet around 800 AD. The royal dynasty of Ladakh introduced Buddhism among them in the tenth century. Projecting chin, Mongolian eyes, large flattened nose and long hair wearing a long wide coat of sheep skin (Shanglak) are some of the key features of people of the Himalayan cold desert. Historians believe that King DPal-gyi-mGon, king of Tibet was the founder of the Ladakhi kingdom and his eldest son, Dpal-gyi-mgon was the first king of Ladakh between 1000 and 1025.

In the present times (Census 2011, Govt reports 2020), the cold desert zone has **762 villages** and **seven towns** which are home to ca, **83,800 households** with **ca. 0.4 million people**. The vast area and small population are thinly spread across the zone. These 762 villages are settled in locations where ever water and some favourable land for farming and animal husbandry are available (Table 12.1).

The cold desert zone has no city, a large town Leh and other few towns and a large number of villages. Unlike UT of Ladakh where there is 24% urban population, the cold arid zone of Himachal Pradesh has 98% rural population. Keylong the district HQ of Lahul, Kaza the Subdivisional HQ of Spiti and Kalpa and Sangla are trading and tourist towns in this belt. However, there are several larger villages with growing

Table 12.1 Area, villages, towns and population of the Himalayan cold desert ecozone

Administrative unit	Geographic area, sq. kms	House holds	Population	Pop, density per sq km	Villages	Towns	% rural population	% urban population	% Literacy rate
Ladakh UT	59,146	57,144	302,766	3	152	2	76	24	77
Leh distt	45,110	31,909	148,171	–	60	1	68	32	-
Kargil distt	14,086	16,338	140,802	–	67	1	88	12	71
Zanskar Tehsil*	7004	2327	13,793	–	25	nil	100	–	60
Lahul, and Spiti distt HP	13,841	6674	31,564	2	369	1	98	2	77
Kinnour distt	6401	19,976	78,334	3	241	2	98	2	80
Himachal part	20,242	26,650	109,898	2–3	610	3	100	2	79
Himalayan cold desert	79,388	83,794	4,12,664	2–3	762	7	88	12	79

Note (Data compiled by the author from census data 2011, as well as updated 2022 census information about UT of Ladakh)

urban amenities. Demographic data also shows a growing trend of urban migration in Leh Ladakh and Kargil.

However, the cold desert community of Himachal presents a contradiction of sorts. Having better incomes from fruit and vegetable farming. There is a higher level of literacy among the tribal cold desert mountain community of Himachal Pradesh. It intrigues many but the matter of fact is that the state has made honest efforts during the seven decades in setting better social infrastructure in this tribal area.

12.3.1 Marginalised Among the Marginal Communities of Cold Desert

Less than a hundred thousand people dispersed over a very large inhospitable area of ca. 79,000 kms. The inhabitants of the cold desert have faced acute environmental constraints and socio-political marginality. **Pastoral nomads** called *Changpas* of Changthang area in Ladakh are the most marginalised highland community among the other marginalised mountain communities of the Himalayan region. They settle in small groups forming a temporary village of yak skin tents. Their Indo-Mongoloid (Tibetan) race dominates. Even during the 2010–2020 decade, ca. 90 per cent population was living a nomadic life, and managing their livelihoods through animal husbandry. The area has a maximum concentration of animal population, pashmina goat, sheep, yak and horses including a fairly large number of wild ass population. Herds of wild ass keep moving back and forth across the Chinese border in Tibet and are very often seen grazing alongside local livestock of yak and goat herds. Their system of livelihood sustenance is further described in more detail under the section on farming and livelihoods.

12.3.2 Highlanders of Zanskar—Another Marginalised Mountain Community

Another marginalized mountain community is an intermediary race between Ladakhis and Baltis of Baltistan, living on subsistent mixed farming. Deep inside the Himalayan mountains is Zanskar Valley, which is over 250 kms east of Kargil. The life and livelihoods of the highland people of Zanskar have been rather cumbersome to manage. Subsistence farming, no cash crops. Living in permanent village settlements, their main occupation is crop livestock mixed farming. Farming is possible only with 100 per cent irrigation so where ever irrigation water source was available human settlement expanded. Every household here maintains a herd of cattle, sheep and goats, some yak and ponies. Since there is only one crop season from April to September, people have to work long hours to accomplish every farming task from seeding to harvesting and storing crops, fodder collection to fuel collection. The main crops are wheat, barley, mustard, peas, black peas and oat and clover are their common fodder crops.

12.3.3 *Those Who Overcame Marginalisation and Entered the Mainstream*

Unlike nomads of Chang thang, highlanders of Lahul and Spiti are settled in villages, own farm land with access to community irrigation and every family maintains large animal herds of cattle, sheep, goats and only some families in villages in the higher reaches of Spiti and Kinnour have yak or hybrids of yak and cow. Economically better-off families of Lahul have emigrated to Kullu Valley during the past five decades. However, these families retain their agricultural land and go back in summer for farming. The winter season from November to March is extremely cold with heavy snowfall (−20–−40 °C). The area receives 90 to 1200 mm of rainfall largely as snow. Rainfall is rather scanty in Spiti in comparison with Lahul and therefore humidity remains low throughout the year. Spiti is higher, drier and colder and is inhabited by over 113 villages, of which 81 are permanent settlements. The inhabitants were largely dependent on agriculture and medicinal plants for their livelihood. In recent decades, one can see expanding apple and pea cultivation that has led to economic well-being among people of Spiti Valley.

Kinnour, the second cold desert district of Himachal, is inhabited by over 78,000 people. It has a high literacy rate of 78% for males and 64% for females. At present Kinnour is known for premium quality apple production. Cold arid climate, abundant sun light and fully irrigated apple orchards are home to red coloured fruits full of aroma, crispiness and sweetness. Supported by state Govt programmes on the promotion of horticulture, fruits and vegetables, in this agroecosystem, farmers are making full benefit of this opportunity. As a result, during the past three decades, the farm economy has seen tremendous improvement. Besides fruits and vegetables, farmers were earlier cultivating the following crops; cereals, maize, wheat, barley, buckwheat, pulses—rajmash, black pea, potatoes, peas, tomato, cabbage, radish, apricot, some dry fruits, kuth and apples. The livestock population in this zone included cross bred cows, local cows, equines and goats including a small number of pashmina goats in Spiti sheep, poultry and yak. There is predominance of sheep and goats in Spiti, cows and cross bred cows in Kinnour and Lahul.

12.4 Cold and Dry Climate Determined Niches and Livelihood System

12.4.1 Nomadic Pastoralism—History and the Trends of Change

The nomads, the cold desert highland inhabitants of Changthang in Ladakh, are called Changpas. They are spread over 21,000 sq kms area, west of Indus River along the high land flat pasture valleys in between 15,000 ft–16000 ft ASL. Changpas graze their animals in a particular area for a specified period, using different valleys for summer, winter and spring. Some of their pastures are nearly 46 kms long and 30 kms wide. The central area of nomads comprises of villages of Rupshu-Kurzok and Kharnak. Kurzok, the HQ of Rupshu served as centre of Central Asian trade until 1947. It is on the western side of Tso Moriri Lake in Rukchen Valley. In 1993 only about 5 acres of farm land was being cultivated by a few families, growing largely barley. Similarly in the nearby village of Rupshu, a few other villagers were cultivating around 10 acres of farm land at an altitude of 15,000 ft AMSL. The farmland of Rupshu village is rated as **the farm land at the highest altitude in India**. Jina, 1995 reported that during the 1990s nearly 16,000 kgs of pashmina was being produced by Changpas of Changthang with annual earnings of ca. Rs six million and that when one kg of pashmina wool was sold by nomadic families for Rs 500/-. As a historical example of socio-political marginality, as per the 1683 treaty, Chang-pas were compelled to sell pashmina wool to Kashmiri traders, and they had the sole right to buy it (Jina, 1995). To gather a guess of the regional scale of the pashmina economy, one male pashmina goat yields 450 gms of pashmina wool annually and the female goat yields 250 gms. The mortality rate of pashmina goat kids used to be nearly 40% until the 1990s but institutional efforts since then have succeeded in reducing it drastically to less than 5% during normal years, i.e. during winter disasters that repeat every 4–5 years and there have been any time on an average 200,000 pashmina goats in Changthang, approximately in the ratio of 40:60 male and female.

Living mostly in tented hamlets/villages, they keep moving every 3–4 months, up and down as determined by the grazing requirement of their animals and weather conditions. From the view point of the source of livelihood, **Chang-pas** are nomads, semi-nomads and semi-pastoral. During the decade 2001–2010, the nomads lived only in the high plateaus of Rupshu, Kharnak and Korzok. In Rupshu and Kharnak 100% population was leading a nomadic life, whereas in Kurzok about 90% were nomadic. Semi-pastoralists were settled along the Indus River such as Kuyul, Damchok, Kery, Chumathang, Nyoma, Ney, Kesar, Linkchey, Henley and Chumur villages. They were more dependent on one summer season farming along with some cattle, sheep goats. They produced wheat, barley, turnips, potatoes, mustard and peas. No other form of property had the same prestige and value as herds of yak and pashmina goats. The income from the sale of pashmina wool and meat was used for purchasing food items; rice, jaggery, wheat flour and Tsampa from the market in Leh.

The Changpas are not bothered as much by rain but wind is their main problem. Wind starts blowing in the afternoon and becomes almost a storm by the evening. Each Changpa tent-shelter consists of two sides of black cloth made from YAK hair. Tents are fixed low so that any of the Changpa family members can stand up and stick his neck out from the smoke chimney of the tent, to keep a watch on his herd outside the tent, without going outside. Within the tent, the space serves as a multipurpose room for cooking, warming, sleeping as well as storing. Changpa food consists of meat, Tsampa (satoo), churpee (dried cheese made from butter milk), curd, butter, tea and chhang (local crude beer).

The Chang-pas, a pastoral community of Changthang highlands of Ladakh, are segregated into two groups, one known as DROK and the other group called RONG. The nomadic pastoralist community "*DROK*" constitutes the original Changpa community and people lead a pure nomadic life style. Key features of the DROK community are that it is a non-caste Buddhist society and speaks the Tibetan dialect *Changkyet*. Among the DROK community women enjoy greater freedom, consumption of meat, including dried meat of yak and crude beer "*chhang*" is an integral part of the culture. Pastoralism and trade form the basis of their livelihoods and economy. **Semi-pastoralist "*RONG*"** community is the second community. These people live in hamlets/villages along the narrow valleys in between high mountains. Probably transformed from nomadism their ancestors decided to settle down along the valleys cultivating small plots of farm land during the summer months and also grazing their animals in the adjoining pasture lands. The RONG also are non-caste Buddhists. Pastoralism, trade and farming are the basis of their economy.

12.4.1.1 A Natural Disaster that Happens Every 4–5 years

Chang thang experiences a very untimely heavy snowfall once every four to five years, leading to a heavy casualty of livestock in each household. For example, it happened in 1982, and 1986 accompanied by a snowstorm in 1990. It also happened in late February 2012 when heavy snowfall and cold conditions for more than 15 days killed half of the livestock (author witnessed it). Even though Govt stockpiles hundreds of tonnes of dry fodder, brought from the plains of Punjab, for distribution to pastoralists it serves a limited purpose. Strangely, there has never been an effort to examine the problem and find other alternative solutions. Livestock casualty is largely due to open penstocks, as animals are kept in open stone-walled enclosures without roofing. *Andean pastoralists faced similar weather problems but have been managing it through underground dwellings for Alpaca herds in the high land Andes—author survey/study, 1988.*

12.4.1.2 Trends of Change

During the past decades, many nomadic families have been settling down in the valleys or migrating to Leh town. Families moving into Leh town are leaving nomadism and semi-nomadism after selling their livestock to fellow nomads. It is a guess that by 2040, there may be a lesser number of "*REBOs*" the nomadic tents in the high lands of Ladakh and those who will be there will have larger herds of pashmina goats, sheep and yak. The next generation of nomads is largely getting educated and they are less inclined to continue to engage in nomadism or semi-nomadism as a way of life. Socioeconomic aspirations are now a stronger cause of change than even the local environment. However, within the cold arid zone, there are subsistent farming communities, as well as those who are experiencing the transformation of traditional farming cultures leading to improved economic well-being of people. The cold desert community of Himachal Pradesh is a good example. Settled along rivulets and river systems as well as on dry mountain landscapes they have innovated ways to harvest glacier waters for irrigating their apple orchards and vegetables.

12.4.2 Agropastoralism of the Indus Valley

In the valley area, below the highlands, villages have mixed families of nomads, semi-nomads and semi-pastoralists. The numbers are reducing faster than expected. By 2016, there were a few semi-nomadic families around, some doing farming and some engaged in trading. **The system of animal movement** with seasons is the key adaptation mechanism these people followed for centuries. As soon as snow melts Chang-pas move their herds from winter pastures in the valley to spring pastures at the height of 16,000 to 17,000 ft ASL Their migration pattern is absolutely weather-determined.

12.4.3 Agropastoralism of Zanskar Valley

Zanskar is a remote hermitage valley between 1200 ft and 15,000 ft, deep inside the mountains south of Leh, east of Kargil, west of Lahul and north of the Chamba district of Himachal. The lower belt of the valley has settlements at 10,500 ft (3500 m) and 13,000 ft (4300 m) is the highest altitude where crops are farmed by the Zanskar farming community. At these altitudes farmers only grow barley crop. The limiting factor at these high altitudes is the reduced length of the growing season, which eliminates the growing of wheat. Below 13,000 ft. peas ripen earlier than barley. Zanskar has no forests. A few trees of juniper, willow and tamarisks grow along the river banks and moist locations.

Zanskar farmers start farming activities by March end, carrying out agricultural work collectively, pooling the village labour force. Several irrigation cycles are repeated depending on the weather and water needs of the crop. Among the crops, peas ripe earlier and the crop is harvested by pulling out plants rather than cutting. Pea harvest is followed by barley and wheat harvest and green fodder for hey making is harvested the last. Animals taken to higher pastures in May, by this time, move down to nearer homes.

Upper reaches of the valley have abundant grass lands and people have evolved mixed farming of a kind to make use of these pastures during the summer months for grazing their livestock. During summer months when they take their cattle to pasture lands for a few months, they **make butter** and ***Chhurpee***. Even today (2010–2015 author's observations) crop threshing is done with the help of animals viz. donkeys, horses, cows, dzos, dzomos and yaks. Just after the crop harvest period, members of a household would divide into two groups, some going for fuelwood—shrubs collection and some would roast and grind wheat and barley to make *stock of satoo* for the family. During the long winter season, all family members live together with their animals. Leisure time in winter is spent celebrating social functions, handicraft making, processing the wool for making woolen clothes for the family members (overcoat, coat, shirt, pyjama, socks and long shoes for snow walking—everything).

Once the winter sets in and Zanskar River is frozen, people would travel to Leh carrying large quantities of stocked butter, for trading in the Leh market. During these several days journey, local people have been walking over the frozen river for centuries (famous Frozen River Walk), using sledges to pull the butter stock, and peg their tents over the frozen river for cooking and sleeping over the night and this happened in the month of January and February. Few tourists join locals for this adventure walk. Year 2024, the frozen river walk has become history, as a road has been constructed linking Padum with Nimu to Leh and Padum to Darcha Lahul in Himachal. The famous **chaddar river** adventure tourism will be only for tourists in the future.

12.4.4 Transforming Cold Desert Farming in Himachal Pradesh

Agriculture in cold arid zones of Himachal Pradesh—Lahul and Spiti and Kinnour has shown a high degree of transformation and has passed through many stages of the transformation process over the decades. Even though crop land is only 0.3% but area presents a success story of vegetables, cash crops-based farm economy in the Indian Himalayas. It has made **off-season mountain vegetables** famous across the north Indian urban markets, namely, potatoes, peas, cabbage, French beans. Some progressive farmers are trying to grow exotic vegetables. Prior to 1962, Kuth was the only cash crop that people grew in Lahul and it was exported to China. Thereafter 1963 to the 1980s Lahul was identified as the niche area for producing disease-free potatoes needed by farmers in other states of India. Cultivation of Hops flourished during the 1990s for a short period until cheap imports from China killed it. Then came peas as off-season cash crop, that spread across the region around the 1980s. *Peas farming in Lahul Spiti and Kinnour became a new niche for producing off season peas (July to September) for the Indian market. It continues to dominate farming and fetches premium prices.* In 2020, farmers of Lahul and Spiti had cash crop options of pea, potato, exotic vegetables and apple farming.

12.5 Driving Forces Behind the Structure and Functioning of Cold Desert Agroecosystem

- Cold temperatures in winter limit agriculture in this zone to one season. Therefore, farmers grow crops only between March to September/October.
- Farming is purely irrigation based, even the pasture lands are irrigated. Irrigation from snow fed water sources.
- There are climate-determined farming cultures in this zone; namely, Nomadic Pastoral, agro-pastoral, Subsistent crop livestock mixed farming, Cash cropping dominated mixed farming of vegetables and fruits mainly apples.
- Crop land and single crop season are limiting factors of farming-based livelihoods.

12.5.1 Dynamics of Limiting Factors of the Himalayan Cold Desert Agricultural Systems

Scholars defined mountain agricultureas a system that includes all land based activities and supportive actions of the farmers to manage a living through farming. If one were to present a combined impact of all the components involved in the functioning of the agroecosystems and then identify one single key factor that can represent the impact of all other factors, it is the energy flow. The ecological understanding behind it is that the energy flow links a farming community with its crops and animals, and therefore it can help discover how the character and magnitude of these flows vary in different farming cultures. Various inputs into an agricultural system—solar radiation, human labour, the work of the machine and the application of fertiliser and herbicide—can all be converted into energy values. However, a complete explanation would require much more than the knowledge of energy flow and how the system functions.

12.5.1.1 *Agroecological Determinants of the Cold Desert Himalayan Agroecosystems*

These factors have over the centuries played a key role in evolving the diversity of agroecosystems within this agro-ecological zone. These are:

i. Solar radiation that determines the solar growing season of the ecozone
ii. Temperature, that determines the thermal growing season of the ecozone
iii. Moisture determines the hydrological growing season of the ecozone
iv. Nutrients, that determine the edaphic potentials of the ecozone
v. Social factors, that observe sum total impact of the above factors to evolve different functional agriculture systems for different agro-ecological conditions.

12.5.2 Solar Radiation, Which is a Determinant of Solar Growing Season of a Cold Desert Agroecosystem

The amount of solar radiation received at a site is the most fundamental of the various factors for agroecosystem productivity. There are substantial geographical variations in the amount of solar radiation that is received in different eco regions. In this context, three factors are important: the altitude of the sun above the horizon, a function of latitude and time of year; day-length, which in high altitude cold arid Himalayan region, especially Ladakh plateau, would determine the proportion of the sun's rays reaching the earth. There is a large seasonal variation in solar energy in the temperate highland Himalayas. Also, the difference between maximum and minimum available solar energy actually widens much more in the cold arid Himalaya region. In temperate areas, the difference is as wide as nine times higher. Short days and lack of sunshine in winter are the contributing factors. Here crops, for example, tomatoes in the greenhouse, can only be grown with the help of artificial sunlight, i.e. by using electric lamps. Further, different types of vegetation vary somewhat in the proportion of the visible light that they can convert into carbohydrates through photosynthesis. These differences in biological capability relate to two different ways of using carbon dioxide, the so-called C3 pathway and C4 pathway. As one proceeds to higher latitudes, the C4 species begin to come into their own. That is why crops like wheat, barley and sugar beet are more suitable crops.

12.5.3 Cold Temperature, that Determines the Thermal Growing Season of a Cold Desert Agroecosystem

Temperature becomes important only at its extremes and in the cold arid Himalayan ecozone, it is extremely low temperatures that act as a determinant for evolving farming practices. It is a crucial limiting factor in this ecozone, which determines whether crops and which crops, can be grown. For example, the cereals of temperate regions, oats, wheat and barley will produce some growth at comparatively low temperatures until a critical minimum is reached (between 0 and 5 °C), while the optimum for these crops is 25–35 °C, max being 37 °C. Beyond this, some hardy grass species grow and dominate the landscape. **It is these low-temperature extremes**

which inhibit the evolution of crop-based agriculture and encourage nomadic pastoralism in the Himalayan cold desert region.

As a general rule, agriculturists reckon that above av. monthly 6 °C temperature, conditions will usually be adequate for the growth of some kind of crops. Livestock can be maintained at this or a little lower temperature. On the basis of this assumption, the thermal growing season can be computed, which means a total number of warm months available for crop growth or for plants in general. For example, the thermal growing season in some villages of Changthang in Ladakh is little less than 5 months, hence crop farming is not possible here and people have adopted total nomadism as a way of livelihood system. However, down in the valley areas of the Indus River, Nubra Valley is below the Siachen glacier, and a thermal growing period of five months is available, hence crop farming is done here.

12.5.4 Rainfall and Soil Moisture the Determinant of the Hydrological Growing Season of Cold Desert Agroecosystem

It is a common ecological limiting factor of both kinds of desert agroecosystems, the hot desert land of Rajasthan and the cold desert of the Himalayan region. Even if the cold arid zone had and will have enough months of solar growing season and adequate months of thermal growing season, some areas of the highlands of Ladakh were and are not suitable for farming because of non-availability of hydrological growing season. The reason is obviously the lack of soil moisture during most part of the year but for different reasons. It is the actual lack of soil moisture content during the summer season months, as there is no monsoon in the summer season over here and there is physiological dryness in the winters, i.e. when water uptake by plants is not possible due to freezing temperatures. Snow and ice conditions are as bad as a hot summer, w.r.t. soil moisture/water availability to plants. In the cold and arid Himalayan ecozone agriculture is supported by 100% irrigation. Farming activities, such as growing crops or maintaining fodder lands or even growing and maintaining Salix and poplar trees for fodder and timber, cannot happen wherever water for irrigation is not available. It is the number one limiting factor that has not allowed farming to expand, whether in the cold desert of the Himalayas or in the hot desert of Rajasthan.

It is because of the limitations imposed by the hydrological growing season, that the nomad community in the highlands of Changthang prefers the greater security of pastoralism as its chief source of livelihood. Actually, pastoralism in the high lands, agropastoralism in high land valley areas of the Indus River and mixed farming in the valley areas down below, can be seen as an adaptation to acute rainfall seasonality and soil moisture scarcity. When a cold winter season is prolonged by unusual climatic conditions then even these adaptations fail, causing disasters.

12.5.5 Nutrient Deficiency in Soil that Determines Edaphic Potentials of an Agroecosystem

Nutrient deficiency can be the fourth category of ecological limiting factor/constraint of agriculture production for farming in cold deserts. In one sense, it is more important in the context of the Himalayan cold desert. The constraints of solar radiation, temperature and rainfall are beyond the farmer's direct control and the mountain communities usually adapt to these features of the environment, through evolving adaptation mechanisms. Local mountain communities of the cold desert zone have usually viewed each of these constraints as a renewable resource more than as a force they would need to alter. Soil nutrients, on the other hand, are not self-renewing in the same inevitable way. The three principal plant nutrients are N (nitrogen), P (phosphorus) and K (potassium). Of this nitrogen is usually the most important agricultural fertiliser. Most land in the cold and dry Himalayas is marginal and it requires a lot of effort to develop crop land. An example of this is of Cold arid zone farmers of Spiti and Kinnour in Himachal Pradesh, who have been transporting fertile soils for over 100 kms to establish their apple orchards.

12.5.6 Social Decision-Making in Agrarian Societies

Community observing the sum total impact of the above four factors and then evolving a functional livelihood system for given agro-ecological conditions.

In fact, the four ecological pressures can be counted as rigid constraints, in the sense that the farmer or the farming community is virtually helpless to modify them. It is true for the Himalayan cold and arid agroecosystems. For other factors, a farming community can exercise choice of suitable crops or develop methods of cultivation that improve yields or reduce farm expenses or increase farm income. The land can be irrigated, mulched, manured, soil can be improved or more remunerative crops can be grown. Therefore, in this case, the exact nature of the response does not depend on the environment, but rather on the society in which the farmer finds himself. In this context, there are four ways in which a society can influence the decisions the farmers make. The influence operates through; population pressure, technological innovation, structures of social organisation and values that farming communities conform to when making decisions.

Together, these social factors actually help determine the reasons underlying the bewildering variety of farming practices that are encountered in the cold and dry Himalayan agroecosystems. These are put into context in the subsequent sections on Perspectives and Challenges.

12.6 Approaches to Assessing Cold Desert Agroecosystem Health

There are a number of approaches available for the analysis of conditions about the Health and Sustainability of Arroecosystems and these can be applied to assessing the health of cold desert Himalayan agroecosystems (Bayliss-Smith, 1982; Costanza et al. 1992; Smith & Smitters, 1994; Smit et al. 1998; Galloping, 1995; 1994; Ramsey, 1997). Agroecosystems are complex and involve interactions between human (including economic) and biophysical components at different levels in hierarchies nested in time and space. These basic elements of agroecosystems provide a generic model for approaches to understanding agroecosystem health.

One of the main features of agroecosystem health for sustainability is that it provides a comprehensive and holistic basis for assessing the condition of rural farming environments and communities (Waltner-Toews, 1996). The underlying assumption is that "human culture is part of the ecosystem".

12.6.1 Human Centred Approach to Assessment of Agroecosystem Health

Accepting that all ecosystems are managed, directly or indirectly, by human intervention (Rapport 1995a, 1995b). This perspective supports the use of human centred approach to the assessment of the health of any ecosystem/agroecosystem, such as the present case of the Himalayan cold desert agroecosystem.

The human-centred approach to the assessment of the health and sustainability of agroecosystems depends on the structural and functional attributes of the system defined as: "the ability to realize its socially desired functions and maintain its functionally and socially needed structure over time" (Xu, 1997). The human-centred perspective of agroecosystem health has been refined further, by some other workers, with the division into two complementary elements: **SYSTEMS–FUNCTIONS** and **SYSTEMS–GOALS**. Systems goals identify the precise state of optimum agroecosystem health, that state in which all of its health-relevant goals are achieved. By contrast, systems–function identifies the modes of functioning by which the system is enabled to achieve these goals. A full account of the state of the system when those goals are realised are an account of agroecosystem health in the primary sense.

Using Community Well-being as a Tool of the assessment of Agroecosystem health and Sustainability, further strengthens this approach. There are instances, where as a matter of national policy, more importance is attached to community health and well-being and in such circumstances, the tool for assessment of agroecosystem health and sustainability has to be different. Here, the interest will be on the role/influence community environment will have on individual life styles, attitudes, well-being, empowerment, and self-efficacy (Coombs, 1991; Dunst et al., 1992; Wallerstein, 1993). Although recent models for human health reflect the incorporation of community or social dimensions, few of them use community-level indicators for their final assessments.

12.6.2 A Perspective Approach to Understand Scenario of Development Imperatives

Using inputs from determinants and perspective tools elaborated in the earlier section, researchers also need to define the scenario of imperatives of sustainability development of cold desert agroecosystems (environment and people). There are a set of conditions inherent to the ecosystem that have to be taken into account in any sustainable development interventions. These set of conditions emerge from the understanding of cold desert mountain specificities of a particular locale that in the present context is the Himalayan cold desert ecosystem (Table 12.2).

To sum up, there are two important aspects of the perspectives on approaches, for assessment of health and sustainability of cold arid agroecosystems, that are highlighted in this section.

First, the five determinants of the evolution of cold desert agroecosystems and **second, the perspective approaches** for assessing the growth, health and sustainability of these cold desert agroecosystems, especially the Mountain Perspective approach.

Table 12.2 Cold desert mountain perspective framework for assessment of agroecosystem sustainability/health

Mountain specificities	Physical dimension	Biological dimension	Socio economic dimension
Inaccessibility	High priority to reduce it, roads, rail, air, telecom	Needs research, what we don't know enhance understanding	Enhance understanding
Fragility	Respect as inherent quality, show sensitivity in R&D	Species/ agroecosystems with narrow ecological amplitude Needs research	Socioeconomically vulnerable, high priority
Marginality	Respect as an inherent quality, *misinterpreted as degradation* needs research	Research needs	High priority vulnerable groups
Diversity	Recognise	Recognise	High priority
Niche	High priority, R&D	High priority, R&D	High priority, R&D
Adaptive mechanisms	Priority, ITK—R&D	Research needs	High priority

These determinants, also known as limiting factors, have much influence on the shape of the Himalayan cold and arid agroecosystem. These factors have played a crucial role in why nomadism is there in the cold desert and there are several categories of agropastoral systems or crop-livestock mixed farming in the Himalayan cold desert. Despite the vast land, what limits the expansion of farming-based livelihoods and the economic well-being of people living here.

In previous sections, there is an explanation of the history and culture of high land people and their livelihoods, highlighting the fact that the solar growing season is a great positive factor of this eco zone but the unique features of the thermal growing season and hydrological growing seasons in the cold and arid zone have forced several limitations w.r.t. choice of farming-based livelihoods for the inhabitant mountain communities. These high-land communities have been for this reason pioneers in evolving a variety of adaptation mechanisms.

The perspective explains that there are some inherent cold mountain desert conditions and for any development interventions, these conditions impose certain development imperatives and if ignored or handled improperly, will result in unsustainable development.

12.7 Elements of Uncertain Future

In this section, an attempt is made to envision the likely scenario of the future of the environment and people of the Himalayan cold and arid zone and the factors behind it. What are the aspirations of local communities? How will climate change reshape the future of agroecosystems? Has the state policy taken cognizance of the imperatives? What are the areas of course correction w.r.t. policies and strategies that will need to be revisited?

To build a certain future, the state needs to frame policies and strategies that give consideration to the following:

One, the local environment is healthy and stable;

Two, well-being of local communities is based on local resources and environment and **three**, the national strategic compulsions do not add to disruptions in the above two.

Winds of change are blowing in the Himalayan cold arid zone, which appears a cocktail of climate change, socioeconomic aspirations and national strategic needs. Indicators of change in ecological, economic and social dimensions can now be seen emerging and what can happen over a period of the next three decades, is a guesstimate, being presented here, as follows:

12.7.1 *Climate Change Will Change the Thermal Growing Season and Hydrological Growing Season, Bringing Unforeseen Changes in Farming Opportunities and Constraints*

In the Himalayan mountains, there is no other agroecosystem than the cold and arid ecozone, which will see a severe impact of climate change. The effect of rising winter temperature is observed in the shrinking of the winter period and the severity of cold. The effect of lesser snowfall and receding glaciers is being felt in different scales by the mountain communities in the form of declining flow of glacier waters, which is so vital for irrigation. Even though there are instances of a few villages shifting their locations. They are not exactly climate refugees but it can right now be considered as climate-led migration. The climate change predictions for the coming three decades, for the cold and arid zone, indicate the following changes:

- Prediction is about warmer winters and lesser snowfall calling for receding glaciers and declining availability of irrigation water in the growing season, i.e. summer months.
- We will see landmark changes in rain fall pattern and the guess is that cold and arid zone will see an increase in summer precipitation. If that happens, the vast dry landscape of the cold and arid ecozone will start becoming green. Let us call, **we may witness a GREENING PROCESS OF COLD DESERT**.
- Authors observations during 2010–2015 and 2018 periods confirm that the greening of grass lands has already started happening. Both, enhanced population density and flourishing growth due to increased hydrological growing season and thermal growing season were leading to greening of brownish pasture lands in the highlands of Changthang. It will be a positive climatic development that the carrying capacity of pasture lands may improve. We may see an increase in the wild population of animals as well as of domestic livestock of nomads, i.e. Cashmere goat and yak.
- Indications are that climate may offer opportunities for pastoral and agro-pastoral systems to flourish. However, it is uncertain to think it will be so because for that to happen much will depend on how the Gen Z of the high land community perceives these opportunities.
- Changing the thermal growing season and hydrological growing season may lead to new opportunities for diversifying agriculture into horticulture and medicinal plants. Farming area may expand, because of the expansion of farming possibilities into flat lands at higher altitudes. (*Author has seen this happening in Tibet high lands, where institutional support services (policy investment R & D services and linking to markets) have led to the evolution of new high value crops and setting up of several agri business enterprises in the Tibet highland).*
- New crops and new farming cultures, emerging from native plants of medicinal value, essential oils, giving very high returns is possible. **However, it is dependent on** positive institutional interventions.

- Improved local economy of people to more economic well-being. It may become a disruptive force, causing rural to urban migration.
- Social transformation with Gen Z, where their aspirations will reflect a generational shift in the way highland people will like to live. The new generation of educated youth is already acting as disruptive force to continuation of traditional livelihood systems and agriculture systems. What will save the old systems from being abandoned is anybody's guess.
- In addition to these three factors, there are other externalities influencing the environment and development in this ecozone.
- For several decades people of Ladakh faced socio-political marginality within J&K. Ladakhi people being strong nationalists, would not fit into the socio-political fabric of J&K. Because of their small population, Ladakhi people faced socio-political marginality, Evidence of this is reported by the social uprising under the banner of Ladakh Buddhist Association (LBA). There have been several agitations for a separate state or at least autonomy within the state. In 1995, the Ladakh Autonomous Hill Development Council (LAHDC) was established in Leh. Later in 2003, Kargil was also brought under LAHDC, Kargil.
- However, declaring Union Territory of Ladakh in 2020 was a land mark event for the locals. A long-aspired ambition fulfilled with the feelings of removing the socio-political marginality.
- The new UT Govt is buttressed by adequate funds for development and is hastening the process of physical infrastructure development. There is enhanced access to social welfare schemes. Government is focussing on development programs that will lead to improving lives of local people. Road and infrastructure development and tourism promotion are a priority.
- Today most villages across Ladakh are connected by road. People get food grains under PDS, fuel wood has been largely replaced by LPG.
- NOMADIC pastoralists now use LPG for cooking food. Although highlanders need fuel wood to keep their houses warm. MANREGA benefits most families w.r.t. generating employment and supplementing family income.

Likely *Potentials for Improving Livelihoods of Cold Desert Communities.*

i. Apricots of Kargil and Leh**:** It is one of the key fruit trees of Kargil and Leh districts and people make use of every part of it. Fruit is eaten fresh or dried with satoo or otherwise, seed kernel is also edible as apricot oil is used in cooking and for massage. Apricot in Kargil and Leh belongs to the HELMUND variety, possibly brought from Gilgit. Every household maintains several trees of apricot and uses fruits, seeds as food and twigs and dry trees for fuel and timber.
ii. Kirkichoo apple of Kargil is a very different variety from those of Kashmir valley. It has a narrow ecological niche. As yet there are no large orchards except that farmers maintain a few trees around their houses. As per a century-old ban order of the King of Kashmir, the fresh apples cannot be exported to Kashmir as it was feared they would carry the prevalent disease and spread in the valley which may destroy its orchard economy.

Kirkichoo apples of Kargil and Ladakh apricots fresh and dried, kernel or oil have great market potential and they can become the leading cash crops of Ladakh. It would though require some policy and R&D institutional support. Similarly, among the likely future opportunities for transforming farming, the climate provides for disease-free seed production of vegetables on a commercial scale.

iii. Felt need for better education for children has made highlanders migrate to near Leh town, selling their livestock to other local households. Therefore, on the one hand, one can notice a trend of declining number of households in the highlands but the number of livestock holdings per family, especially pashmina goats, is expanding. Economically it makes sense to make the animal husbandry vocation also viable.
iv. Tourism of all kinds: It is a unique land and therefore tourism development offers good hope for strengthening non-farm economic development**.** There is promising scope fo**r** high land travel tourism/ pastoral life tourism/ecotourism/ Agro-tourism with apricots/kirkichoo apple/pashmina goats husbandry. Ladakh is already on the tourist map of India. However, to widen the scale of the tourism economy, these other niches may also be included in the tourism sector list. Each kind of tourism activity will need to be developed within this agro-ecozone. The good thing about it will be that not only local population will be involved in it, but tourism flow will see increased tourist movement along the border belt which, lead to help in enhanced **human presence for better border** security.

12.7.2 Increased Geopolitical Significance of the Himalayan Cold Desert

Presence of defence forces in large numbers across the Ladakh Himalayas has its own positive lateral impact. It contributes to helping locals by providing work and jobs on a substantial scale. Local people get ample opportunities for employment and contractual work, travel rides in military vehicles, and opening tea shops, Dhabas and homestays even in remote areas like GALWAN valley (Galwan Dhaba). The army needs labour and other logistic support and locals get non-farm income opportunities. It is a positive development of proto-cooperation, where both parties are benefited. Locals are a big help to defence forces in the borders and the presence of defence forces contributes to the local economy and helps make the lives of local people comfortable. There is a renewed effort to build an extensive road network by the defence forces and it is equally benefiting local highlanders. One can see nomads using pickups to transport their tents and food, and getting gas cylinders delivered to highland camps. There is an increasing trend of movement of motorcycle-borne tourists who stop over with the nomads for tea, relaxing for a while, food and overnight stays.

12.8 Conclusion

i. How the climate change in cold desert may impact the agroecosystem?

Signs of climate change impact in cold arid zones are visible and these can be both positive and negative. Looking at the negative signs, glaciers are receding faster than expected, snowfall is declining in terms of annual snowfall quantity as well as a widening cycle of years when heavy snowfall is experienced. These events may become severe. In Zanskar, instances were recorded during 2010–2015 when the whole village migrated due to the drying up of the village water source. Where to go? Ladakh still has the opportunity to migrate to some other area because of less population density which may not be the case elsewhere. In Leh district villagers in Indus Valley report declining water availability but so far it has not turned into a serious issue.

Summers are warmer now and if the trend continues, that will mean the availability of longer periods for plant growth—good for agriculture. In the future, Ladakh may see upwards movement in agriculture and opportunities for growing new crops. Horticulture and vegetable farming will stand to gain from warming of weather. Possibilities of winter cropping of some of the crops may also open up—it will directly impact cropping intensity positively. Highland pasture land—the nomad land stands a good chance of starting farming of high value cash crops.

In Himachal Pradesh, warming of local weather has been not only observed but strategies to benefit from it in farming have already been put in place. For example, the apple belt is moving upwards into the cold arid zone. The premium apple is today grown in Kinnour, it is full of orchards. From Kinnour apple has already moved into Spiti and Lahul, and it is hoped that during the next 25 years the vast area of Spiti will have favourable climate for growing fruits. If it happens in Spiti and Lahul in Himachal the practice may be replicated elsewhere in Ladakh. The people of Zanskar and the new political dispensation in Ladakh are already working out strategies for replicating the Himachal experience. There is a new road linking Zanskar with Keylong replacing the need to go via Kargil. It will change the whole picture of farming by the Zanskar farmers. There is a high possibility of high value cash crops, viz vegetables, vegetable seeds, floriculture and medicinal plants.

In nutshell, both climate change and economic change, i.e. access to markets, will open up opportunities for farmers of this area to introduce new crops and receive premium value for their produce.

ii. What lessons can be learnt from the success story of transformation of farming and economic well-being in the cold desert areas of Himachal Pradesh?

The success story of vegetable and fruit (apple) farming in the cold and dry zone of Himachal highlights the fact that with sincere institutional efforts, the farm economy of the cold arid zone can be turned into a profitable proposition. Focus on road infrastructure, horticulture R&D support services, planting material/seeds and facilitating trading, each one of these initiatives helped contribute to building the success story of Kinnour, Lahul and Spiti. The cold and arid climate also is a niche, it needs to be understood and strategies worked out to harness its uniqueness as a strength. It is an imperative that must be seen as a strength rather than a limitation. It has been the thinking and development strategy, behind the development philosophy of Himachal Pradesh for the cold arid zone. Continuing the same perspective thinking works well for the whole Himalayan cold arid zone in the future also.

iii. There is a China factor that makes a positive contribution to the development of Himalayan cold desert area, what is the weight in this argument?

The cold and arid area was a remote area and the highland community lived in isolation evolving their own social institutions and mechanisms that were necessary to live in this ecozone. The national attention is drawn to this zone only when there has been Chinese aggression, whether in 1962 or now in May 2020. With these incidents, one thing that has happened is the need felt for robust road infrastructure across the cold desert area, because of being a border area. One mountain scholar had commented, "*Will it not be right to state that cold arid area and people, of Ladakh and Himachal, have been beneficiaries of aggressive posturing by the Chinese on the borders. It made the central Govt realise the need for good road infrastructure for the use of defence forces and the value of strategies encouraging local people stay put*".

"It is development efforts that encourage local people to continue staying in this inhospitable environment". There is some truth in this statement. Govt of India has now realised the strategic defence significance of strengthening road and rail connectivity to this area. Promoting tourism movement in this border belt is *an important policy shift by the national Govt, that has been made*. The increased access through roads and railways and the subsequent flow of tourists is going to make a significant positive impact on the infrastructure development process, agriculture transformation and improving the lives of local people.

iv. Is there a problem of out siders and insiders perspectives to development thinking for cold desert zone, what is the message to policy planners?

It is painful to agree that yes in some ways there is a problem of perspective thinking. For outsiders, it is a fragile and marginal ecosystem and a place for national parks or biosphere reserves for the benefit of conserving biodiversity. The fact is that it is inhabited by human communities that possess tremendous traditional knowledge and skills to stay on under these harsh climatic conditions and serve as uncared warriors on the borders. For insiders, it is their homeland and they nurture dreams of living a better life. For the local institutions development focused on making life of local people comfortable is the priority but for the outsiders conserving nature remains a priority. Can it be made possible to put in place policies and launch development programs that take care of both nature and people? That may require a balancing act in formulating development strategies for the Himalayan cold desert zone.

v. The Himalayan cold desert by 2050—what may it look like?

Based on weather data predictions, we imagine there will be warmer winters with less snowfall and more wet summers that may make the desert greener. Increased growth period and greening of the cold desert will have unleashed new human activities, like farming may expand laterally and vertically. Road, rail and air connectivity and the educated society of the future will have brought changes in everything old and traditional about farming and livelihoods. Whether pastoralism and related practices survive or not will depend on how much of these prove useful and can be improved upon to serve the changed circumstances. Many traditional pastoral practices will remain show pieces for tourism or they will be transformed to fit into the cash economy of the new times. "*There will be no harm in being a nomad by 2050 but* ***a nomad of the year 2050*** *may like idea of* ***living in a refined yak tent****, having a vehicle, machines for processing of milk and meat and all other amenities necessary to make life comfortable*". The annual income of the family will have to be better than the families who migrated and will be living elsewhere. Such a scenario of positive change is not impossible but difficult to achieve.

The inherent strength of cold arid mountains—highlands and valleys, may offer other kinds of opportunities to the nation, in the future. The technological advancements may create other kind of opportunities to the nation, most notably valued silica source for high tech use, for harnessing the abundant bright sun light and clean environment.

vi. Winding up the conclusion

It is reiterated that the Himalayan cold arid zone has **unique environment** not available elsewhere in India. It is inhabited by high mountain communities, largely of mongoloid Buddhist origin, and they have over the centuries evolved several farming cultures and practices to manage livelihoods to live under these ecological conditions. Life had been tough here for the inhabitants, however, in the past few decades one finds increasing national attention being given to these areas for various reasons. Consequently, there is enhanced institutional support in all spheres be it physical infrastructure, social infrastructure, human well fare schemes and flow of outside population into the cold arid hinterlands. A change process in the economy and ecology, both is visible. Observing the trend of the changes in ecological, economic and socio-political conditions it seems that the impact of these driving forces of change will leave a new mark on economic prosperity and the natural environment. The changed ecological and economic conditions may offer new development opportunities and it is for the national, provincial and local institutions to have a better understanding of the new normal of the ecological and economic environment for designing better development policies for this zone.

As uncertainty continues to prevail about the level of understanding of the new normal of ecology and economy and of the social dimension, it is also not yet clear what strategies will be put in place for the governance and management of changed ecological, economic and social circumstances, that is why the uncertain future remains.

References

Bayliss-Smith, T. P. (1982). *The ecology of agricultural systems*. Cambridge University Press.

Coombs, J. (1991). The report of the national task force on health information—A summary. *Health Reports, 3*(9), 349–361.

Costanza, R., Norton, B., & Haskell, B. (Eds.). (1992). *Ecosystem health. New goals for environmental mnagement*. Island Press.

CSKHPKV, & CAZRI. (2011). *Status report on cold arid areas of Himachal Pradesh*. CSK HP.

Dunst, C., Trivette, C., & LaPointe, N. (1992). Towards clarification of the meaning and key elements of empowerment. *The Family Science Review, 5*, 113–130.

Galloping, G. (1994). *Impoverishment and sustainable development: A systems approach*. International Institute for Sustainable Development.

Galloping, G. (1995). The potential of agroecosystem health as a guiding concept for agricultural research. *Ecosystem Health, 30*, 129–140.

Jina, P.S. (1995). *High pasture lands of Ladakh Himalaya* (p. 141). Indus Pub Co.

Ramsey, D. (1997). *The dynamics of rural community wellbeing: The case of the tobacco belt*. Univ of Geulph.

Smit, B., Waltner-Toews, D., Rapport, D., Wall, E., Wichert, G., Gwyn, E., & Wandel, J. (1998). *Agroecosystem health analysis and assessment*. University of Guelph.

Smit, B., & Smithers, J. (1994). Sustainable agriculture and agroecosystem health. In N. O. Nielsen (ed.), *Agroecosystem health: Proceedings of an International Workshop*. University of Guelph.

Waltner-Toews, D. (1996). Ecosystem health-A framework for implementing sustainability in agriculture. *BioScience, 46*(9), 686–689.

Chapter 13
Climate Change, Farmers Adaptation Strategies, and Technological Interventions Led Agricultural Development Process in the Himalayan Cold Desert

Anurag Saxena, M. S. Raghuvanshi, Hardev Ram, Changchuk Lamo, Stanzin Landol, and Kamal Garg

Climate change, is a critical global issue, with the Intergovernmental Panel on Climate Change (IPCC) predicting warmer years ahead. The global average temperature has been increasing by 0.15 °C every decade, adversely affecting ecologically fragile regions. Communities in these areas, which rely exclusively on scarce natural resources, are particularly vulnerable.

Ladakh, located in the high-altitude Trans-Himalayan region, covers an area of 96,701 km^2. It is one of the highest regions in the country, with elevations ranging from 2,900 m to 5,900 m above mean sea level, and one of the coldest, with temperatures dropping as low as −30 °C. The region receives very scant precipitation, less than 150 mm annually, of which rainfall constitutes less than 50 mm (Raghuvanshi et al., 2019a, 2019b, 2019c). Over 90% of the terrain is barren. Approximately, 87.4% of India's cold desert region is located within the union territory of Ladakh, with the remainder in Himachal Pradesh and Uttarakhand. The soils here are predominantly gravelly and sandy loams, characterised by poor fertility. Other major agricultural challenges in Ladakh include barren topography, a short growing season and poor availability of high-quality seeds. The region's agricultural production systems struggle with low biomass and frequent crop failures, attributed to several factors

A. Saxena (✉) · H. Ram · K. Garg
ICAR-National Dairy Research Institute, Karnal, Haryana, India
e-mail: anurag.saxena@icar.gov.in

M. S. Raghuvanshi
ICAR-National Bureau Soil Survey and Land Use Planning Nagpur, Maharashtra, India

C. Lamo · S. Landol
ICAR-Central Arid Zone Research Institute RRS, Leh, Ladakh, India

M. S. Rathore et al. (eds.), *Desert Ecosystems*,
https://doi.org/10.1007/978-3-032-03237-9_13

including a hostile climate, fragile ecosystem, limited food and fodder supplies, low precipitation (mostly as snowfall), extremely low temperatures, and high wind speeds reaching 20–40 km/hr (Raghuvanshi et al., 2019a, 2019b, 2019c).

Ladakh, referred to as the 'cold desert' due to its altitude and cold arid climate, is a region that has experienced extreme weather and climatic events in the recent past. Recent research and data indicate rapid climatic changes in Ladakh. There have been shifts in the patterns of rainfall and snowfall, and the glaciers and permanent snowfields are receding, which affects water runoff in the rivers and streams. Additionally, a rise in temperature is creating conditions that are more favourable for agricultural crops (Shaheen et al., 2013).

13.1 Ladakh Farmers and Farming

Nearly 70% of Ladakh's population is directly or indirectly dependent on farming and livestock rearing. With small land holdings—62% of households possess less than 1 hectare of cultivable land—the local population relies on limited cropping and heavily depends on natural resources to meet their basic needs. Barley, wheat, and alfalfa are the main crops, while apples and apricots are the major fruits cultivated. Control of weeds, insects, and other pests is often neglected due to cultural reasons (Raghuvanshi et al., 2016), resulting in very low production of all fruits and crops. Additionally, suitable technologies have not yet reached many of the region's farmers. During the winter months, the region faces shortages of food, fresh vegetables, and fodder. Livestock rearing forms the backbone of tribal livelihood and is crucial for economic and nutritional security. However, an acute shortage of green and nutritious fodder for most of the year, due to the extremely cold and arid climate, severely limits animal productivity (Tewari et al., 2016). Animal husbandry and perennial vegetation provide stability in sustainability. To address these challenges, there is a need for diversified farming and site-specific technological packages implemented in a holistic manner.

In Ladakh, farmers can plant and harvest only a single crop each year, typically wheat, barley, or vegetables. The region receives very little rainfall, leading farmers to rely heavily on natural glacier melt for irrigation. Only 10–15% of agricultural farming benefits from the waters of the Indus and Shayok rivers in Ladakh, while the remaining 85% of farming irrigation depends entirely on streams formed from snow melt and a traditional water management system. This system has been prevalent and practiced for centuries by farmers across all the watershed areas in the cold arid desert of Ladakh (Saxena et al., 2021). Consequently, there is minimal opportunity for the expansion of land holdings, as there is no scope to allocate water to new areas for plantation or vegetation (Jigmet et al., 2019a, 2019b).

Ladakh region faces prime challenges of adapting to changing climate and climate shift. The traditional subsistence agricultural systems and cropping patterns are increasingly under threat (Meena et al., 2015). One of the climate change-related features is Frost Heaving. The region is a good habitat for pasture lands under wet conditions, but the degradation of grass or pasture lands occurs when they dry up due to deficit water conditions (Giya, Saboo, Sakti, etc.). The shift in seasons has been observed with the warm season. Sometimes prolonged winter creates a shortage of irrigation water and crop sowing is delayed. Apple cultivation was not possible earlier in higher reaches of Ladakh, now it is now possible but incidence of new insect pests and severe outbreaks of codling moth, brown-tail moth and aphids also increased in the past few years.

Problems associated with existing practices

S. no	Existing practices	Problems associated
1	Sowing by broadcasting method	High competition, no weed control and low yield
2	Use of higher seed rate	
3	No weed control	
4	No insect pest management	Poor quality of produce. Low yield
5	Primitive cellars	Damage to produce during storage
6	Improper animal sheds	Low productivity of animals
7	Fodder shortage in winters	Deterioration in quality of stored fodder
8	Winter season	Low availability of fresh vegetables
9	Open drying of vegetables/apricot	Poor quality of dried product with limited export potential

To enable and encourage smallholder farmers to adopt new technology and practices under more unpredictable and extreme climatic circumstances for a resilient agricultural production system, new techniques and policy interventions are critically needed. Over time, Indian farmers have developed a variety of coping strategies, but these have not been sufficient to address the frequent and severe extreme weather events as well as the slow but steady changes in the climate, such as temperature increases, altered rainfall patterns, declining soil moisture levels, and altered evapotranspiration rates. In the Ladakh region, agriculture and livestock productivity are suffering because of the unpredictable weather.

Serious risks, such as insufficient precipitation and the devastation of wetlands, marshlands, and springs, lead Ladakh to experience drought-like conditions during the early planting season in settlements located in the valleys. In the near term, melting glaciers cause surplus water, but in the long run, it causes shortages. According to data from the India Meteorological Department, minimum temperatures in Ladakh have increased by 0.5 degrees Celsius in the summer and by about 1 degree Celsius in the winter over the previous 35 years. The peak winter months of November, December, and January have mean low temperatures of −7.8 °C, −12 °C, and −14.2 °C, respectively. The precipitation amount decreased from November to March between 1973 and 2008. Ladakh is sheltered from the south-west monsoon by the trans-Himalayan range, which is located on its leeward side. The yearly precipitation in the desert district is only 102 mm. In addition to the overall trend of decreasing precipitation, the shift in precipitation from winter to summer can occasionally result in extremely heavy and intense rainfall that can cause cloud bursts, flash floods, and other phenomena. These occurrences cause rich soil to erode, and agricultural lands and forests in areas with little vegetation to be destroyed, turning rich lands into unproductive and desolate areas (Singh et al., 2019). Because there is less snowfall, streams have less water in them and frequently go dry. Between 2013 and 2017, there was a 50% to 80% yearly precipitation shortfall, with 2016 setting a record low for precipitation. In some way, customary wisdom from the area helps the mountain inhabitants adjust to their precarious circumstances.

In Ladakh, a vulnerability assessment of the cold, dry terrain was done. Strongly resistant hard rocks to the majority of brittle slopes, river dissection, and freeze–thaw erosion are vulnerability concerns. Other associated characteristics include: (a) Frost shattering, which primarily occurs at higher elevations (Tangtse, Ulitopo, Basgo villages) and is a phenomenon of rock erosion, slope failure, and land slide. (b) Frost Heaving: Provides excellent cover for grazing areas during rainy seasons, but damages grass when dry due to low water levels (Giya, Saboo, Sakti etc.). In Ladakh, the effects of climate change on agriculture are being witnessed.

13.2 Impact of Climate Change on Agriculture in Ladakh

Agriculture's productivity is reliant on the availability of glacial meltwater. Agriculture, and by extension the economy and food security, are seriously threatened by any alteration in the glacial-melt water supply. A rise in temperature may have a beneficial or negative impact on agricultural output. Depending on the options accessible to farmers, the effects on agriculture and the ensuing ramifications on society are expected to vary. While warmer temperatures may boost wheat in higher altitude areas, they may decrease wheat yield in other areas. Although many crops may grow more quickly in warmer climates, yields may also decrease. Trees that are apricots now grow larger than they did thirty years ago. One benefit of global warming is the ability to grow a variety of foods and vegetables in Ladakh, something that was previously unfeasible. These days, farmers cultivate watermelon, tomatoes, okra, cucumbers, cauliflower, and capsicum. This is conceivable here because of the rise in temperature.

Furthermore, cattle in the area are susceptible to the effects of climate change. Climate change is particularly felt by the Changpas, a nomadic group in the Changthang region that depend on raising Pashmina goats for their wool. In addition to the loss of several animals in the area because of unpredictable snowfall, Changpas have recently had to alter their migration paths due to a decrease in pasture space (Raghuvanshi et al., 2020). In addition, a rise in the frequency of weeds and pest infestations might devastate the area.

In Ladakh, about 90% of farmers rely on snowmelt water for irrigation. The area's water crisis is mostly caused by an average temperature increase of three degrees Celsius during the previous forty years, which has led to less snowfall and earlier melting in the upper elevations. There is an existential threat to some of the glaciers in the western Himalayas, where the glacier cover has decreased by over 20%. Farmers' crop productivity has suffered greatly. A few crops have had a 30% to 50% decrease in production in recent years. Lack of water has caused crops like potatoes, barley, turnips, radish, and peas to suffer. It's hard to grow even grass these days for cattle feed. Arid climates in cold deserts make farming challenging when glaciers begin to melt more quickly and as a result of reduced snowfall and rain due to global warming. Farmers are left with few choices. Seasonal changes were noted. Extended winters can occasionally result in a scarcity of irrigation water, delaying the planting of crops. Because of the warm weather, agricultural innovations, and the introduction of new crops, there are now more veggies available in the summer and winter. It is now feasible to cultivate apples in Ladakh, something that was previously impossible.

The increasing risk of plant pest invasion worldwide is a result of the globalisation of trade in plants and plant products. The fast expansion of the agricultural and horticultural industries is probably going to cause the threats to get worse. A number of invasive pests have been introduced into the High Himalayas in recent years as a result of climate change. These include various species of aphids on cereals, vegetables, and horticultural crops; yellow, brown, and black rust in wheat and barley; locust outbreaks in the Zanskar Valley and Nyoma block; outbreaks of brown tail moth in Kargil; and outbreaks in the lower belt of district Leh (Dha, Bheema, Hanuthang, and Achinathang Villages) and Nubra Valley (Turtuk) (Stobdan et al., 2016).

13.3 Suggested Practices/Technologies

(A). **Strengthening Input Supply**

1. Ensuring reliable and timely availability of quality seeds and planting material is the primary factor for increasing productivity (Providing good quality seeds/planting material). Seed is procured during the month of November for the next cultivation season to avoid delays due to blocked road connectivity.
2. Appropriate mechanisation of agriculture aimed at improving efficiency and reducing drudgery (Providing agricultural/horticultural tools). Small hand driven improved equipment to be introduced to facilitate cultivation.
3. Increasing profitability from Farm Enterprise through post-harvest processing.
4. Suitable veterinary interventions to improve the health status of livestock. Improved breeds along with feed and medicines to be provided at the doorstep of farmers.

(B). **Resource Management for Sustainable Agriculture**

1. Protected agriculture in high mountain sloping areas with low-cost poly house technologies for growing vegetables and quality planting materials.

2. Underground cellars, low tunnels, black polythene mulch, etc. to be used for enhancing diversity in crops and their productivity.
3. Seed material including waste decomposers and bio-fertilisers to be demonstrated for enhanced productivity.

(C). **Need-Based Training for the Farmers**

1. Field crops production technologies
2. Vegetable cultivation and home stead farming
3. Hill horticulture
4. Fodder production technologies
5. Soil health management
6. Organic farming, organic manure, and vermi-compost production.

(D). **Improvement of the Livelihood and Nutritional Security of Tribal Farmers**

- Mushroom cultivation
- Poultry rearing
- De-worming and vaccination.

13.4 Technology Dissemination Accomplishments of Institutions in Ladakh, Cold Arid Region

On-farm technologies disseminated in different adopted villages with the concept of seeing is believing and learning by doing. Resultantly, many technologies, i.e. using good quality seeds of different crops including new crops, line sowing, use of appropriate quantity of seed, earthen up in potato, weed control, preservation of local seeds, prevention/control against insect-pest, adopting organic farming, etc. were disseminated among the farmers (Raghuvanshi et al., 2019a, 2019b, 2019c). Poultry intervention was initiated specifically in Khardung and Kukshow villages to enhance livelihood.

In spite of facing several challenges, farmers were successful in producing more than what they had been producing through traditional cropping in the past. Farmers engaged in vegetable growing have become an important economic activity of this region. Line sowing of field crops was introduced in Ladakh, now it is showing better results compared to a traditional broadcasting method of sowing.

Crop productivity in Ladakh

Crop	Productivity (kg ha^{-1})	Crop	Productivity (kg ha^{-1})
Barley	1487	Apple	6276
Wheat	1469	Apricot	4048
Pulses	585	Potato	6000
Oilseed	620	Onion	1870

(continued)

(continued)

Crop	Productivity (kg ha^{-1})	Crop	Productivity (kg ha^{-1})
Millets	200		

Solar drying of vegetables and fruits (apricot/apple) also added to the value of their products. Today the market remains a challenge for farmers in far-off areas. The timely availability of seeds is a major concern in this remote area. Farmers were trained for the community seed bank concept, its importance and future scope. This concept helped the farmers very much and now farmers are getting timely seeds. Demonstration of promising genotypes of fodder crops (barley, alfalfa) was conducted to improve the fodder availability.

Encouraging results were obtained by introducing poultry chicks in Ladakh to improve the livelihood and nutritional security of local dwellers. The beneficiaries were earning by selling the organic eggs. Good quality pashmina wool was obtained with chain-link fencing due to better air circulation besides saving their livestock from wild animals.

Productivity Enhancement

1. Line sowing and optimum seed rate improved grain production by 1.5 times besides saving seeds.
2. Weed management improved yields by more than 50%, especially in staple crops, onion and potato (Raghuvanshi et al., 2018a, 2018b).
3. Orchard sanitation in Apricot enhanced yield by 40% in Nimoo and Basgo villages.
4. Seeds and other planting materials are timely available for timely sowing.
5. Better quality apricot and other vegetables was obtained with solar dryers besides reducing drying time compared to open air drying of these commodities.

Improved yield achieved

Crop/vegetables	Yield with existing practice (kg ha^{-1})	Yield after adoption of improved technologies (kg ha^{-1})	Yield increase* (%)
Thiksey Village			
Wheat	700–1200	1600	More than 50%
Chushot Village			
Wheat	800	1250	More than 50%
Stakmo Village			
Barley	900–1600	1800	38
Potato	6000	10,000	60%
Ranbirpur and Gopuk Villages*			
Onion	1870	6000–12,000	50 and more

*Vegetable production farm

13.5 Adaptation Measures to Cope with Climate Change

The methods through which societies strengthen their ability to deal with an uncertain future are known as adaptation. In order to adapt to climate change, one must take the necessary steps to minimise its bad consequences or to take advantage of its beneficial effects. This involves making the necessary modifications and adjustments. Adoption opportunities and options abound. These include everything from individual behaviour modification to technological possibilities. The community's adaptation and mitigation efforts to counteract the externalities of climate change are identified, and their wider consequences are evaluated. In the environmentally severe cold arid desert, where nothing grows naturally without the correct application of water, water storage, and water use efficiency are crucial for adaptation (Saha et al., 2020). The main method for adapting and mitigating the effects of climate change was the construction of artificial glaciers to counteract the retreat of natural glaciers, which would increase water storage and its availability in the summer and the production of vegetables in the winter through enhanced solar greenhouses that lessen the effects of CO_2 emissions.

- **Water Management Practices**

Due to the excessive strain on natural resources, the Ladakh region's prime sectors—agriculture and other—are particularly vulnerable to the effects of climate change. As a result, it is critical to manageably use the natural resources that are already available and to advance agriculture in order to sustain agricultural productivity and production.

Insufficient precipitation (less than 150 mm) and limited water resources are insufficient to meet the agricultural plants' typical requirements. The glaciers at high altitudes, which melt and flow in the summer, are the only sources of usable irrigation water. The amount of snowfall in previous winter receipts determines the availability of this water, making it difficult to anticipate. The primary irrigation supply for maintaining agricultural output under effective management is water melted from glaciers. The main source of irrigation for maintaining agricultural output under an efficient use management pattern is water melted from glaciers. In this case, additional irrigation is necessary to boost the crop, and the only significant source of irrigation is water melted from glaciers. Different kinds of ice reservoirs, also referred to as 'artificial glaciers', have been introduced and pushed as suitable adaptation solutions to deal with changes in the cryosphere in Ladakh in order to address the issue of recurring water scarcity in the Leh district, the first artificial glacier was tested at Poktse Pho hamlet in 1987. After proving to be effective, the technology was expanded to neighbouring communities. It raises the amount of meltwater available in the spring when water scarcity is most crucial. The community irrigation system is a local institution that helps people manage their uncertain resources. This technique rotates the supply of water to farmers for irrigation. Aside from this, farmers are also using micro-irrigation, which is common in cold, arid regions, to address the problem of water management. In vegetables, micro-irrigation led to a notable increase in yield, quality, and fertiliser utilisation efficiency.

Rich dividends from artificial glaciers.

13.6 Practicing Creation of Artificial Glaciers

In order to battle the water crisis, farmers are diverting the water stream and slowing it down during the winter so that it would freeze due to the low temperature. This creates fake glaciers. This is accomplished by creating stone embankments around the slope's shallow ponds and water streams. This clever method, which creates an artificial glacier, aids in the conservation of water as ice. Ice begins to melt in the spring, releasing water suitable for irrigation.

Photo credit (Ice stupa): Padma Thinles at Gya (left one).

- **Soil and Moisture Conservation Through Snow Harvesting**

By using snow harvesting techniques like snow fences, trenches, and pits, dry wastelands have been transformed into cultivated areas fit for farming or grazing. A gabion wall is built at the bottom of the depression, and snow barriers 8.5 feet high are erected along its edge for a distance of 100 feet. By catching blowing snow and compacting the settled snow, the system helps the snow melt over a longer period of time.

- **Low Carbon Solar Greenhouse Vegetable Production**

The region, which is isolated from the rest of the country during the winter, has few choices for family food security due to small-scale agriculture and a short growing season. In Ladakh, summertime is when most fresh veggies are grown. Since most households do not have access to fresh vegetable supply during the winter, they are more likely to experience dietary inadequacies.

A solar greenhouse that utilises the passive solar principles of heat release, insulation, ventilation, and solar gain presents a significant opportunity to help small-scale farmers in arid, cold climates enhance their standard of living. Even in cold weather when the outside temperature might drop as low as −25 °C vegetables can be grown in greenhouses because they are fully heated by sunshine. In addition to generating revenue, it encourages the growth of vegetables (such as spinach, coriander, onions, or garlic) that enhance the dietary intake of people residing in rural locations (Shaheen et al., 2013).

- **Livelihood Diversification**

i. Traditional Agroforestry System

The region's traditional agroforestry system combines border plantations of popular (*Populus* spp.) and willow (*Salix* spp.) species with agricultural crops. These are a major supply of fuel wood, feed, and lumber. An estimate, states that these plants contribute 400 tonnes of leaf litter per ha area yearly to the ground. As a result, they are excellent sources of organic carbon and are in charge of sequestering over 75,000 tonnes of carbon (Butola et al., 2012).

Traditional agroforestry system in Ladakh.

ii. Dairy Farming

Due to the paucity of grazing pastures, their short lifespan, and a scarcity of green fodder, dairy farming is restricted in the area. Because the area is a popular tourist destination, there is a significant influx of strangers, which increases the need for milk and milk products. Dairy farming might therefore become the foundation of the local population's family economy. Numerous studies have demonstrated that dairy farming can be more profitable than arable farming. Dairy production will advance with the inclusion of cattle in the agroforestry system.

iii. Goat Farming

Already, goat farming has a significant impact on the local economy. Since the pashmina goat is indigenous to the Himalayas, it is well-suited to the frigid temperature and high altitudes of Ladakh. In addition, it provides the best wool and is utilised as a pack animal. During the winter, it is lucrative to raise pashmina goats primarily for their meat and wool. If pastures and fodder supplies are improved, this business can grow and give the locals a great deal of economic stability.

iv. Poultry Farming

Poultry in the cold, dry regions of the Himalayas is a backyard endeavour with very few birds. The main barriers to the growth and acceptance of poultry farming in this specific area are the lack of commercial hatcheries, issues with the transportation of day-old chicks, and the scarcity of chicken feed. There is a huge demand for chicken meat and eggs due to increased tourism, changing lifestyles, and eating habits. Additionally, farmers have the chance to engage in poultry farming throughout the winter months when typical agricultural activities come to an end. Consequently, there is a significant chance that the poultry sector will grow in this area despite a number of obstacles.

13.7 Conclusion

From the discussion above, it can be inferred that further research is needed, primarily on agro-climatic components. This includes developing more straightforward methods for simulating adaptation responses, such as improvements in agro-technology that allow farmers to reduce risks and/or maximise profits in the face of changing climates. Researchers are working to create new seeds, crops, and germplasm that can withstand novel environments. Research centres are striving to create kinds that can withstand climate change and are also developing water-efficient technology.

References

Butola, J. S., Malik, A. R., & Baba, J. A. (2012). Livelihood diversification in cold arid desert of Indian Himalaya: Urgent need of transforming traditional agroforestry system with emphasis on adoption of herbal farming. *International Journal of Medicinal and Aromatic Plants, 2*(3), 544–578.

Jigmet, Y., Dawa, S., Raguwanshi, M. S., Tundup, P., & Gupta, V. (2019). Impact of agriculture and land use on ground water quality: A case study of Ladakh cold arid region. *International Journal of Current Microbiology and Applied Sciences, 8*(9), 1447–1455. https://doi.org/10.20546/ijcmas.2019.809.166

Jigmet, Y., Raghuvanshi, M. S., Kumar, A., & Arya, C. K. (2019b). Climate change-induced impact on water resource of Ladakh cold arid region. *International Journal of Current Microbiology and Applied Sciences, 8*(5), 1996–2009.

Meena, H. M., Tewari, J. C., Raghuvanshi, M. S., Pandey, C. B., & Ahmed, L. (2015). Influence of weather variation on cropping pattern of Leh district of Ladakh region. *Current World Environment, 10*(2), 489–493.

Raghuvanshi, M. S., Dorjay, N., Singh, R. K., Manjunatha, B. L., Moharana, P. C., Spalbar, E., Stanzin, L., & Saxena, A. (2019). Ladakh traditional farming: An approach to resource utilization under changing climate. *International Journal of Current Microbiology and Applied Sciences, 8*(9), 654–666. https://doi.org/10.20546/ijcmas.2019.809.079

Raghuvanshi, M. S., Gupta, V., Stanzin, J., Dorjey, N., & Chauhan, S. K. (2016). Outbreak of insect-pests on apricot and its management efforts for better livelihood in cold arid region of Ladakh. *Indian Journal of Ecology, 43*(2), 590–592.

Raghuvanshi, M. S., Moharana, P. C., Enoch, S., Dorjay, N., Singh, R. K., Saxena, A., Gaur, M. K., & Arunachalam, A. (2019b). Spatial distribution of nutrients in soil profile under different land use systems of cold arid region of Ladakh. *Indian Journal of Hill Farming, 33*(2), 394–399.

Raghuvanshi, M. S., Moharana, P. C., Enoch, S., Dorjay, N., Singh, R. K., Saxena, A., Pandey, L., Rastogi, A., & Arunachalam, A. (2019c). Organic farming enlivening the Ladakh subsistence agricultural system. *International Journal of Current Microbiology and Applied Sciences, 8*(8), 897–906. https://doi.org/10.20546/ijcmas.2019.808.103

Raghuvanshi, M. S., Moharana, P. C., Saxena, A., & Saha, D. (2020). Pasture and land degradation by weedy invasion on frost heaves in Changthang-Ladakh: Retrospect and Prospects. *Food and Scientific Reports, 1*(4), 46–49.

Raghuvanshi, M. S., Tewari, J. C., Bhatt, R. K., Landol, S., Spalbar, E., Dorje, N., & Raza, M. (2018a). Importance of weedy endemic in unique subsistence agriculture and their traditional management in Ladakh Region. *Indian Farming, 68*(11), 53–57.

Raghuvanshi, M. S., Tewari, J. C., Pareek, K., Landol, S., Raza, M., & Stanzin, J. (2018b). Energy budget of crops and weed management to enhance crop productivity in cold arid Ladakh Region. *Defense Life Science Journal, 3*(2), 157–161.

Saha, D., Raghuvanshi, M. S., & Saxena, A. (2020). Algae as a biorefinery: Generic applicable resonance of bio-ecological engineering in exploring possibilities for desert landscape. *International Journal of Current Microbiology and Applied Sciences, 9*(3), 3201–3212.

Saxena, A., Raghuvanshi, M. S., & Suna, T. (2021). Traditional water management of subsistence agriculture system in cold arid Ladakh: A review. *International Journal of Scientific & Engineering Research, 12*(4), 1019–1028.

Singh, R. K., Moharana, P. C., Goyal, R. K., Bhatt, R. K., Raghuvanshi, M. S., & Meena, H. M. (2019). Soil erosion- status, causes and preventive measures in Trans-Himalayan region at Leh-Ladakh of J&K. *Journal of Soil and Water Conservation, 18*(3), 235–240.

Shaheen, F. A., Wani, M. H., Wani, S. A., & Norphel, C. (2013). Climate change impact in cold arid desert of North–Western Himalaya: Community based adaptations and mitigations. InS. Nautiyal, K. Rao, H. Kaechele, K. Raju, & R. Schaldach (eds.), *Knowledge Systems of Societies for Adaptation and Mitigation of Impacts of Climate Change*. Environmental Science and Engineering. Springer, Berlin, Heidelberg. https://doi.org/10.1007/978-3-642-36143-2_15

Stobdan, T., Deen, M., Gupta, V., Raghuvanshi, M. S., & Phunchok, T. (2016). Biology, damage and integrated management of apricot defoliator and aphid in trans-Himalayan Leh District, Ladakh. Ladakh Autonomous Hill Development Council, Leh. 20 pp.

Tewari, J. C., Pareek, K., Raghuvanshi, M. S., & Roy, M. M. (2016). Fodder production system-a major challenge in cold arid region of Ladakh, India. *MOJ Ecology & Environmental Science, 1*(1), 22–28. https://doi.org/10.15406/mojes.2016.01.00005

Chapter 14
Indicators of Changing Climate, Natural Landscape and Agriculture-Dependent Livelihoods in the Desert Region of Ladakh

M. S. Raghuvanshi, Anurag Saxena, R. Naitam, Nirmal Kumar, Changchuk Lamo, Stanzin Landol, Amrita Daripa, and N. G. Patil

14.1 Introduction

Agriculture is the mainstay of the Ladakh economy and Ladakhis are self-supporting farmers, living in small settlements scattered across the high desert and have developed small-scale farming systems. Nearly 70% population is directly or indirectly dependent on this sector (LAHDC, 2019). The farmers of Ladakh region have access to their agricultural land only for 5 months, while it remains closed from all directions due to the harsh winter for about 7 months a year. This region is experiencing recent land use changes and shifts which are alarming. The crisis arising due to the shift of land to unproductive uses is further compounded by the increasing fragmentation of holdings (Baba, 2018). Expansion of agriculture in time and space has been made possible by rising atmospheric temperatures, and organically grown cash crops like the variety of vegetables and fruits. Although the major sources of irrigation are glacier-based and perennial rivers with fairly large drainage basins, on the contrary, in the last few decades, it has experienced receding snowline, depleting glaciers and drying of springs. As a result, many glaciers have either shrunk or disappeared. In this regard, scientific reports have claimed that 87 glaciers near the Pangong, Changthang region, have retreated 6.7 per cent since 1990 (Ahmed, 2020). This is attributed to atmospheric warming and an increase in the anthropogenic footprint in the glacial

M. S. Raghuvanshi (✉) · R. Naitam · N. Kumar · A. Daripa · N. G. Patil
ICAR-National Bureau of Soil Survey and Land Use Planning, Nagpur, India
e-mail: omsai.msr@gmail.com

A. Saxena
ICAR-National Dairy Research Institute, Karnal, India

C. Lamo · S. Landol
Regional Research Station, ICAR-Central Arid Zone Research Institute, Leh-Ladakh, India

M. S. Rathore et al. (eds.), *Desert Ecosystems*,
https://doi.org/10.1007/978-3-032-03237-9_14

and periglacial environments. The pace of glacial recession among glaciers is in the range of 0.24% to 15% (Ashiq, 2022).

In view of the above, the Himalayas being a dynamic complex mountainous range thus appears more vulnerable to global warming and increasing anthrophonic activities (Bandyopadhyay & Gyawali, 1994). Under this threat of glacial retreat and water stress, studies have predicted a 30 per cent decrease in crop yields and associated floral diversity (UNDP, 2006). In Ladakh, climate change may worsen the rate and magnitude of several ongoing land degradation processes with changed degradation patterns. It may further exacerbate continuing land degradation through increasing floods, drought frequency, frost shattering/heaving, soil erosion, vegetative processes and land management (Olsson et al., 2019). It is consequently exacerbating the soil salinity due to frost-heaving, resulting in novel degradation in a natural ecosystem. This process of climate change has influenced exotic species to enter the region, particularly invasive weedy species viz. *Cirsium arvense* causes a noticeable loss to native biodiversity due to its allelopathic impact enhancing vegetative degradation. Anthropogenic activities are another factor that influences the entry of exotic species helping them to adapt and establish themselves in a new ecosystem (Hellmann et al., 2008; Olsson et al., 2019; Raghuvanshi et al., 2020a, 2020b).

14.2 Cold Desert of Ladakh

Ladakh region is situated between Karakoram (North), the Himalayan ranges (Southwest) and the Tibetan plateau (East) with an altitude ranging from 2300 to 7672 m. The geographical area of Ladakh (Leh and Kargil Districts) is 96701 square kilometers. The terrain in the entire Ladakh represents the barren Rocky Mountains with steep slope sand as such the total reporting area for the region is only 64,626 ha. Out of that, 30,654 ha is under cultivation and fallow lands. It remains inaccessible to the outside world for nearly five months of the year due to snowfall. Subsistence agriculture is the predominant characteristic of the Ladakh. It continues to play a key role in the economy of the region. Due to small land holdings, the local population subsists on limited cropping and largely depends on natural resources to meet their basic needs. The agrarian settlements and villages are mostly located around banks and terraces of rivers and streams of glacial water flow.

The region receives an annual precipitation of 115 mm on average (Thayyen et al., 2013) as it is situated in the rain shadow area of the Trans-Himalayan region. The average summer rainfall between July and September reaches 37.5 mm and average winter precipitation between January and March amounts to 27.3 mm and falls almost entirely as snow (Archer & Fowler, 2004). The mean annual temperature is −5.6 °C, and the thermal range is characterised by high seasonal variation. During January, the coldest month, the mean temperature drops to −7.2 °C (IMD, 2011). Despite these low temperatures, only the upper tracts of the mountains receive a continuous winter snow cover, whereas the lower valleys get snow for a few days only. During August—the warmest month—the mean temperature rises to 17.5 °C (IMD, 2011).

The factors determining water supply in summer are the high-altitude glaciers and some seasonal snow fields. Air is very dry and relative humidity ranges 6–24%.

Major challenges of the cold desert ecosystem of Ladakh are land degradation, frost heaving, soil salinity, infiltration of invasive weedy species, glacier retreat and increasing temperature leading to cloud bursts, flash floods, etc. Almost 80 per cent of the glaciers in the Ladakh region are very small (less than one square kilometer) but are critical as local communities are entirely dependent on glaciers and snow melt. Degradation of land, which leads to the process of desertification, is not limited to arid and semi-arid areas but is also visible in high-altitude regions that get very little rainfall. Some reports attribute that for the last three decades, Ladakh has been the victim of modern influences brought largely by tourism and other economic reasons. Its agriculture seems uneconomic, and many Ladakhis are abandoning their farming in pursuit of paid jobs in or outside Ladakh. Some reports predict that agriculture is gradually declining under climate change (Tewari et al., 2014).

In the cold desert region of India, the expanse of deserts appears to be growing. The data of the India Meteorological Department shows a shift in precipitation from winter to summer which sometimes leads to very high and intense rainfall causing flash floods, cloud bursts, etc. leading to the erosion of fertile soil, and destruction of agricultural lands. Inadequate precipitation forces the region to face drought-like conditions during the early cropping season. The melting of glaciers leads to excess water in the intermediate time period, while in the long term, it causes shortage. Local traditional knowledge and practices somehow support the local communities to adapt to changing conditions. Ladakh village communities are already taking initiatives to build ice reservoirs to provide water for irrigation and households during periods of scarcity.

14.3 Challenges to Agriculture in Ladakh

Agriculture still is one of the major economic drivers of the Ladakh region. Subsistence farming by smallholder farmers is conducted on assured irrigation through glacier-melt water, located in deep valley areas that are characterised by less favourable agro-climatic conditions, limited basic infrastructure, difficult accessibility and limited access to agricultural services besides few of the villages are facing retreat of glacier recharge, while villagers of the Indus plains get their irrigation either from canals or rivers. But these small farmers face marginalisation and susceptibility to environmental changes and degradation in the process of protecting natural resources. With assured but limited access to irrigation, farmers in Ladakh obtain low productivity. Small farmers inevitably turn to overexploitation and unsustainable agricultural practices just to maintain a livelihood.

As a result of this, land degradation becomes a key challenge for Ladakh due to the poor fertility of agricultural lands. The Ladakh province experiences a very harsh climate for about 3/4th part of a year and the farmers have access to their agricultural land only for a few months in a year. Kargil and Leh districts fall under land use class III which has a cropping intensity of less than 125% due to variations in the geographical, climatic and other factors.

Leh's and Kargil's total area reported are 45.17 and 19.46 ('000 hectares), respectively. Out of that, they have 10.18 and 9.05 ('000 hectares) area under net sown with more than 80 per cent area under irrigation, respectively (Wani et al., 2009).

The traditional system of agriculture in Ladakh is classified by the Food and Agriculture Organization of the United Nations among the possible "Globally Important Agricultural Heritage Systems (GIAHS)" (Tewari et al., 2015). The agricultural economy is based on the production of barley, wheat, peas and rearing livestock especially, yak, cows, dzos (yak—cow cross breed), sheep and goats. In addition to these crops, a range of vegetables including potato, cabbage, cauliflower, peas, onion and green leafy vegetables has grown in the region. Tethys agriculture (Zanskar Valley) is more diversified and richer than that of high-mountain trans-Himalayan agriculture (Leh and beyond). Thus, essentially the farming systems are mixed livestock-crop-based. It appears that the area under staple crops is gradually decreasing as farmers prefer to grow cash crops, namely potatoes, peas and vegetables, and the market is readily available for these commodities (Tewari et al., 2015).

Rearing of Pashmina goats or Changthangi Goat (*Capra aegagrushircus*) provides cashmere wool, and ecotourism activities are other livelihood income-generating activities of the people in Ladakh. Table 14.1 shows a declining trend in the total cultivated area recorded in Leh district (12,364 ha in 1996–97 and 10,347 ha in 2017–18) (LAHDC, 2003, 2018).

The vegetable cultivation has increased by 62 per cent from 2001 till 2018 (Table 14.1). While fruit orchard did not expand significantly, wheat has kept its space as compared to barley, grim and other food crops. Naked barley is known as

Table 14.1 Growth of agriculture in Ladakh, 1950–2018 (Area in terms of percentage to total cropped area)

Year	Wheat	Barley	Other food grains	Fruits & vegetables	Oilseeds	Non-Food Crop	TCA (hectare)
1950–51	16.93	5.84	54.94	0.72	1.45	18.21	16,238
1955–56	17.80	7.94	51.53	2.60	1.26	18.82	16,858
1959–60	17.27	5.94	54.79	3.03	1.23	17.68	17,012
1967–68	18.86	4.86	53.34	2.92	0.54	19.51	17,460
1969–70	18.24	4.52	55.20	2.19	0.33	19.51	17,807
1996–97*	30.17	29.62	4.60	2.36	0.68	49.40	12,364
2001–02*	24.74	45.00	6.71	3.21	0.7	40.00	10,523
2009–10*	25.36	43.79	6.02	4.22	0.81	40.30	10,608
2010–11*	22.06	37.81	1.64	3.07	0.74	16.65	11,692
2015–16*	22.40	44.65	7.95	4.29	0.80	40.48	10,542
2017–18*	21.94	44.13	7.86	4.77	1.01	41.48	10,347

* Fodder area in Non-food crops; TCA—Total Cropped Area (Source: LAHDC, 2019)

Greenhouse Poly-tunnel

Fig. 14.1 Polyhouse utilisation during winter in Ladakh

Grim in Ladakh and increased many folds. Under non-food crops, fodder crop has expanded and other food crops area has declined significantly. Vegetables are mainly cultivated by every household under protected structures (Fig. 14.1). After 1960s, more than 30 leafy and other vegetables have been introduced in Leh–Ladakh. The DRDO research laboratory—DIHAR has developed various types of poly- and green houses for winter season (Mishra et al., 2010).

In Ladakh, protected cultivation of vegetables has offered quality produce and favourable market prices to the growers.

Soils in Ladakh region (Leh and Kargil districts) are highly variable in their characteristics due to differences in climate, topography, vegetation and rocks (Gupta & Arora, 2017), ranging from sandy to sandy loam in texture with common occurrence of gravels and pebbles, medium to medium–high (0.75%) in organic matter with poor water holding capacity; pH is neutral to alkaline in reaction (5.65 to 10.12 in Leh and pH of Kargil varied ranging 6.57–9.47) (Acharya et al., 2012; Gupta & Arora, 2017). Due to low rainfall, dryness and low temperature during most part of the year accompanied by scarce natural vegetation and steep topography-resistant nature of the rocks towards weathering result in the formation of skeletal type of soils with poor profile development, i.e. shallow to very shallow in depth.

Terraces and broad valleys have deep soils with excessively drained, yellowish brown to dark grayish brown in colour, calcareous, occurring on gentle to moderate slopes with neutral to slightly alkaline, moderate to high in organic carbon (0.5 to 1.7 per cent) depending upon the altitude (Gupta & Arora, 2017; Walia et al., 1999). The majority of the soils in cold deserts are alkaline in nature, low in phosphorus and high in potassium.

14.4 Land Degradation in Cold Desert: Intensity and Cause

Ladakh is a vulnerable region where various processes are involved in land degradation. These are barren rock/stony waste, frost heaving, frost shattering, gullies, mass movement, stabilised dunes, ravines-extremes, ravines-very severe, rills, saline-sodic sheet erosion, unstabilised dunes, wind erosion, vegetation, etc. Wind erosion and glacial processes (frost shattering and heaving) are predominant in cold arid regions and occur in the cold desert region of Ladakh, Himachal Pradesh, Uttarakhand, Sikkim and Arunachal Pradesh. In the Cold Arid region, the processes involved in desertification/land degradation are Vegetation Degradation, Water Erosion, Wind Erosion, Frost Shattering, Mass Movement, Barren/Rocky and Settlement. The dominant degradation type is frost shattering, followed by frost heaving. The frost heaving occurs in the valley regions where the subsurface is frozen, making it difficult for useful plants to grow. Most of these lands remain as pastures and cannot support any crops.

The intensity of processes involved in soil degradation in Ladakh has been estimated as follows. Frost shattering (13.36% in 2011–13) followed by vegetation degradation (8.78% in 2011–13), wind erosion (7.52% in 2011–13) and mass movement (4.18% in 2011–13 (SAC, 2016). These processes have fastened their pace in due course of time.

The causes of degradation due to direct/indirect human interventions are (a) deforestation and removal of natural vegetation, (b) overgrazing, (c) agriculture-related activities and (d) overexploitation of the vegetation for domestic purposes. These processes, in turn, reduce agricultural productivity leading to social insecurity.

14.5 Vulnerability of Farm Land to Degradation

The major processes involved in degradation are frost shattering, vegetative shift and mass movement. In addition, in highland pastures, the major processes involved also include frost heaving, anthropological activities, infiltration of invasive weedy species and erosion processes. More than half of the degraded land is either rain-fed farmland (responsible for food security), or forest land (that offers the best defence against climate change) (SAC, 2016). In the cold arid region of Ladakh, approximately 10,000 hectares of cultivated land area is most vulnerable. It faced a cloud burst impact in 2010 when a large area of agricultural land was washed away through erosion, followed by wind erosion, salinity/alkalinity, etc. In cold arid regions, in recent times, anthropogenic pressures are reducing both the quantity and quality of biophysical environment. The drivers of degradation include grazing, lack of natural regeneration, short growing season prolonged winter, frost shattering/heaving and erodibility.

14.6 Socio-Economic Vulnerability and Ecological Consequences

Severely degraded lands are mostly inhabited by marginal farmers and tribal populations and are devoid of land-based amenities and infrastructure (Aulakh and Sidhu, 2015). This degradation has led to nutrient depletion thus reducing soil quality. Poverty and natural resources degradation coerces individuals to search for more land for food, with the estimation of six factors, i.e. number of households, occupation, literacy status, assets, food sufficiency and public awareness to natural disaster.

Almost all the village farmers/nomads had agricultural land ownership, indicating high dependency on agriculture and livestock to sustain their livelihood. Vulnerability Indices (VI) were classified into three categories using "Three Categorized Ranking Method" (TCR) assigning scores of 1 to 3, being the least vulnerable (Lama & Devkota, 2009).

Socio-economic vulnerability of a settlement was calculated by combining these six VI, and ranking of each VI is as interpreted below (Table 14.2).

- Higher household number in a settlement is associated with higher vulnerability.
- People with diversified occupation are considered less vulnerable and people involved only in agriculture are considered highly vulnerable.
- Lower level of literacy is associated with higher vulnerability.
- People having higher property investment are considered more vulnerable.
- Higher food deficiency is associated with higher vulnerability.

Table 14.2 Socio-economic vulnerability assessment of the five settlements

Villages altitude	Situation	VI1	VI2	VI3	VI4	VI5	VI6	VI Combined	Vulnerability
Saboo 11,627ft	Irrigated glacier-fed	2	2	1	2	1	1	1.50	L
Stakmo 12,675ft Semi-irrigated glacier-fed		2	2	2	2	2	2	2.00	M
Nang 11,755ft	Dry and rain-fed glacier-fed	3	3	2	2	3	3	2.67	H
Chushot 11,755ft	River bank salt-affected	2	1	1	2	1	1	1.33	L
Debring 16,100ft	Changthang glacier-fed	3	3	3	2	3	3	2.83	H
Tangtse 13,501ft	Changthang East glacier-fed	2	3	3	2	2	3	2.50	H

VI1—No. of HHs, VI2—occupation, VI3—education, VI4—property value, VI5—food sufficiency and VI6—awareness; vulnerability: L = low, M = medium and H = high

- Lower level of awareness to natural disaster and climate change and its adaptation and mitigation options are associated with higher vulnerability.

14.7 Climate Change Impact on Natural Resources

Farming communities of this cold desert hardly have access to a wide range of natural resource base as compared to those plains do. Subsistent farming communities respect every inch of the valuable cultivated lands and water from snowmelt water to grow enough food only during one cropping season to last for the whole year. Farming was almost a common practice with neighbours and relatives helping each other through the sowing and harvesting. Nomad communities follow migration plans and routes discovered every year for their livestock depending on the availability of pastures.

14.8 Loss of Carbon Stock and Entry of Invasive Species in Ladakh

Present climate trend and variability indicated that in India, trend in temperature hike is to the tune of 0.68 °C increase per century, i.e. increasing trends in annual mean temperature, warming more pronounced during post-monsoon and winter but in Himalayas, it is 0.09 °C per year, more in winter (Anbumozhi et al., 2012). The approximate area under Changthang is about 22,000 sq km which is now undergoing a climate and biotic crisis, although highland pastures are at lower altitude with meagre and threatened vegetative biomass (Tashi, 2008). Half a degree hike in temperature will make pine vegetation vulnerable (Callaghan & Jonasson, 1995; Holten & Carey, 1992; Körner, 1995, 1999) wherein mobile species are running out of the area to migrate to newer areas and cause upward shift of species (Klanderud & Birks, 2003). *Kobresia* turf holds its dominance by 90 per cent (approx.) among the vegetative cover in Changthang pastures and is a key component of the C-stocks and cycle in these pastures (Miehe et al., 2019).

Similarly native species are disappearing gradually and invasive species are adapting to a complex ecosystem. It is stated that spread of alien plant species is confined to the high altitude but towards eastern Ladakh, the road development might have driven the upward spreading by the alien plant species by disseminating their seeds/plant parts through anthropogenic activities. This might be the reason for altitudinal movement of alien species as observed in case of *Cirsium arvense* which is adapted up to Pangong lake area. Changthang is a sensitive ecosystem, where native species have become vulnerable to alien species, like creeping thistle. As a consequence to invasive weedy species of plants (*Cirsium arvense, Phragmites australis,* etc.) and wild ass in Ladakh, it has put the sensitive cold arid ecosystems of Ladakh at risk of loss of native species in the Eastern Changthang pastures.

Miehe et al. (2019) documented the major consequences of native vegetation cover and reported the following release of large amounts of carbon, earlier diurnal formation of clouds and decreased surface temperatures. These effects decrease the recovery potential of *Kobresia*-dominated pastures and make them more vulnerable to anthropogenic pressure and climate change. Traditional migratory range land management was sustainable over millennia and offers the best strategy to conserve and possibly increase C-stocks in the *Kobresia* turf.

14.9 Reduction in Vegetation Cover: A Case Study of Changthang Pastoral System

In Changthang region, the pastures are dominated by one of the smallest Cyperaceae endemic growing not taller than 2 to 3 cm covering more than 90%, and consist of only 8 to 10 mostly tiny rosette species, sedges and other cohesive group of vegetation such as *Kobresia* spp, *Carex, Leontopodium pusillum, Astragalus strictus, Helerpestus* spp, *Triglochin* spp., *Puccinellia* spp. and *Glaux maritime, Cirsium arvense* and spread in an area of 3743.86 km^2 (Singh et al., 2018). Among all, the main constituent is endemic dwarf sedge, *Kobresia* species occurring at 3000-nearly 6000 masl, existing since the onset of pastoralism. Its plants are mostly monoecious, with androgynous spikes—upper flowers male, lower female (Dortch et al., 2011; Zhang & Noltie, 2010). A reduction of vegetation cover in the Changthang region is alarming as it is leading to the acceleration of extreme surface runoff. This reduction in transpirational processes results in accelerated convective precipitation processes due to an increase in sensible heat fluxes or local water recycling is reduced by a loss of vegetation cover such as due to overstocking (Babel et al., 2014; Lehnert et al., 2016).

14.10 Salinity Stress in the Pasture Lands of Ladakh

Pastures in East Ladakh, i.e. Changthang are significantly vulnerable to the destructive impacts of frost heaving, rising soluble salts and invasive weedy species, thereby ruining the non-competitive natural vegetation. Infiltration of invasive weedy species like *Cirsium arvense* has been established up to highlands due to anthropogenic activities and these are now replacing native species. High salt concentrations are causing various physiological disturbances in plants, such as leaf burn and water uptake inhibition and uptake of certain essential elements like calcium. Na^+ and Cl^- are toxic to cellular metabolism as they affect enzyme activity and osmotic regulation of the cell inhibiting sodium–potassium pump activity (Figs. 14.2 and 14.3).

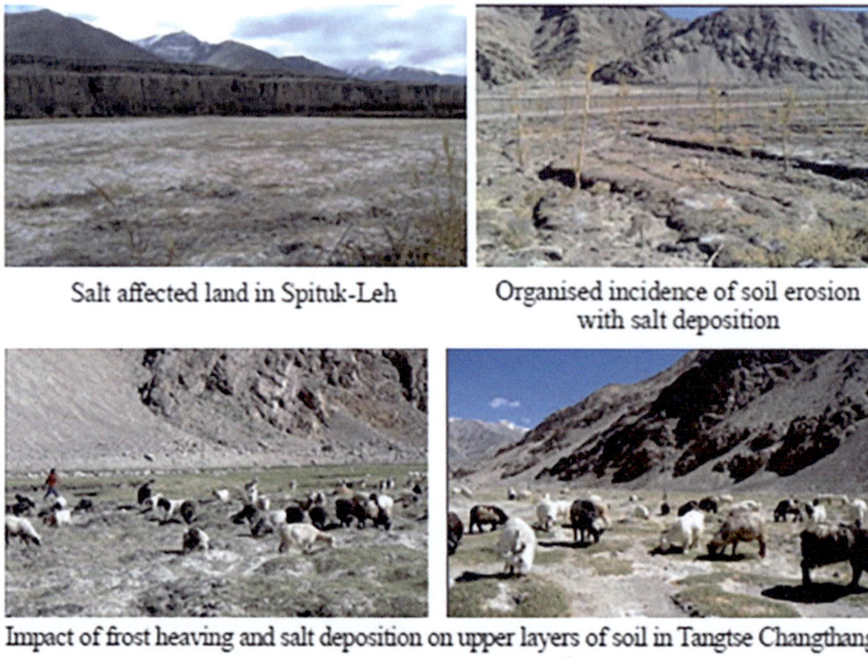

Fig. 14.2 Impact of frost heaving on soil salinity

14.11 Upward Movement of Species in the Himalayan Cold Desert of Ladakh

Ladakh is undergoing various biotic (BS) and abiotic stresses (AS), which are associated with each other. Abiotic stresses such as drought, high and low temperature and salinity play an important role in influencing the occurrence and spread of pathogens, insects and weeds (Coakley et al., 1999; McDonald et al., 2009; Peterset al., 2014; Scherm & Coakley, 2003; Ziska et al., 2010). Consequent to climate changes, as stated in the above para, Ladakh is experiencing infiltration of exotic floral and faunal species and significantly influencing the native scarce vegetation. Under the influence of climatic variability, one of the worst weeds, i.e. creeping thistle (*Cirsium arvense*) has entered in Ladakh region and adapted to the altitude at Pangong lake, Eastern Ladakh and Changthang and growing as an ornamental shrub. This shrub has not only reached the highland but also fast-replacing the native species through its own allelopathic potential directly and indirectly reducing the biomass availability for the livestock/flocks/other animals in and around pastoral areas. This leads to soil degradation, which further attracts minor pests as potential threats in the future (Duveiller et al., 2007). These stress conditions also directly affect plant–pest interactions by altering plant physiology and defence responses (Scherm & Coakley, 2003).

Fig. 14.3 Leh 2010 Cloudburst triggered flash floods damaged agricultural fields (Photo source-https://en.wikipedia.org/wiki/2010_Ladakh_floods) and mass movement

Additionally, abiotic stress conditions such as drought enhance competitive interactions of weeds on crops as several weeds exhibit enhanced water use efficiency than crops (Patterson, 1995; Valerio et al., 2013; Ziska et al., 2010).

In Eastern Ladakh, another emerging conflict with Trans-Himalayan pastoral communities in Ladakh's Changthang Plateau is the increasing population of wild ass "Kiang". They are overstocking at low-productivity Trans-Himalayan pastures (moist sedge meadows) and resulting in forage competition with livestock. They are degrading the pastures (Fig. 14.4).

Ladakh region is basically an agri-horticultural/agri-pastoral-based economy confined to limited crops and vegetables. During the 1960s, changes in the economy had a significant impact on the dominant economic activities of Ladakh, i.e. agriculture and pastoralism, and there was a predominance of naked barley locally called Grim, in the region, followed by non-food crops and wheat before 1970 (Dolker, 2018). The cropping pattern includes Grim (Naked Barley), wheat, peas, other grains [Garas and Bakla (Broad Beans) and Trumba (Buckwheat) (Singh, 1978)] and nonfood crops, which mainly consist of fodder crops, especially alfalfa. The introduction of various cash crops such as apples, apricots, potatoes, peas, mustard and other vegetables occurred in the late 1980s.

Fig. 14.4 Invasive weed entry at high altitude

References

Acharya, S., Singh, N., Katiyar, A., Maurya, K. S. B., & Srivastava, R. B. (2012). Improving Soil Health Status Of Cold Desert Ladakh Region Defence Institute of High Altitude Research C\O56 APO. Extension Bulletin No.-24.

Ahmed, N. (2020). Shrinking glaciers of Ladakh—A Clarion Call. www.mountaininitiative.in 18 September 2020.

Archer, D. R., & Fowler, H. J. (2004). Spatial and temporal variations in precipitation in the Upper Indus Basin, global teleconnections and hydrological implications. *Hydrology and Earth System Sciences, 8*(1), 47–61.

Ashiq, P. (2022). Ladakh's glaciers threatened by growing vehicular emissions. *The Hindu, 9*, 2022.

Aulakh, M. S., & Sidhu, G. S. (2015). Soil degradation in India: Causes, major threats, and management options. *In:* MARCO Symposium 2015 - Next Challenges of Agro- Environmental research in Monsoon Asia. National Institute for Agro- Environmental Sciences (NIAES), Tsukuba, Japan.

Baba, S. H. (2018). S&T intervention in agricultural and allied sector for strengthening livelihood security. Final Technical Report of Department of Sciences and Technology (DST) sponsored research project, Shere-Kashmir University of Agricultural Sciences & Technology of Kashmir, Shalimar Campus, Srinagar, Jammu & Kashmir.

Babel, W., Biermann, T., Coners, H., Falge, E., Seeber, E., Ingrisch, J., Schleuß, P.-M., Gerken, T., Leonbacher, J., Leipold, T., Willinghöfer, S., Schützenmeister, K., Shibistova, O., Becker, L., Hafner, S., Spielvogel, S., Li, X., Xu, X., Sun, Y., … Foken, T. (2014). Pasture degradation modifies the water and carbon cycles of the Tibetan highlands. *Biogeosciences, 11*, 6633–6656. https://doi.org/10.5194/bg-11-6633-2014

Bandyopadhyay, J., & Gyawali, D. (1994). Himalayan water resources: ecological and political aspects of management. *Mountain Research and Development, 14*(1), 1–24

Business Insider Indian Bureau. (2019). Most of the glaciers of the Himalayan mountain range surrounding Mount Everest will melt by the end of the 21st century: Report. www.businessinsider.in/most-of-the-himalayan-glaciers-will-melt-by-the-year-2100

Central Ground Water Board. (2012). District Ground Water Information Brochure, Leh district. 2012 (http://cgwb.gov.in/District_Profile/JandK/Leh.pdf)

Coakley, S. M., Scherm, H., & Chakraborty, S. (1999). Climate change and plant disease management. *Annual Review of Phytopathology, 37*, 399–426. https://doi.org/10.1146/annurev.phyto.37.1.399

Dolker, P. (2018). An overview of transition in traditional agriculture of Ladakh. *J. Himalayan Ecology Sustainability Development, 13*(2018), 26–48.

Dortch, J. M., Owen, L. A., Caffee, M. W., & Kamp, U. (2011). Catastrophic partial drainage of Pangong Tso, northern India and Tibet. *Geomorphology, 125*, 109–121. From: International Encyclopedia of the Social & Behavioral Sciences.

Duveiller, E., Singh, R. P., & Nicol, J. M. (2007). The challenges of maintaining wheat productivity: Pests, diseases and potential epidemics. *Euphytica, 157*, 417–430. https://doi.org/10.1007/s10681-007-9380-z

Gupta, R. D., & Arora, S. (2017). Characteristics of the soils of Ladakh Region of Jammu and Kashmir. *Journal of Soil and Water Conservation, 16*(3), 260–266.

Hellmann, J. J., Byers, J. E., Bierwagen, B. G., & Dukes, J. S. (2008). Five potential consequences of climate change for invasive species. *Conservation Biology, 22*, 534–543. https://doi.org/10.1111/j.1523-1739.2008.00951.x

IMD. (2011). Climatological Table. Accessed on 2 February 2011. Available from: http://www.mausam.gov.in/WEBIMD/Climatological Action. do? function5get Station Details & action Param51.

LAHDC. (2019). Mission Organic Development Initiative of Ladakh- Policy, Strategy and Action Plan. 18 pp.LAHDC (2003). Statistical hand Book-District Leh-2002–03.

LAHDC. (2018). Statistical hand Book-District Leh-2017–18.

Lehnert, L. W., et al. (2016). Climate variability rather than overstocking causes recent large scale cover changes of Tibetan pastures. *Science and Reports, 6*, 24367. https://doi.org/10.1038/srep24367

Mcdonald, A., Riha, S., ditommasob, A., & Degaetanoa, A. (2009). Climate change and the geography of weed damage: Analysis of U.S. maize systems suggests the potential for significant range transformations. *Agriculture Ecosystem Environment, 130*, 131–140. https://doi.org/10.1016/j.agee.2008.12.007

Miehe, G., Schleuss, P.-M., Seeber, E., Babel, W., Biermann, T., Braendle, M., Chen, F., Coners, H., Foken, T., Gerken, T., Graf, Hans-F., Guggenberger, G. … & Wesche, K. (2019). The Kobresia pygmaea ecosystem of the Tibetan highlands – Origin, functioning and degradation of the world's largest pastoral alpine ecosystem: Kobresia pastures of Tibet. *Science of The Total Environment, 648*, 754–771, https://doi.org/10.1016/j.scitotenv.2018.08.164.

Mishra, G. P., Singh, Kumar, N., Singh, H., & Bala, S. (2010). Protected cultivation for food and nutritional security at Ladakh. *Defence Science Journal, 61*(2), 219–225.

Olsson, L., Barbosa, H., Bhadwal, S., Cowie, A., Delusca, K. D., Flores-Renteria, D., Hermans, K., Jobbagy, E., Kurz, W., Li, D., Sonwa, D. J., & Stringer, L. (2019). Land degradation. In Climate Change and Land: An IPCC special report on climate change, desertification, land degradation, sustainable land management, food security, and greenhouse gas fluxes in terrestrial ecosystems [P.R. Shukla et al. In press. https://www.ipcc.ch/site/assets/uploads/sites/4

Patterson, D. T. (1995). Effects of environmental stress on weed/crop interaction. *Weed Science, 43*, 483–490.

Peters, K., Breitsameter, L., & Gerowitt, B. (2014). Impact of climate change on weeds in agriculture: A review. *Agriculture Sustainability Development, 34*, 707–721. https://doi.org/10.1007/s13593-014-0245-2

Raghuvanshi, M. S., Moharana, P. C., Saxena, A., & Saha, D. (2020a). Pasture and land degradation by weedy invasion on frost heaves in Changthang-Ladakh: Retrospect and prospects. *Food and Scientific Reports, 1*(4), 46–49.

Raghuvanshi, M. S., Moharana, P. C., Spalbar, E., Dorjay, N., Singh, R. K., Saxena, A., Gaur, M. K., & Arunachalam, A. (2020b). Spatial distribution of nutrients in soil profile under different land use systems of cold arid region of Ladakh. *Indian Journal of Hill Farming, 33*(2), 394–399.

SAC-Space Application Centre. (2016). Desertification and Land Degradation Atlas of India. Space Applications Centre Indian Space Research Organisation Government of India

Scherm, H., & Coakley, S. M. (2003). Plant pathogens in a changing world. *Australasian Plant Pathology, 32*, 157–165. https://doi.org/10.1071/AP03015

Singh, H. (1978). Ladakh: Problems of Regional Development in the Context of Growth Point Strategy. Unpublished PhD thesis, Centre for Study of Regional Development, Jawaharlal Nehru University, New Delhi.

Singh, J. P., Ahmad, S., Radotra, S., Dev, I., Mir, N. H., Deb, D., & Chaurasia, R. S. (2018). Extent, mapping and utilization of grassland resources of Jammu and Kashmir in western Himalaya: A case study. *Range Mgmt. & Agroforestry, 39*(2), 138–146.

Tewari, J. C., Pareek, K., Tewari, P., & Raghuvanshi, M. S. (2015). Productivity and energetics of crops in farming systems in Ladakh: A case study of a village in Leh. Paper presented in National Symposium on Sustaining Agricultural Productivity in Arid Ecosystems: Challenges & Opportunities, 19–22 August 2015. pp. 31–39.

Tewari, J. C., Raghuvanshi, M. S., & Kamlesh, P. (2014). Indian cold desert. *Den News July-December, 2014*, 1–3.

Thayyen, R. J., Dimri, A. P., Kumar, P., & Agnihotri, G. (2013). Study of cloudburst and flash floods around Leh, India, during August 4–6, 2010. *Natural Hazards, 65*(3), 2175–2204. https://doi.org/10.1007/s11069-012-0464-2

UNDP. (2006). *Human Development Report: Beyond Scarcity: Power, Poverty and the Global Water Crisis*. United Nations Development Programme.

Valerio, M., Lovelli, S., Perniola, M., Di Tommaso T., & Ziska, L. (2013). The role of water availability on weed–crop interactions in processing tomato for southern Italy. Acta Agriculture Scand. Sect., B, *63*, 62–68. https://doi.org/10.1080/09064710.2012.715184

Walia, C. S., Rana, K. P. C., Sidhu, G. S., Mahapatra, S. K., & Tarsem, L. (1999). Characterisation and classification of some soils of Ladakh region for land use. *Agropedology, 9*, 16–21.

Wani, M. H., Baba, S. H., & Shahid, Y. (2009). Land-use dynamics in Jammu and Kashmir. *Agricultural Economics Research Review, 22*, 145–154.

Zhang, S., & Noltie, H. J. (2010). *Kobresia* Willdenow, Sp.Pl.4:205.1805. *Flora of China, 23*, 269–285.

Ziska, L. H., Tomecek, M. B., & Gealy, D. R. (2010). Evaluation of competitive ability between cultivated and red weedy rice as a function of recent and projected increases in atmospheric CO_2. *Agronomy Journal, 102*, 118–123. https://doi.org/10.2134/agronj2009.0205

Chapter 15
Land Resources and Agricultural Economy of Leh District of Ladakh: A Case Study

Mahesh K. Gaur

15.1 Introduction

Land-use change is perhaps the most significant driver of environmental change as it is deeply entwined with many issues like loss of biodiversity, greenhouse gas emissions, soil degradation, and alteration of hydrological cycles (Chan et al., 2017) as well as leading to human migration. The term "land use" refers to human activities or economic functions associated with a specific area of land, while "land cover" pertains to the physical features present on the Earth's surface (Gaur & Goyal, 2022). Insights into land use and land cover enhance our understanding of land utilisation aspects such as cropping patterns, fallow lands, forests, pasturelands, wastelands, and surface water bodies, all of which are essential for effective development planning (Lillesand & Kiefer, 2000). Globally, land-use change is driven by a variety of human activities, encompassing everything from large-scale alterations in land cover to shifts in cropping intensity and types, as well as transitions from crop production to infrastructure development and conservation efforts (Gaur & Squires, 2021). The evolution of farming technologies and practices varies significantly, ranging from subsistence methods to mechanised approaches (Squires & Gaur, 2020). In India, most farms are operated by resource-poor smallholders. While food security issues are primarily local and influenced by both local and regional factors, land-use changes are also rooted in local contexts. However, many driving forces are regional in nature, impacting food systems based on different land-use types and their transformations. Additionally, measures aimed at enhancing food security can further influence land use and its changes (Chan et al., 2017).

Historically, agriculture has been the backbone of the Ladakhi economy, with most Ladakhis relying on subsistence farming in small settlements scattered throughout

M. K. Gaur (✉)
Regional Research Station, ICAR-Central Arid Zone Research Institute Stakna, (The U.T. of Ladakh), Leh, India
e-mail: geo.maheshgaur@gmail.com

M. S. Rathore et al. (eds.), *Desert Ecosystems*,
https://doi.org/10.1007/978-3-032-03237-9_15

the high-altitude cold desert. The region's economy is predominantly based on small-scale farming and livestock. Pastoralism plays a crucial role, particularly in raising sheep, yaks, goats, and cows. Yaks provide essential resources: milk for butter, hair for making ropes and tents, skins for boots, aprons, and baskets, horns for agricultural tools, dung for fuel, and meat. Goats are especially valued for their fine, long-staple fleece, which is renowned for producing luxurious pashmina shawls. Ladakh's economy is largely biomass-based, with crops like barley, wheat, and peas being cultivated. However, the region has also opened up to tourism, which now accounts for 50% of Ladakh's GDP.

This paper analyses the dynamic spatial interactions between land use and food security, focusing on local changes and their impacts on climate, ecosystems, and social systems, along with their economic and cultural dimensions.

15.2 Arid Lands of India

The country's arid lands include both cold and hot dry zones (Gaur & Squires, 2016). It covers over half of the country and is home to 43% of the country's people. Water availability, soil conditions, and the duration of the growing season all reveal significant differences in arid lands. Over 60% of the arid lands are in Rajasthan, Maharashtra, Gujarat, Andhra Pradesh, Ladakh, Himachal Pradesh, Uttarakhand, Karnataka, and Tamil Nadu. In these arid lands, the annual rainfall of cold arid zone ranges from 70 to 170 mm. Soil types vary from gravelly and sandy loams found on alluvial fans to sandy and silty clay loams in the Indus floodplains. The soil of the cold desert region of Ladakh is described as skeletal, calcareous, and alkaline. Soil is often coarse and sandy, with differing proportions of pebbles (Gupta & Arora, 2017). The following are the most notable arid areas in India:

1. Western Himalayas: Cold arid region with rainfall <150 mm;
2. Western Rajasthan, Kutch, and northern part of Kathiawar peninsula: Hot arid region with rainfall <300 mm;
3. Rajasthan uplands (Aravali) and Chambal districts of Madhya Pradesh: Semi-arid region with extensive land degradation leading to ravines;
4. Central Highlands (including Gujarat plains and Malwa region of Madhya Pradesh): Semi-arid region with rainfall of 500–1000 mm;
5. Deccan Plateau (including Maharashtra and northern Karnataka): Semi-arid region with rainfall of 600–1000 mm;
6. Interior Andhra Pradesh (Telangana): Semi-arid region with 600–1000 mm rainfall;
7. Tamil Nadu Uplands and western Karnataka;
8. Sub-humid Eastern Plateau (Chhattisgarh): rainfall of 1000 to 1600 mm;
9. Sub-humid Eastern Chhotanagpur Plateau including Jharkhand, western Orissa, and northern Andhra Pradesh: rainfall of 1200–1600 mm.

Dryland areas in different regions of India

States	% of arid land area
Ladakh, Uttarakhand, and Himachal Pradesh	60 to 81
Rajasthan and Gujarat	66 to 88
Madhya Pradesh, Maharashtra, Andhra Pradesh, Karnataka, and Tamil Nadu	76 to 82
Eastern Uttar Pradesh, Bihar, Jharkhand, Orissa, and West Bengal	33 to 73

15.3 Cold Arid Region of Ladakh—Leh District

The Ladakh region consists of two districts, Leh and Kargil. Leh, spanning an area of 45,110 km^2, is the largest district in the country. It lies between 32° and 36° north latitude and between 75° and 80° east longitude. The population density of Leh is about three people per km^2 (Table 15.1). This district includes Leh City and 112 inhabited villages. Even with the harsh terrain and unpredictable cold desert climate, farmers have effectively managed the environment for approximately 8,800 years (Miehe et al., 2009). Ladakh is a high-altitude cold desert. The southwest monsoon keeps the area dry. The great Himalayan Range separates the Vale of Kashmir from Ladakh, running northwest to southeast through the region. The Zanskar Mountains encircle the Zanskar River Valley to the east and they run in the same direction. The Ladakh Range lies further east, forming the Indus River Valley, while the Karakorum Mountains lie towards the northeast, forming the eastern boundary of Nubra Valley. Rain clouds cannot cross these ranges, so Ladakh only gets about 2 inches (5 cm) of rain per year. Therefore, the weather pattern in the Ladakh Desert is unusually cold and dry. The air at this altitude is so thin that solar radiation remains very high. Global solar radiation is up to 6–7 Kw/mm. The region experiences arctic cold in the winter when temperatures drop to −23 °C. It is frigid with temperatures below minus 40 degrees Celsius for most of the winter. There is little rainfall in this area (less than 100 mm per year). The area is exposed to strong winds and scorching sun. There is occasional snowfall in April, but only at higher elevations (above 4,000 or 4,500 m).

According to the 2011 population census, the district has a total population of 133,487, comprising 78,971 males and 54,516 females. The population density is approximately 3 individuals per square kilometer, and the district has experienced a decadal growth rate of 13.86%. There are 16 blocks within the district. The sex ratio stands at 690 females for every 1,000 males, while the literacy rate is 77.2%.

The historic city of Leh is built on an alluvial fan. The alluvial plain stretches on both sides of Leh with a width of about 30 km. From above, the alluvial plain appears to be a simple succession of golden grasslands and barren snow-capped mountains. Agriculture is primarily practiced along the alluvial plains, especially along the Indus River and its tributary plateaus, such as Zanskar, Nyoma, Doras, Sulu, Waha, Kanj, and Wangla. However, irrigation facilities are limited, and water

Table 15.1 Changes in land-use pattern in Leh district (1985–86 to 2018–19)

Year	Area (in ha.)	Area not available for cultivation*	Cultivable wasteland excluding fallow*	Fallow land*	Net area sown*	Area sown more than once*	Total cropped area*
1985–86	44,479	60.2	16.3	1.5	21.9	NA	NA
1990–91	45,167	62.3	14.7	1.1	21.9	NA	NA
1995–96	45,167	62.1	14.6	1.2	22.0	NA	NA
2000–01	45,167	62.1	12.3	0.9	24.7	NA	NA
2005–06	51,352	65.4	13.85	0.89	19.84	0.78	20.61
2010–11	51,684	66.26	13.77	0.8	19.73	2.89	22.62
2015–16	45,167	67.42	9.51	1.04	21.97	1.37	23.34
2018–19	45,611	67.69	28.1	0.11	22.00	0.71	22.71

* Percentage to Total Reporting Area (PTRA)
Source (1) LAHDC Statistical Handbook, Leh; and (2) Financial Commissioner (Rev.) J&K

supplies are unstable due to low temperatures that hinder plant growth for most of the year, small and scattered farms, and poor soil development. Owing to the poorly formed soil profile, soils in the Ladakh region differ not only in morphological and physicochemical properties but also in occurrence, taxonomy, and microbiological properties. Temperature and humidity are the two main factors responsible for the sparse and reduced vegetation in this area (Kachroo et al., 1977). South of the Leh district, in alpine areas, fine-flowered herbs from *Aconitum, Astragalus, Potentilla, Primula, Saxifrage*, and several genera of the *Asteraceae* family are found.

15.4 Land Use in Leh District

The study of land utilisation in the Leh district encompasses economic, geographic, and demographic dimensions. Land use in this region is influenced by various physical and human factors, including terrain, soil, climate, population density, occupations, and the technological advancements of the local population (Mandal, 1990). Situated in a rain-shadow area, Leh is one of the driest places on Earth, making agriculture particularly challenging (Gaur et al., 2021). Despite these harsh conditions, local residents have developed innovative practices, such as channelling glacial-fed rivers into stone-built terraces, capturing soil through sedimentation, and enriching the land with organic manure. As a result, they have successfully cultivated staple crops. Thus, traditional Ladakhi agriculture is not only unique but also reflective of the broader agricultural practices found on the Tibetan plateau.

The two land-use categories—*forests* and *permanent pastures* and *other grazing lands* classified under *cultivable wasteland* (*excluding fallow land*)—are either absent or occupy a negligible area. Consequently, they have not been included here.

Table 15.1 illustrates the relative changes in land-use patterns in the Leh district. During these decades, the largest proportion of land use was classified as "*area not available for cultivation*," which accounted for 67.7% of the reporting area in 2018–19, while *fallow land* represented a mere 0.11% during the same period. Each land-use category exhibited varying degrees of significance within the total reporting area. (i) *Area not available for cultivation*: This category peaked at 68.13% in 2012–13 and reached a minimum of 60.2% in 1985–86. (ii) Additionally, it was found that *cultivable wasteland* (excluding fallow land) had a maximum share of 28.1% in 2018–19, compared to just 9.5% in 2015–16.

In the years 1985–86, *fallow land* constituted a stake of 1.5% of the total reporting area. However, by 2013–14, there was no land left *fallow*. The area designated as *Net Sown* reached its highest level in 2000–01 at 24.7%, but this figure declined to 14.9% by 2012–13. Notably, throughout the period from 1985–86 to 2018–19, the category of *land not available for cultivation* consistently represented the largest proportion of the total reporting area (Fig. 15.1).

Area Not Available for Cultivation: This land-use class includes *area under non-agricultural uses* and *barren and uncultivable lands*. The area under non-agricultural

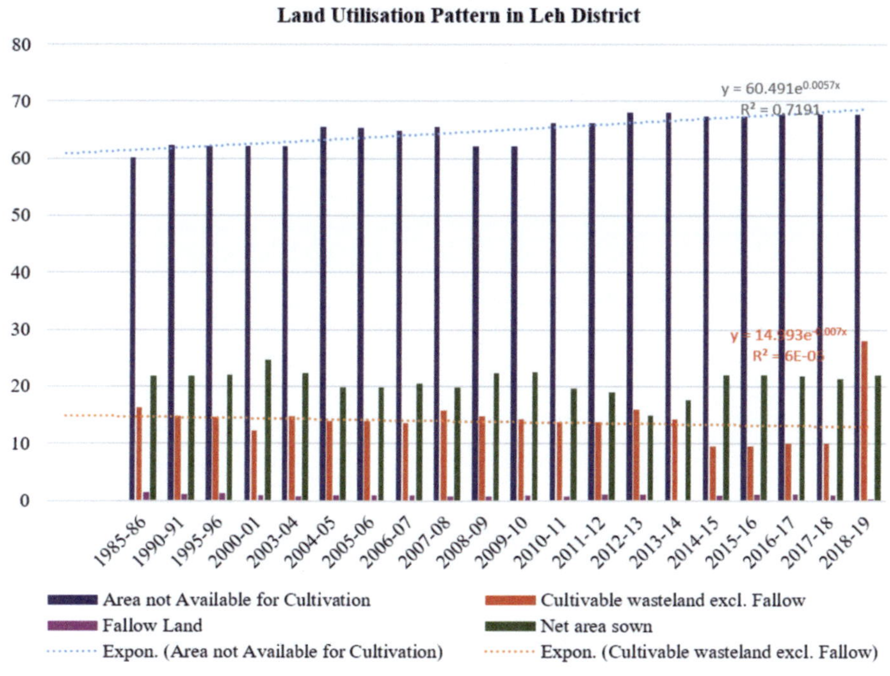

Fig. 15.1 Land utilisation pattern in Leh District

Table 15.2 Area not available for cultivation

Year	Area (in ha.)	Land put to non-agricultural uses	PTRA	Relative change	Barren and uncultivable land	PTRA	Relative change
1985–86	44,479	1806	4.1	–	24,983	56.2	–
1990–91	45,167	2280	5.0	26.2	25,255	55.9	1.1
1995–96	45,167	2902	6.4	27.3	25,168	55.7	−0.3
2000–01	45,167	2908	6.4	0.2	25,163	55.7	0.0
2005–06	51,352	7066	13.8	0.9	26,520	51.6	0.0
2010–11	51,684	7073	13.7	143.2	27,169	52.6	8.0
2015–16	45,167	2978	6.6	0.0	27,474	60.8	0.0
2018–19	45,611	7381	16.2	0.0	23,492	51.5	0.0

* PTRA: Percentage to Total Reporting Area
Source (1) LAHDC Statistical Handbook, Leh; and (2) Financial Commissioner (Rev.) J&K

uses includes all land occupied by buildings, roads or underway, and all land put to non-agricultural activities, whereas barren and uncultivable land covers all land occupied by mountains and deserts. Such lands require a huge amount of money and labour to bring them under agricultural practices, if at all that will be required at any point of time in the future (Gaur & Goyal, 2021).

Table 15.2 presents data on the PTRA and relative changes in the *area not available for cultivation*. The PTRA indicates the highest proportion of land designated for *non-agricultural uses* occurred between 2016 and 2018, reaching 16.2%. In contrast, the lowest proportion was recorded in 1985–86 at just 4.1%. However, the relative change indicates a remarkable increase of 147.9% in land designated for *non-agricultural use*, shifting from 0.0% in 2003–04 to −59.0% in 2012–13 within the same category.

The PTRA data presented in Fig. 15.2 shows that the proportion of *barren and uncultivable land* within the total reporting area reached its highest point in 2012–13 and 2013–14, at 61.7%. Conversely, the lowest representation was noted in 2006–07, at 51.2%. Additionally, the data on relative change reveals a slight increase of 8% in this category, moving from 0.0% in 2009–10 to 8% in 2010–11. The trend according to the analysis in the case of *barren and uncultivated land* is somewhat less and lacking any substantial transformation.

15.5 Cultivable Wasteland Excluding Fallow

Cultivable wasteland excluding fallow includes the land-use category like permanent pasture and other grazing land, land under miscellaneous trees and grooves, and cultivable wasteland. The first category mentioned above is not analysed since there was no separate information available about them besides one year. Land under

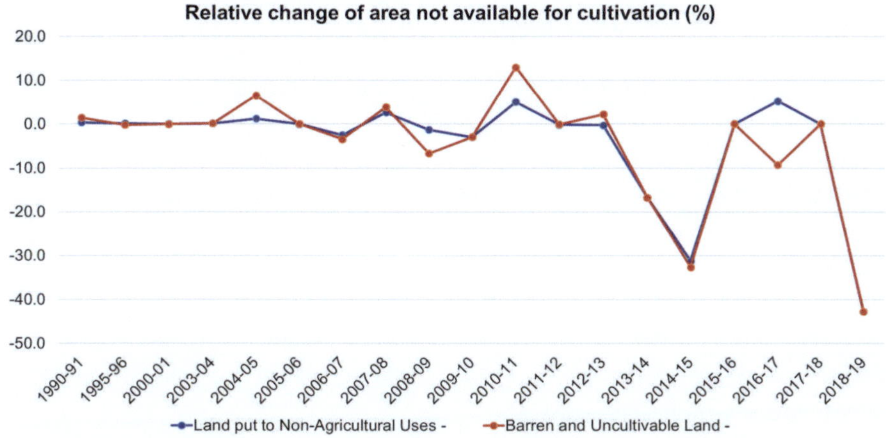

Fig. 15.2 Relative change of area not available for cultivation (%)

miscellaneous trees and grooves comprises land that is used for cultivation but has not been included as a part of net area sown. The land not included in orchards and which is covered by trees, grasses, bushes, and grooves for fuelwood is classified under this category.

Cultivable wasteland excluding fallow comprises land that is cultivated once or can be cultivated if tried but has not been cultivated for five successive years including the current year due to a lack of sufficient water availability. They may be accessible or inaccessible and could be covered by shrubs and bushes that may be found in isolated patches or within some cultivated lands.

Table 15.3 depicts PTRA and relative change regarding *cultivable wastelands excluding fallow.* The PTRA indicates that the maximum stake of the land-use class of *land under miscellaneous trees and grooves* was during 1985–86 (6.4%) which continuously fluctuated between 2.4% (2009–10) and 6.0% (2012–13). The data on the relative change shows the highest increase of 178.3% in the class of *land under miscellaneous trees and grooves* between 2000–01 (−47.7%) and 2010–11 (138.3%). In addition, there was a decline of 238.3% between 2010–11 (138.3%) and 2013–14 (−100%) (Figs. 15.3 and 15.4).

Fallow Land

Fallow land as a whole comprises *current fallow and long fallow*. The land-use class of long fallow includes all land that is once cultivated and is left uncultivated for a period of 3 to 5 years, whereas current fallow are those land which is kept uncultivated for a short duration of 1 to 2 years for the rejuvenation of fertility (Table 15.4).

Table 15.4 deals with the PTRA and relative change regarding fallow land in the Leh district. The PTRA data indicate that the highest stake of the long fallow (0.7%) remained for five long years 1985–86, 2000–01, 2003–04, 2008–09, and 2009–2010, whereas the least stake was found in 2004–05 to 2007–08, 2010–11 to 2016–17, and 2018–19 (0.0 to 0.1%). The data on the relative change shows two major declines in

Table 15.3 Cultivable wasteland excluding fallow (in ha.)

Year	Area (in ha.)	Land under misc. trees & Grooves	PTRA	Relative change	Cultivable wasteland	PTRA	Relative change
1985–86	44,479	2859	6.4	–	4416	9.9	–
1990–91	45,167	2243	5.0	21.5	4400	9.7	0.4
1995–96	45,167	2196	4.9	−2.1	4406	9.8	0.1
2000–01	45,167	1148	2.5	−47.7	4406	9.8	0.0
2005–06	51,352	1594	3.1	−4.2	4465	8.7	0.0
2010–11	51,684	2621	5.1	138.3	4498	8.7	5.1
2015–16	45,167	1732	3.8	0.0	2567	5.7	0.0
2018–19	45,611	1911	4.2	8.6	1546	3.4	−42.8

* PTRA: Percentage to Total Reporting Area
Source (1) LAHDC Statistical Handbook, Leh; and (2) Financial Commissioner (Rev.) J&K

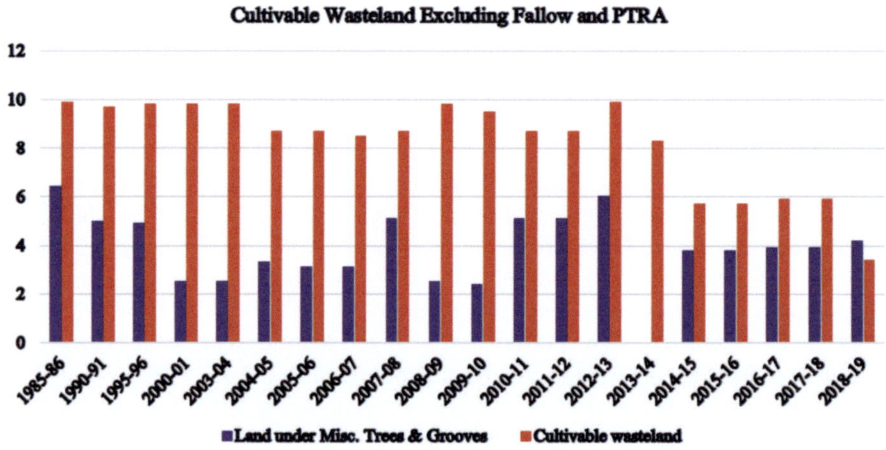

Fig. 15.3 Cultivable wasteland excluding fallow and PTRA

the stake of the land-use category; the first being a fall of 79.2% between 2000–01 (31.8%) and 2005–06 (−47.4%), and the second drop of 113.5% between 2013–14 (−100%) and 2015–16 (−13.5%) (Figs. 15.5 and 15.6).

Net Area Sown

This category of land refers to the total land area where crops are cultivated once annually. The net area sown, combined with the area sown more than once in a year, constitutes the gross cultivated area. All of this cultivated land is irrigated and primarily supports mono-cropping, with canal and khul irrigation being the main sources. In certain regions, such as Khaltsi and Nubra blocks, double cropping is also practiced.

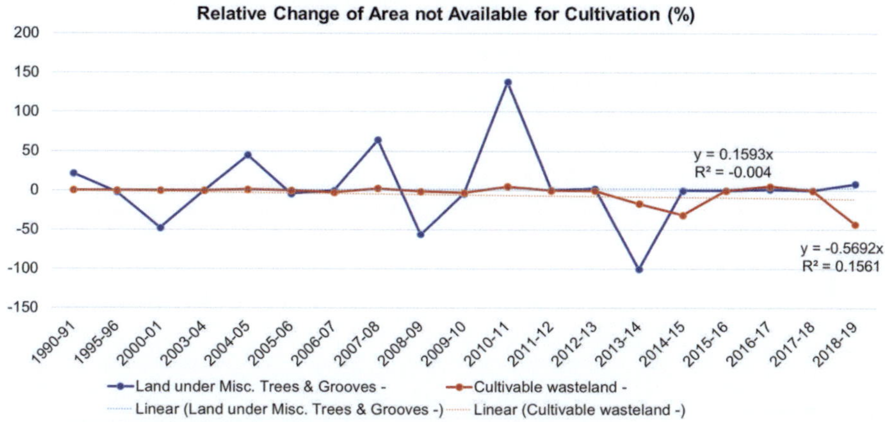

Fig. 15.4 Relative change of cultivable wasteland excluding fallow

Table 15.4 Fallow land (in hectares)

Year	Area (in ha.)	Long fallow	PTRA	Relative change	Current fallow	PTRA	Relative change
1985–86	44,479	106	0.2	–	563	1.3	–
1990–91	45,167	–	–	–	514	1.1	−8.7
1995–96	45,167	88	0.2	–	469	1.0	−8.8
2000–01	45,167	116	0.3	31.8	282	0.6	−39.9
2005–06	51,352	61	0.1	−3.2	397	0.8	2.6
2010–11	51,684	59	0.1	−82.0	358	0.7	258.0
2015–16	45,167	32	0.1	−13.5	439	1.0	16.8
2018–19	45,611	51.6	0.1	−65.6	0	0.0	−100.0

* PTRA: Percentage to Total Reporting Area
Source (1) LAHDC Statistical Handbook, Leh; and (2) Financial Commissioner (Rev.) J&K

Table 15.5 represents the PTRA and relative change in net area sown and it suggests that the highest stake of this class was during 2000–01 (24.7%), whereas the lowest stake was observed in 2012–13 (14.9%). The maximum relative change was 11.5% between 1995–96 (0.6%) and 2001–02 (12.1%). A decline of 27.7% has also been observed between 2011–12 (−3.6%) and 2012–13 (−31.4%) (Fig. 15.8).

There are vast stretches of land available but irrigation facilities are not there, therefore, these lands cannot be brought under cultivation. Even the future plan of reclamation of these desert areas, the creation of sources of irrigation will be uneconomical.

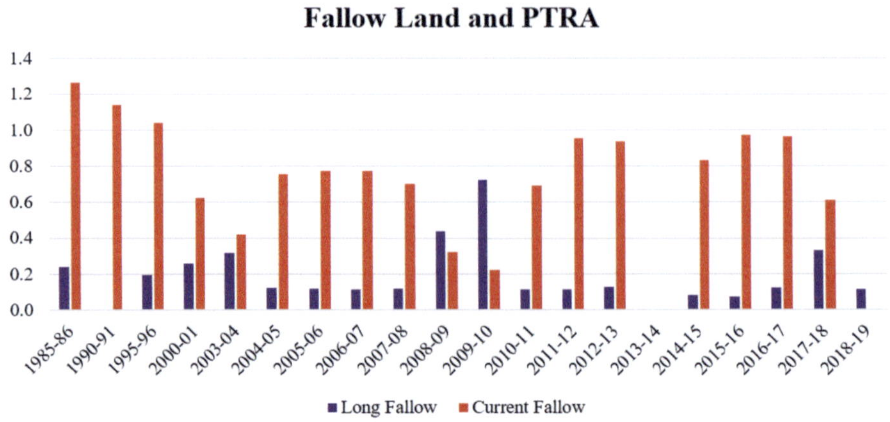

Fig. 15.5 Fallow land and PTRA

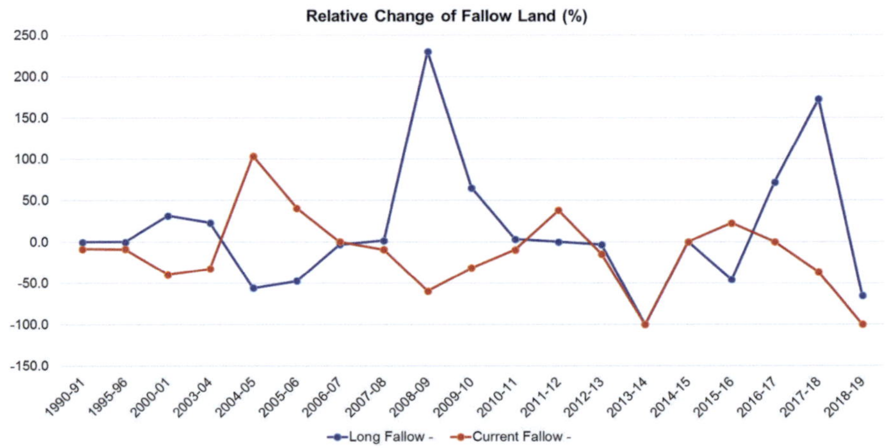

Fig. 15.6 Relative change of fallow land (in %)

Landholding Size

In the Leh district, the number of marginal, semi-medium, medium, and large agricultural landholdings are termed operational holdings and as per the tenth agricultural census these holdings have registered a slight increase in number during the year 2015–16 compared to 2010–11 (Table 15.6). This indicates a growing number of people owning small parcels of agricultural land. Out of a total of 21,542 operational holdings, only 14,698.2 hectares are actively cultivated. Additionally, there is a trend towards the consolidation of landholdings in the semi-medium, medium, and large categories within the district. Notably, the proportion of female landholders has risen, reflecting an increase in women's participation in the management and operation of agricultural lands (Fig. 15.9).

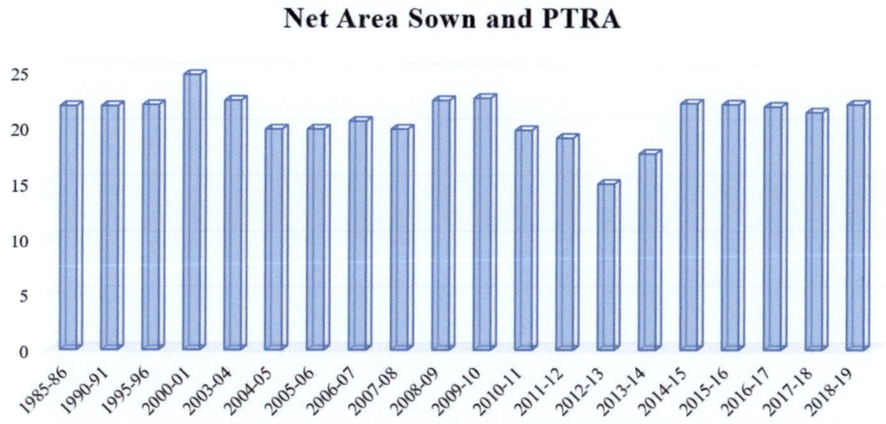

Fig. 15.7 Net area sown and PTRA

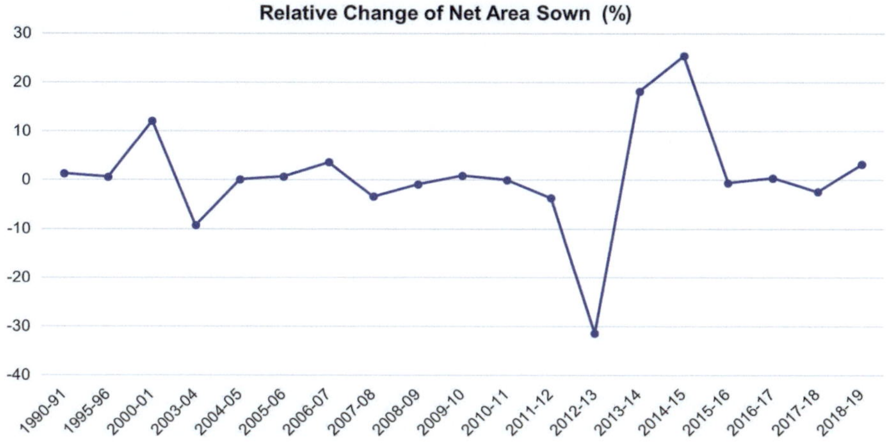

Fig. 15.8 Relative change of net area sown (in % age)

It is significant that marginal, small, and medium land holdings constitute the vast majority of the operational area, with large landholdings comprising just 0.22% of the total operational area (Table 15.7). The average size of operational holdings in the Leh district has decreased from 1.38 hectares in 1995–96 to 0.68 hectares in 2015–16. The small and marginal holdings account for 93% of the operational holdings in this dryland agriculture. The future growth of agriculture in these arid regions is heavily dependent on the performance of these small and marginal farmers. In such environments, having a large landholding holds little significance (Gaur et al., 2014). This distribution of landholdings reflects broader agricultural practices, with large holdings often indicative of the challenging environmental conditions faced (Gaur & Squires, 2018; Gaur et al., 2020).

Table 15.5 Net area sown (in hectares)

Year	Area (in ha.)	Net Area Sown (ha.)	PTRA	Relative change
1985–86	44,479	9746	21.9	–
1990–91	45,167	9875	21.9	1.3
1995–96	45,167	9938	22.0	0.6
2000–01	45,167	11,144	24.7	12.1
2005–06	51,352	10,186	19.8	0.7
2010–11	51,684	10,197	19.7	0.0
2015–16	45,167	9924	22.0	−0.6
2018–19	45,611	10,034	22.0	3.2

* PTRA: Percentage to Total Reporting Area
Source (1) LAHDC Statistical Handbook, Leh; and (2) Financial Commissioner (Rev.) J&K

Due to the subdivision of holdings, a good amount of land (about 3–5%) is being wasted for drawing boundaries and planting hedges between huge numbers of tiny plots. Furthermore, due to subdivision, the plots become so small that it is quite uneconomic to cultivate such land. Fragmentation of agricultural holding creates difficulties for the farmers to manage the agricultural operation smoothly (Gaur et al., 2019; Squires et al., 2020). In addition, the smaller size of holdings cannot provide full-time jobs to all the members of the farmer's family. Thus, in the absence of alternative occupations, chronic (often hidden) unemployment started to occur in rural areas (Gaur and Squires, 2020).

Animal Husbandry

The importance of livestock in the fragile ecosystem zone of Ladakh goes beyond its traditional role in terms of source of income. Since the area remains isolated for about half of a year from the rest of the world with limited access to food items, animal products particularly meat and milk play a pivotal role in meeting the nutritional security of local people. However, the "traditional" agro-pastoral land-use system of Ladakh has seen manifold changes in recent years due to the household income diversification, which is similar to other high mountain regions of South Asia (Kreutzmann, 2006; Nüsser & Gerwin, 2008). Animal husbandry is an important sector in the allied sector of agriculture. Animal husbandry plays a pivotal role in the rural economy by providing manure and dairy products to the district and generating substantial income for the rural population. The livestock population in the district as per the 17th, 18th and 19th livestock censuses is shown hereunder (Table 15.8).

Livestock contributes significantly to total household income, ranging from 16.32% to 86.59%. Milk alone accounts for approximately one-third of the dietary fat needs for farmers, with the exception of those in the Nyoma block, where it fulfils over 80% of their requirements. A similar trend has also been observed in a milk-based diet in fulfilling the family requirement of Chuchot blocks. FYM obtained by the farmers has been highest in Chuchot (1.99 quintals/year) and lowest in Nyoma (0.67 quintals/year). However, employment (man-days/year) generated by livestock

Table 15.6 Number of land holdings by different sizes and classes in Leh District

Class and size		2005–06				2010–11					2015–16		
		Area (ha.)	% age area	No. of holdings	% age holdings	Area (ha.)	% age area	No. of holdings	% age holdings	Area (ha.)	% age area	No. of holdings	% age holdings
Marginal	Up to 1 ha	6105.0	34.1	19,410	77.89	5281.91	36.5	17,306	80.47	5312.35	36.1	17,320	80.4
Small	1–2 ha	4988.4	27.9	3563	14.3	4010.62	27.7	2875	13.37	3921.85	26.7	2818	13.08
Semi-medium	2–4 ha	4278.6	23.9	1620	6.5	2967.77	20.5	1124	5.22	3038.7	20.7	1141	5.3
Medium	4–10 ha	1596.8	8.9	280	1.12	876.88	6.1	160	0.75	1202.8	8.2	216	1.0
Large	10 & above	935.7	5.2	48	0.19	1328.41	9.2	40	0.19	1222.5	8.3	47	0.22
Total		17,904.5	100.0	24,921	100.0	14,466.6	100.0	21,505	100.0	14,698.2	100.0	21,542	100.0

Source Agriculture Census of different years and publications of LAHDC

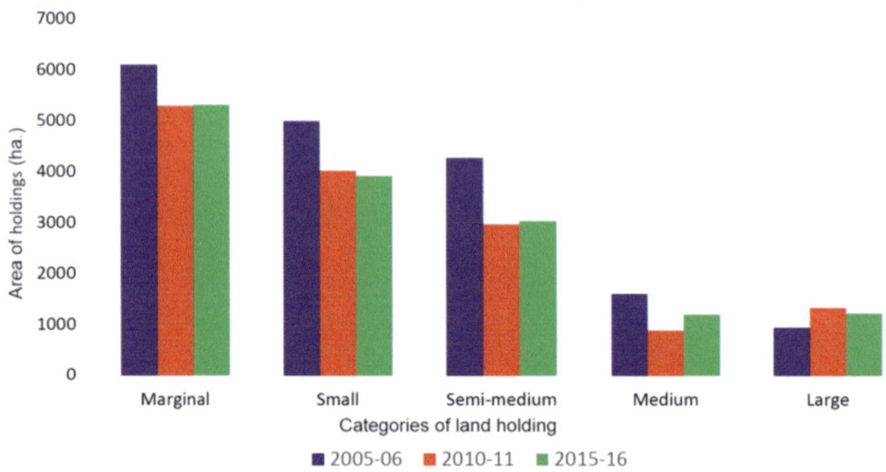

Fig. 15.9 Land holding pattern in Leh district

Table 15.7 Number and area of operational holdings in Leh District

S. no	Year	Number of holdings (000 Nos.)	Area (000 ha.)	Average holding size (ha.)
1	1995–96	12.04	16.359	1.38
2	2000–01	16.91	12.76	0.75
3	2005–06	24.92	17.90	0.72
4	2010–11	21.51	14.47	0.67
5	2015–16	21.54	14.70	0.68

Source Agriculture Census of different years

is more in Nyoma (561.91) as compared to the farmers of other blocks ranging from 294.11 to 326.67 per annum per family. The native livestock breeds are multipurpose with unique genetic characteristics and exhibit a distinct superiority over other breeds of India which include adaptation to harsh or extremely cold arid dry climatic conditions, utilisation of poor quality feed, and better resistance to tropical diseases.

Sheep: Malluk and Merino-Malluk hybrid breed are the main sheep breeds of this region. As they provide better quality clothing wool, the Carpet industry wool demand is met by the Changluk breed mostly prevalent in the Changthang area. Ladakh Region contributes 2.697 lakh kg to the total wool production. Nyoma and Durbuk are the two regions where sheep and goat rearing is mostly concentrated in the district.

Goats: Small ruminants are preferred because of their small size, adaptive feeding behaviour, and easy management. Goats are a viable option in improving the household cash flow and food security of rural tribal communities in the Leh district. Apart from cash income from the sales of pashmina wool products, goats could also be a valuable source of milk, meat, manure, and skin. *Changthangi* goat is primarily

Table 15.8 Livestock population of Leh district (from 2003 to 2012)

S. no	Type of Livestock	Population					
		2003	%	2007	%	2012	%
1	Cattle/ Dzo-Dzomo	31,800	7.97	35,370	10.72	46,807	11.28
2	Sheep	102,400	25.66	60,799	18.43	81,192	19.57
3	Goats	219,000	54.87	185,856	56.34	222,028	53.53
4	Horses & Ponies	7900	1.98	5418	1.64	5066	1.22
5	Yak–Demo	29,600	7.42	21,874	6.63	16,357	3.94
6	Donkey/ Mules	8300	2.08	7936	2.41	5709	1.38
7	Camel	100	0.03	141	0.04	214	0.05
8	Poultry	DNA*	DNA*	9159	2.78	34,486	8.31
9	Dog	DNA*	DNA*	3310	1.00	2949	0.71
	Total	399,100	100.0	329,863	100.0	414,808	100.0

* DNA: Data not available
Source Animal Husbandry Department, Leh and https://jkanimalhusbandry.net/wp-content/uploads/2017/09/BiReport-1415.pdf

raised for its valuable pashmina wool. There are other breeds of goats, mostly raised for meat and wool production. Each and every part of goats is valuable, such as, hair used for making ropes, baskets, and coarse blankets, and the skin is used for garments.

Yak: The yak is a distinctive bovine species of significant economic value in the high-altitude, snow-covered regions of the Indian Himalayas. This versatile animal serves as a vital source of livelihood for communities living in challenging terrains. Yaks provide a variety of products and services, including meat, milk, wool, and leather, as well as draught power and soil nutrients. Additionally, they serve as a financial asset and form of security for the highlanders.

Cattle: The Leh block boasts the highest number of exotic cow breeds. Despite having the least cultivable land, they have higher milk production in the district. In contrast, Durbuk has the fewest exotic cow breeds. Khaltsi, Kharu, and Leh have the highest populations of indigenous cow breeds. Dairy production is a key sector of the rural economy in Leh, meeting domestic needs and allowing for surplus sales in the market (Akand et al. 2017). Notably, 84.38% of the total livestock consists of milk-producing animals, underscoring the vital role of dairy production in the arid region of Ladakh.

15.6 Agricultural Productivity

Agricultural productivity is a key driver for the well-being of farmers, the agro-based industry, and the general economy. It affects food security, food prices, and poverty alleviation efforts (Darku & Malla, 2010). In addition, agricultural productivity helps in meeting the challenges of the increasing population, changes in income, and the resultant changes in food habits (Bruinsma, 2009).

Agriculture serves as the backbone of the district's economy, employing over 70% of the workforce primarily as cultivators, agricultural labourers, and livestock rearers (Darku et al., 2016). In this context of limited agricultural resources, the local population has sustained their livelihoods through a combination of mountain agriculture (Ehlers & Kreutzmann, 2000). The core pillars of livelihoods in these villages consist of crop farming, animal husbandry, and gathering activities, with millet and wheat being the most dominant crops. Other cultivated crops include barley, pulses, and fruits. The sowing season is brief and often begins before the majority of glacial meltwater flows into the region (Table 15.9).

Barley and wheat are the primary staple crops grown on irrigated terraces during the brief agricultural season from May to September. These cereals are rotated with peas and, occasionally, mustard. A variety of vegetables are also produced in the area which include turnips, cabbage, potatoes, onions, carrots, and green leafy vegetables. Mono-cropping predominates because of the agro-climatic conditions, as double-cropping is only feasible at altitudes below approximately 3,000 m. Wheat cultivation is suited at elevations ranging from about 3,600 m (in Saboo village, according to Labbal, 2001) to 3,800 m (in Stongde, according to Osmaston, 1994), while barley is grown at altitudes exceeding 4,000 m to about 4,400 m (Osmaston, 1994).

The region is mostly suitable for agri-horticulture and agri-pastoral farming systems. However, given the climatic conditions, people have evolved mono-cropping as the predominant practice. However, double cropping is carried out in certain areas of the Sham belt and Nubra Valley. Wheat and barley are the primary crops cultivated in the district. Though there is significant potential for vegetable production in Ladakh, advancements in technology have extended the vegetable growing season, allowing for the cultivation of various new vegetable crops. The largest share of vegetable cultivation is found in Leh, accounting for 52.6% of the total, followed by the Sham belt at 34.2% (Table 15.10). In total, vegetable production in the Leh district reached 11,867 tonnes, with Leh leading at 55.4%, followed

Table 15.9 Cropping season and duration at different altitudes

Altitude (m asl)	Cropping season	Duration (in months)
2400–2700	April–October	7.0
2700–3000	May–mid-October	5.5
3000–3300	Mid-May–September	4.5
3300–3600	June–mid-September	3.5
3600–4000	Mid-June–August	2.5

by Sham at 29.1%, Nubra at 13.4%, and Changthang at 2.1%. Major crops grown in each block of the district are detailed in Table 15.11.

Of the total 2414.5 ha under wheat in Leh district, this cereal crop is primarily grown in Chuchot (740.75 ha), Thiksey (453.1 ha), Leh (408.1 ha), Diskit (263.1 ha), and Nimoo (176 ha). Barley occupies a 3631.4 ha area and of this, the largest acreage is found in Karu (803.7 ha), Chuchot (392 ha), Skurbuchan (388.8 ha), Nimoo (362.5 ha), Khaltsi (357.4 ha), and Saspol (249 ha). Of the 118.2 ha under pulses, the largest area has been in Panamik (19 ha), Khaltsi (16.2 ha), Nimoo (14.1 ha), Diskit (14 ha), Skurbuchan (12.6 ha), and Karu (11.6 ha).

In Ladakh, approximately 33% of the total land is put under agricultural activities. The annual food grain production is around 100,000 quintals. The major crops of the districts are wheat, barley, and rice, and are mainly cultivated in the Indus, Zanskar, and Shyok river basins. Ladakh has significant potential for growing various vegetable crops. Advances in technology have extended the vegetable growing season, enabling the cultivation of a variety of new vegetable crops in the region.

As the farms in the region are relatively small, averaging about 0.2 acres is more suitable for vegetable cultivation. Presently the area under vegetable cultivation is 5.5% of the total 10,319 hectares of agricultural land in the Leh district. The vegetables grown include potatoes (75.6%), peas (10.7%), onions (3.4%), cabbage (3.3%), carrots (2.0%), and cauliflower (1.8%). The emphasis on potatoes, onions, cabbage, and carrots is suitable because of suitable long-term storage which facilitates consumption during the landlocked winter months. The rapid growth in vegetable production in the region is primarily because of the increasing demand from the army and the tourism industry. Vegetables are harvested from June to September to meet the severe scarcity of fresh produce during the winter months.

Mixed crop-livestock farming systems are being practiced throughout Leh-Ladakh, characterised by the interaction of five key components:

i. Field crops
ii. Fruit and multipurpose trees around the crop fields
iii. Community woodlots and pastures
iv. Livestock composition, and
v. Humans.

For centuries, agriculture in Ladakh provided sufficient food grain; however, in recent decades, the region has faced a significant food deficit. This is because over 60% of its food needs are being met through imports from other parts of the country. Research indicates that agricultural production in the region is not energy-efficient, and there is a concerning trend of declining area allotted to cereal crops. The energy inputs for cash crops are considerably higher because their economic viability may be more favourable. Today, farmers in Ladakh are able to cultivate a variety of crops and vegetables that were previously unfeasible, such as capsicum, cucumber, cauliflower, tomatoes, okra, and even watermelon.

Maintaining a diverse array of crops serves the purpose of fulfilling the growers' requirements and the diversified requirements of the consumers. Additionally, crop diversity enables farmers to better handle unforeseen climatic circumstances that may

Table 15.10 Area sown under crops (in hectares)

Year	Wheat	Barley/ Grim	Other millets	Pulses	Fruits	Vegetables	Total food crops	Oil seeds	Fodder	Total non-food crops	Total Area sown
2001–02	2604	4734	436	270	98	240	8382	73	2068	2141	10,523
2002–03	2653	4702	377	272	85	246	8335	73	2070	2143	10,478
2003–04	2894	4504	349	349	85	245	8327	73	2024	2097	10,424
2004–05	2894	4480	384	249	35	299	8341	67	2020	2087	10,428
2005–06	2973	4463	375	272	38	310	8431	65	2089	2154	10,585
2006–07	2973	4463	375	272	38	310	8431	65	2089	2154	10,585
2007–08	2968	4452	359	286	39	313	8436	74	2089	2163	10,599
2008–09	2634	4639	379	306	91	348	8402	86	2028	2114	10,516
2009–10	2690	4646	342	297	90	358	8427	86	2095	2181	10,608
2010–11	2579	4421	–	192	131	229	7552	86	1947	–	11,692
2011–12	2595	4488	97	243	40	280	7743	90	1991	2081	9824
2012–13	1092	2869	556	258	88	321	5188	86	2093	2179	7367
2013–14	1077	3934	622	251	134	393	6415	80	2100	2180	8595
2014–15	2776	4288	564	258	104	388	8364	89	2161	2250	10,614
2015–16	2362	4708	575	263	452		8366	84	2092	,2176	10,542
2016–17	2713	4390	548	265	79	360	8361	97	2082	2179	10,540
2017–18	2271	4566	549	265	494		8149	104	2094	2198	10,347
2018–19	2509	4491	318	267	495		8141	105	2112	2217	10,358
2019–20	2414.5	3631.4	246.5	118.2	822.3	629.0	7861.9	70.80	2289.8	2360.6	10,222.6

Source (1) Deputy Commissioner's Office Leh; and (2) Financial Commissioner (Rev.) J&K

Table 15.11 Block-wise major crops grown in Leh district

S. no	Block name	Major crops grown
1	Chuchot	Wheat, Barley, Pulses, Vegetables, Oilseeds
2	Durbak	Barley, Vegetables
3	Karu	Wheat, Barley, Vegetables, Pulses
4	Khaltsi	Wheat, Buckwheat, Barley, Pulses, Vegetables, Fruits
5	Leh	Wheat, Barley, Pulses, Vegetables, Fruits
6	Nimoo	Wheat, Barley, Pulses, Vegetables, Fruits, Oilseeds
7	Nubra	Wheat, Barley, Vegetables, Pulses
8	Nyoma	Barley, Vegetables
9	Panamic	Wheat, Barley, Pulses, Vegetables, Oilseeds, Fruits
10	Rong Chumathang	Barley, Pulses, Vegetables, Fruits
11	Rupsho Puga	Barley, Fruits
12	Saspol	Wheat, Buckwheat, Barley, Pulses, Vegetables, Oilseeds, Fruits
13	Singaylalok	Wheat, Barley, Pulses, Vegetables, Fruits
14	Skurbuchan	Wheat, Buckwheat, Barley, Pulses, Vegetables, Fruits
15	Thiksey	Wheat, Barley, Vegetables, Fruits, Pulses, Oilseeds
16	Turtuk	Wheat, Barley, Vegetables, Fruits

impact specific vegetables. The preference for crops like potatoes, onions, cabbage, and carrots is largely due to their capacity for long-term storage during the winter months when the region is isolated and open-field cultivation is unfeasible.

With the economic development in the region, it has been observed that people are trying to move out of farming. It is evident from the facts that there is a change in the cropping pattern and an increasing number of fields kept fallow (Hinz et al., 2020). This abandonment of fields is largely driven by income diversification. Small farmers, characterised by average land holdings of less than 2 hectares, increasingly view traditional cropping practices as economically unviable.

15.7 Horticulture

Khaltsi, Leh, and Nubra, with Khaltsi blocks, are best suitable for horticultural production in the district. Apricots, apples, and walnuts are the main horticultural products, while in the lower areas of the Khaltsi block almonds and grapes are also cultivated (Sharma & Mir, 2000). There is limited scope of fruit marketing within the district. Fruits are either sold in Leh Town and other markets or supplied to the defence forces stationed in the region. Apricots, being highly perishable, often suffer significant losses during transportation and marketing. In order to take care of the perishable nature of fruits, the cultivators are increasingly using osmotic dehydration techniques to process the fruit before sale to maximise their profits. In recent years, horticultural crops have become a major source of supplementary income for farmers, particularly gaining substantial importance in the Leh district. Details on the area allocated to fruit cultivation, fruit plant nurseries, and plant distribution in different years and blocks are given in Table 15.12.

There are relatively new enterprises in Leh, which are based on collection and processing of wild fruit of sea buckthorn for its juice, seeds, and pulp. It is quite rich in protein, amino acids, Vitamin C, etc. and seed oil has medicinal value and is being used for the treatment of cancer. Local farmers are trained for the processing and marketing of its products.

Table 15.12 Area under horticultural activities

Year	Area (Ha) under Fruits		
	Apricot	Apple	Others
2013–14	785.42	792.46	9.89
2014–15	790.62	796.24	60.84
2015–16	791.57	797.67	60.81
2016–17	791.57	797.67	60.81
2017–18	571.732	294.17	169.246
Block-wise figure for the year 2018–19			
Leh	17.6	28.844	1.256
Nimoo	36.272	27.226	2.083
Saspol	44.82	25.27	6.68
Khaltsi	150.37	59.73	5.448
Skurbuchan	158.72	29.91	7.66
Lingshed/Singelalok HQ at Wanla	20.868	9.724	0.404
Diskit	27.444	40.442	0.344
Turtuk	101.968	23.854	13.144
Panamic	52.6904	64.276	2.78
RongChugut HQ at Chumathang	8.932	4.34	0.07
Durbuk	0.6016	0	0.06
Karu	8.91	9.364	0.01
Chuchot	14.388	20.472	1.253
Thiksey	10.876	15.208	0.05
Total	654.46	358.66	41.242

15.8 Food Security Analysis

Meanwhile, the local food systems in the Leh district have experienced significant changes over the past two decades, largely driven by an influx of tourists. While subsistence agriculture continues to be the backbone of the farming economy (Dame & Nüsser, 2011), the unique characteristics of the Leh-Ladakh region significantly influence its food systems. Key factors include remoteness due to challenging topography, limited accessibility, restrictions on market access, a constrained agrarian resource base, and reduced income opportunities (Mora et al., 2020). Understanding the dynamics and challenges of food security in Ladakh from an integrative and multiscale perspective is essential for grasping the complex social-ecological interactions at play.

Internal and external migration has become a major feature of hills in the country. It is largely governed by livelihood diversification and off-farm employment opportunities made available because of the economic development of hilly regions. The pattern of migration also includes rural-to-urban migration to the district capital, Leh (Goodall, 2004; Nüsser et al., 2018), as well as internal migration within the region and outside (Squires and Gaur, 2020). It is for these reasons that the urban population has surged dramatically, growing from 28,639 in 2001 to 45,671 in 2011, while the rural population has remained relatively stable (LAHDC, 2011).

The changing migration pattern and the Central Government's Public Distribution System (PDS) had a significant influence on the food habits of the local population. Barley being an important part of the diet of the people is declining, though it remains the main staple crop due to dietary preferences influenced by sociocultural factors. Barley products, such as *tsampa* (roasted barley flour) and *chang* (beer), hold significant cultural value and are yet not market products. As a result, families in urban Leh often rely on their villages of origin for these essential products. Wheat farming has become less common than barley farming; however, wheat-based meals, such as local noodle dishes and bread, are still staples in the diet. Many households choose not to cultivate wheat because the Central Government's Public Distribution System (PDS) offers fine wheat flour (*atta*), rice, and sugar at highly subsidised rates (Tiwari, 2000). Though the PDS primarily targets those below the poverty line, market prices of rice, wheat, and sugar in Leh tend to be higher even for those above the poverty line. Most households benefit from these PDS supplies and there is direct availability of goods in their villages.

15.9 Challenges and Opportunities of Agriculture in Leh District

Given a very limited agricultural season, i.e. from April–May to September–October months farmers have adjusted their cropping pattern accordingly. In the winter season because of the freezing harsh cold, crops cannot be grown. Mono-cropping is predominant in this region. The area available for cultivation is also limited because of snow-covered high mountains and inhospitable environmental conditions. The cultivated area is mostly irrigated and monocropped. The main source of irrigation is canals/khuls. Double-cropping is undertaken in some parts of the Saspol, Turtuk, Skurbuchan, Khaltsi and Nubra blocks.

Vegetable cultivation is becoming a major activity in the district. Protected cultivation technologies like greenhouses, shade-nets, trenches cultivation, and black plastic mulching are promoted to overcome weather constraints. While fruit production is concentrated mainly in Khaltsi, Leh and Nubra blocks, the main fruits produced are apricots and apples, and in part of Khaltsi block fruits like almonds, grapes, etc. are grown. Marketing of fruits and vegetables is a major challenge.

Rapid urban growth, scarce water resources, and a high risk of natural disasters pose serious ***challenges*** for the agricultural sector of Leh. These are as follows:

(i) *Harsh climatic condition*: There is high variability in rainfall patterns that directly influences the snowfall. The fast melting of glaciers has been observed. These changes have influenced the farming pattern in recent times. As an example, there was a 50% to 80% deficit in annual rainfall between 2013 and 2017 and Leh recorded the lowest rain in 2016. Very often, too early prevalence of frost in the month of August causes extensive damage to the crops. A shift in precipitation patterns from winter to summer, a clear indication of climate change, occasionally results in heavy and intense rainfall, leading to flash floods and cloudbursts. These events cause erosion of fertile soil and the destruction of agricultural lands, as well as trees and forests, in a region where vegetation is already sparse. Consequently, fertile lands are transformed into barren, non-productive areas. Approximately 80% of the glaciers in the Ladakh region are small, measuring less than one square kilometer, yet they are vital since local communities rely entirely on glacier and snowmelt water for irrigation. Over the past two decades, a negative mass balance of these glaciers has been observed, primarily due to insufficient winter precipitation. As a result, the region experiences drought-like conditions during the early cropping season.

(ii) *Continuing dominance of native crop varieties in farming*: Farmers are still dependent on traditional seed varieties of wheat, barley, and pea. The old varieties of these crops are low yielding and prone to insect, pests, and diseases. Location-specific high-yielding hybrid seeds of new varieties may perform better and become a game-changer of the agricultural economy in this region.

(iii) *Scope for post-harvest value addition facilities:* Because of inaccessible conditions, it is hard to export surplus produce outside the region as it remains cut-off by road in winter due to weather conditions. Therefore, there is a need for establishing value-added processing facilities in different parts of the district.

(iv) *Soil erosion, land degradation, and desertification landscape:* The rugged, sloping, and undulating topography of the region makes it highly susceptible to soil erosion, particularly during the summer months when snowmelt occurs. This process leads to the loss of the top fertile soil from agricultural lands. Approximately 24% of the district's area is degraded due to wind and water erosion (Gaur, 2022). Additionally, about 58% of the land is categorised as sandy, rocky, or wasteland, while 11% is covered by snow and glaciers. The limited development of soil profiles in the Ladakh region results in significant variability in the morphological, physico-chemical properties, and genesis of the soils. There is clear evidence, both direct and indirect, of increasing desertification in this trans-Himalayan area.

(v) The area under land degradation/desertification in the Nubra sub-basin and Shyok sub-basin has decreased by about 0.11% and 0.04% respectively since 2003–05. The significant process of land degradation and/or desertification in the district is through Frost Shattering (36.21% during 2011–13 and 2003–05) followed by Vegetation Degradation (e.g. 3.04% during 2011–13 and 2003–05). The most significant process of land degradation/desertification in the Shyok sub-basin is Frost Shattering (54.97% during 2011–13 and 2003–05) followed by Wind Erosion (7.37% during 2011–13 and 7.38% during 2003–05).

(vi) *Tourism*: Tourism is adding to the process of depletion of the water table and destruction of marsh lands, springs, and wetlands. Growing tourism has added pressure on its natural resources. Because of the occupation shift to tourism, locals are also gradually giving up farming—their traditional source of livelihood. In fact, many villages in the valley, such as Kulum near Leh and Kumik in Zanskar, were abandoned due to water shortage. However, this may be the initial signs of climate change impact-driven migration. While attempts are being made to rehabilitate ghost villages by solving the water crisis, the problem will not vanish away. In addition, Leh town used to get its supply of water from glacial meltwater and springs. However, over the years, the water demand skyrocketed so much that the hotels and guesthouses had to dig their bore wells to extract the groundwater, which further depleted the water table. Now, the majority of springs in the town have dried up, and thousands of water tankers are required daily to fulfil the needs of the tourists for a 24x7 water supply. It is recommended that there has to be an assessment of the carrying capacity of Leh town and surrounding areas w.r.t. number of tourists it can accommodate every season. It may help in policy formulation about the requirements of sustainable tourism in Leh Ladakh (Fig. 15.10).

(vii) *Urban development*: In Leh, there has been a rapid spatial expansion of housing settlements, both on former agricultural land and barren land, accompanied by an increase in the density of built-up areas. The pace of construction has more than doubled, with data showing that 9,400 new buildings were

Fig. 15.10 Tourists Arrival in Leh District

erected between 2003 and 2017. This figure is comparable to the number of buildings constructed in the previous 34 years, from 1969 to 2003. Furthermore, the total building area surged from 36 hectares in 1969 to 196 hectares by 2017. Consequently, the loss of agricultural land due to construction activities rose from 1% in 1969 to 8% in 2017. Given the limited availability of fertile land in Leh, urban expansion has also encroached upon the surrounding barren areas. Analysing the spatio-temporal data of urban development in Leh reveals three main trends: (a) the expansion of urban areas into barren land, (b) the encroachment of urban areas onto agricultural land, and (c) the densification of existing built-up areas.

(viii) *Fodder shortage:* Given the agroecological status of the Leh district, there is limited scope for the development of fodder for animals. The carrying capacity of grazing lands is also constrained by snow cover. On an average, there is a 40–45% shortfall in fodder production in the region. It is a major constraint farmers face in livestock farming. The problem is compounded by the depleting area under cereal crops and it is posing a threat to the indigenous system of agro-pastoral and crop-livestock mixed farming systems of Leh valley (Kumar et al., 2016).

15.10 The Potential Opportunities for Growth of the Agricultural Economy

Living in snow-bound hilly areas is a big challenge for humans and livestock. In the above sections, attempt has been made to describe the livelihood conditions in the district by listing challenges, such as lack of soil cover, water scarcity due to low precipitation, and low temperatures impacting agriculture (Singh, 1992). Till now the agricultural sector in this region has successfully sustained both the local population and livestock (Osmaston, 1989). The introduction of modern technologies has significantly helped people to diversify their occupations and also adopt diversified agriculture. The progress in the field of horticulture and expanding livelihood opportunities in the off-farm agri-food system further improved the economic status of people. Advancements in exploring the potential of organic farming to enhance the region's self-reliance and sustainability offer a promising vision for the future. Additionally, Ladakh possesses three major untapped resources—solar energy, geothermal energy, and agro-based industries—that could also contribute to the faster growth of the Leh district.

(i) *Protected Cultivation of off-season vegetables*: The harsh climatic conditions are the major constraint in agriculture development in the district. The new advancements in agriculture have led to the development of various protected cultivation practices tailored to specific agro-climatic zones can be the solution to address the constraints. The technology known as greenhouses and polyhouses are particularly beneficial for year-round vegetable cultivation in the Ladakh region. Using Polycarbonate Greenhouse technology, these structures enable the production of fresh organic vegetables even during the harsh winter months, helping Leh-Ladakh achieve self-sufficiency in vegetable production and potentially doubling farmers' incomes. This innovation also paves the way for young agri-entrepreneurs to engage in commercial farming, generating significant employment opportunities.

The Central Arid Zone Research Institute (CAZRI), the Regional Research Station of SKUAST Kashmir, and the Defense Research and Development Organization (DRDO) have developed technologies to enhance local vegetable production, benefiting farmers, local markets, and the armed forces. During the summer months, there is substantial production of brinjal, capsicum, tomatoes, and other cucurbits. Meanwhile, various green leafy vegetables are cultivated during the long, frozen winter months, when temperatures can drop to −40 °C. Unique practices, such as underground greenhouses and soil trenches, are widely adopted by farmers in the Leh district. They cultivate a range of vegetables, including cabbage, cauliflower, and *knol-khol*, to meet the demands of the local market and the defence forces stationed in the region.

(ii) *Cultivation of high value crops*: In the Ladakh region during winter, temperatures and solar radiation levels are not conducive to growing vegetables.

However, leafy greens such as Swiss chard, spinach, coriander, and fenugreek can be cultivated in greenhouses during the extreme winter months from October to February. Additionally, high-value crops like asparagus, broccoli, leeks, tomatoes, cucumbers, and capsicum are crucial for off-season production during this period, as they command premium prices.

(iii) *Cultivation of medicinal and aromatic plants*: This region is abundant in medicinal plants, with approximately 1,100 species identified and documented. Also, valuable ethnobotanical information on the medicinal uses of these phyto-resources is available. Among these, at least 500 species are recognised for their significant medicinal properties. Some of the most commonly used medicinal plants from high-altitude areas include *Aconitum heterophyllum, Angelica glauca, Bergenia stracheyi, Capparis spinosa, Caragana brevifolia, chamomile, Chenopodium botrys, Dactylorhiza hatagirea, Dracocephalum, Ephedra gerardiana, Gentiana aligida, Hippophae rhamnoides, Hippophae salicifolia, Hippophae tibetana, Inula racemosa, Meconopsis aculeata, Origanum vulgare, Podophyllum hexandrum, Rheum webbianum, Rhodiola imbricata, Rhodiola tibetica, Salsola collina, Saussurea lappa*, and *Thymus serpyllum*, among others. However, ethno-medico-botanical studies have indicated that many trans-Himalayan medicinal plants fall in the categories of rare, endangered, or threatened (Saxena et al., 2018), therefore, need for proper planning to extract them to save and conserve those species.

Lack of infrastructure development and inaccessible areas make the provision of healthcare services to human and livestock populations a big challenge. Therefore, people in this region are largely dependent on the traditional system of medicine, locally known as the *amchi* system, that supports approximately 60% of public health in Ladakh (Chaurasia & Singh, 1996). The *amchi* system is based on the local medicinal plants. that have therapeutic benefits with minimal side effects. Most medicinal plants are found growing in the wild. In the Suru Valley of Ladakh, several medicinal plants are incorporated into the *Sowa Rigpa* system of medicine (Stanzin, 2019). Ladakh has a huge scope in medicinal plants because of its climate and topography. Because of the exceptional medicinal value of these wild plant resources, there is a need for R&D investment to develop a new range of high-value crops in this region. It will encourage farmers to grow medicinal plants as commercial crops.

(iv) *Mushroom production*: Ladakh has a large number of people who are vegetarian and mushrooms can be one of the main sources of nutrition for them. Mushroom cultivation is an effective method of "bioconversion" of wastes and woods into valuable food. It can be a very significant contribution to the sustainable agriculture economy of Leh-Ladakh. *Oyster mushroom* (Dhingri) is a famous variety for cold arid zones. In order to increase the shelf life by 4 to 6 months mushrooms can be dried using solar dryers. Solar-dried mushrooms exhibit a natural brown colour and retain their nutritional value and moisture content, making them ideal for use in soups. Mushroom growers in Ladakh can have a sustainable livelihood as there is a high demand for mushrooms from tourists, the army, and local people. Further, mushroom cultivation can offer a promising opportunity for local people to supplement their income both during harsh winters and summers.

(v) *Floriculture*: Floriculture is a fast-emerging agro-enterprise of Leh-Ladakh. Several promotional schemes are introduced to encourage farmers to adopt floriculture. At the same time, financial assistance in the form of subsidy has been provided to flower growers as part of the promotion of floriculture. While farmers can be motivated to grow flowers in the greenhouses the limiting factor will be high airfreight charges necessary for accessing the markets in Delhi and other places and that will limit the promotion of wider cultivation of the flowers.

(vi) *Buckwheat cultivation*: Buckwheat was a staple food for the Ladakhi people until a few decades ago and remains an important part of their diet. Buckwheat (*Fagopyrum esculentum*) is a native crop cultivated in several villages of Leh-Ladakh. Locally buckwheat is known by various names, viz. *dyat, dro, bro, fafar,* etc. One of the most popular dishes made from buckwheat flour is *kiseer*, or *giziri*, which resembles a plain dosa. This crop is nutritionally beneficial, providing high-quality proteins and seeds rich in carbohydrates. Additionally, it offers several advantages for local communities, including the hardiness of the plants, a short growing season, and the use of its foliage as a green vegetable.

Nowadays, its cultivation and consumption have reduced drastically. This may be attributed to changing food habits, increasing demand for land for fodder and wheat, and growing competition between old and new crops like French beans, turnip, and green peas. Keeping in view the nutritional quality of its grain and the early maturity and suitability of the crop for marginal and degraded lands, there is a scope for reviving its cultivation in Ladakh.

(vii) *Dairy industry*: Leh district has high milk production in rural areas but there is a lack of government-sponsored programmes, an absence of cattle feed, high-yielding varieties of cows, and poor water supply that hamper the development of the dairy industry. The cultivable area in livestock-dominant villages does not match the needs of livestock, especially cows. There is a large number of areas, which are uncultivated due to a lack of assured water supply and lack of fertile soils. If proper humus and inorganic salts are maintained, these soils can offer better agricultural production, especially fodder crops. Alluvial fans of sandy loam can boost agriculture by increasing the water-retaining capacity of the soil. The presence of glaciers on the elevated peaks can also be used for irrigation to increase the production of crops and vegetation. Attention is also needed to identify new cold-tolerant fodder crops for augmenting the fodder availability in the region.

(viii) *Utilisation of solar energy*: Energy is key to economic development. Cold arid regions of Leh-Ladakh are deprived of electric energy despite having a large availability of solar irradiation to the tune of 7–7.5 kWh m^{-2} day^{-1}. Ladakh region is capable of generating up to 1,200 watts of solar power. Solar energy holds significant potential for ensuring the future energy security of Leh-Ladakh, whether through solar thermal or solar photovoltaic technologies.

Various solar-based systems can be implemented for harnessing solar energy at the domestic level, in agricultural fields, and even on an industrial scale.

For irrigation, solar PV pumping systems can efficiently lift water against gravity, making them ideal for the region's undulating terrain. As investigated in the PM-KUSUM scheme, solar farming can enable simultaneous crop production and photovoltaic electricity generation on a single plot of land. Solar air heating can create a comfortable indoor environment during the winter months, while solar greenhouses offer a viable solution for growing vegetables in harsh climatic conditions. Rooftop PV systems can help meet the energy needs of office buildings, hospitals, guesthouses, and hotels. Furthermore, solar lanterns have the potential to replace traditional kerosene lighting systems.

Other promising solar technologies for Leh-Ladakh include inclined solar dryers, solar PV winnower cum dryers, and solar water heaters. The adoption of these solar-based technologies could significantly enhance the renewable energy landscape in the region (Santra, 2016). Ministry of New and Renewable Energy aims to install facilities that will generate 7,500 MW of solar energy by 2023 in Leh-Ladakh. The Government says that the solar potential of Ladakh is as high as 23,000 MW (23 GW).

(ix) *Hydro Energy*: Hydro energy is a significant and promising resource for Leh-Ladakh, given the region's abundant water sources and rugged terrain. Hydropower has become an increasingly popular as well as a clean source for generating electricity. The numerous rivers and streams, primarily fed by glacial melt, provide ideal conditions for hydroelectric power generation. Indus River offers exciting prospects for hydropower projects in Ladakh due to the presence of several rivers and run-off streams. There is vast scope for hydro-power projects in Ladakh, the most recent being by the National Hydro-Power Corporation (NHPC) and for setting-up new "Small Hydro" projects by providing incentives and subsidies.

(x) *Food Processing and Horticulture*: Food processing and horticulture are vital sectors in Leh-Ladakh, contributing to the region's economy, food security, and the livelihoods of local communities. Ladakh produces approximately 15,789 metric tons (MTs) of apricots out of which, only four to five varieties meet the export standards. Sea buckthorn has emerged as one of the superfoods of Ladakh; its fruit has unique potential for processing into food, medicines and nutraceuticals. Food processing plays a crucial role in adding value to raw agricultural products. Local processing units produce jams, juices, dried fruits, and snacks, which can enhance shelf life and marketability.

(xi) *Potential of Apricot-based agri-enterprises:* Apricots hold significant potential as a key agri-enterprise in Ladakh, given the region's unique climatic conditions and the fruit's nutritional value. It is a very versatile fruit and it gives a range of opportunities for value addition in the food processing industry in the form of products like apricot. Promoting it through the Farmer Producer Organization (FPO) may help promote value addition and marketing of apricots in the form of jams, dried apricots, and apricot kernel oil. These products have longer shelf lives and can reach broader markets. These are utilised in several other products as lip balms, facial creams, essential oils, and protein isolates, and these have good quality attributes as well as great consumer applicability.

15.11 Conclusion

The agricultural economy of Leh District in Ladakh is intricately tied to its unique land resources and environmental conditions. This case study highlights the region's rich potential for agricultural development, driven by its diverse landscapes, traditional farming practices, and the resilience of its communities. The analysis of land resources and the agricultural economy in Ladakh shows the ongoing shift from a production-oriented to a supply-oriented food system in the context of socioeconomic change. The current transitions affect the three interrelated food system components: availability, access to, and utilisation of food. Changes in agricultural land use have led to a remarkable decline in regional food production. The trends indicate that the agriculture economy of Leh in particular and Ladakh in general will continue to face challenges of climate change, shrinking land resources and increasing demands of the economically viable agricultural economy. Future efforts should focus on strengthening infrastructure, improving water management, and fostering market access to fully realise the potential of its land resources. With continued support from governmental and non-governmental organisations, Leh can pave the way for a robust and resilient agricultural economy that benefits both its people and the environment.

References

Akand, A. H., Singh, B. P., & Chander, M. (2017). Contribution of livestock in livelihood: Ladakh region. *The Indian Journal of Animal Sciences, 87*(5), 649–652. Accessed Oct 15 2022.

Bruinsma, J. (2009). The resource outlook to 2050: How much do land, water and crop yield need to increase by 2050? In *Expert Meeting on how to Feed the World* (Vol. 2050, pp. 1–33), Rome, Italy.

Chan, C., Sipes, B., Ayman, A., Zhang, X., LaPorte, P., Fernandes, F., Pradhan, A., Chan-Dentoni, J., & Roul, P. (2017). Efficiency of conservation agriculture production systems for smallholders in rain-fed uplands of India: A transformative approach to food security. *Land, 6*(3), 58. https://doi.org/10.3390/land6030058

Chaurasia, O. P., & Singh, B. (1996). *Cold Desert Plants of Leh* (pp. 1–4). Field Research Laboratory, DRDO.

Dame, J., & Nüsser, M. (2011). Food security in high mountain regions: Agricultural production and the impact of food subsidies in Ladakh, Northern India. *Food Security, 3*, 179–194. https://doi.org/10.1007/s12571-011-0127-2

Darku, A., & Malla, S. (2010). Agricultural productivity growth in Canada: Concepts and evidences. Canadian Agricultural Innovation Research Network (CAIRN) Policy Briefs, Saskatoon, SK. 2010: 21. http://www.ag-innovation.usask.ca/. Accessed September 10, 2012.

Darku, A. B., Malla, S., & Tran, K. C. (2016). Agricultural Productivity: What Does The Evidence Say? https://www.researchgate.net/publication/301550946_Agricultural_Productivity_What_Does_The_Evidence_Say. Accessed Oct 06 2022.

Ehlers, E., & Kreutzmann, H. (2000). High mountain ecology and economy: Potential and constraints. In: E. Ehlers & H. Kreutzmann (Eds.), *High Mountain Pastoralism in Northern Pakistan* (Vol. 132, Erdkundliches Wissen) (pp. 9–36). Stuttgart: Steiner Verlag.

Gaur, M. K., & Goyal, R. K. (2022). Achieving natural resource resilience through people's science—A case study of hot arid zone of India. *International Journal of Agriculture Sciences, 14*(4), 11221–11227.

Gaur, M. K., Goyal, R. K., Saha, D., Singh, N., Shekhar, S., Ajai, & Chauhan, J. S. (2021). The estimation of snow cover distribution using satellite data in the cold arid Leh Region of Indian Himalaya. *Polish Journal of Environmental Studies, 31*(1), 63–73. https://doi.org/10.15244/pjoes/135606

Gaur, M. K., & Goyal, R. K. (2021). Optimizing productivity through watershed management and improving livelihoods in Rainfed areas. In *Perspectives in Natural Resources Management: Watershed based Approach* (Editors R.K. Goyal & Mahesh K. Gaur). Central West Publishing, Australia. pp. 207–234.

Gaur, M. K., & Squires, V. R. (2020). Changes in agricultural land use and food security: Challenges and opportunities for western drylands of India. In V. R. Squires and M. K. Gaur (Eds.), Food Security and Land Use Change under Conditions of climate variability: A multidimensional perspective. Springer, pp. 257–280.

Gaur, M. K., Goyal, R. K., Raghuvanshi, M. S., Bhatt, R. K., Pandian, M., Mishra, A., & Sheikh, S. J. (2019). Geospatially extracting snow and ice cover distribution in the cold arid zone of India. Internat. *Journal System Assurance Engineering and Management, 11*(1), 84–99. https://doi.org/10.1007/s13198-019-00883-wGaur, M. K., & V. R. Squires (2018). Climate Variability, Land Use change and Impact on Livelihoods in the Arid Lands. Springer. ISBN: 978-3-319-56680-1.

Gaur, M. K., & Mohran, P. C. (2014). *Environment, people and development: Experiences from desert ecosystems.* New India Publishing Agency, New Delhi. ISBN: 978-93-81450-33-8.

Goodall, S. K. (2004). Rural-to-urban migration and urbanisation in Leh, Ladakh. *Mountain Research and Development, 24*(3), 220–227. https://doi.org/10.1659/0276-4741

Gupta, R. D., & Arora, S. (2017). Characteristics of the soils of Ladakh region of Jammu. https://doi.org/10.5958/2455-7145.2017.00037.6. Accessed Sep 29, 2022.

Hinz, R., Sulser, T. B., Huefner, R., Mason D'Croz, D., Dunston, S., Nautiyal, S., Ringler, C., Schuengel, J., Tikhile, P., Wimmer, F., & Schaldach, R. (2020). Agricultural development and land use change in India: A scenario analysis of trade-offs between UN Sustainable Development Goals (SDGs). *Earth's Future, 8*, e2019EF001287. https://doi.org/10.1029/2019EF001287.

Kachroo, P., Sapru, B. L., & Dhar, U. (1977). Flora of Ladakh: An ecological and taxonomical appraisal. *Bishen Singh Mahendra Pal Singh.*

Kreutzmann, H. (ed.) (2006). Karakoram in transition. Culture, Development, and Ecology in the Hunza Valley. Oxford

Kumar, P., Pareek, K., Raghuvanshi, M. S., Roy, M. M., & Tewari, J. C. (2016). Fodder production system—a major challenge in arid region of Ladakh, India. *MOJ Ecology and Environmental Science. 1*(1), 1–7

Labbal, V. (2001). Travail de la terre, travail de la pierre. Des modes de mise en valeur des milieux arides par les sociétés himalayennes. L'exemple du Ladakh. Unpublished PhD thesis, Université Aix-Marseille 1 - Université de Provence, Marseille.

LAHDC-Leh (Ladakh Autonomous Hill Development Council – Leh) (2011). District Profile. Online, available at http://leh.gov.in/pages/leh.pdf

LAHDC-Leh (2012). Statistical Handbook. Online, available at http://leh.gov.in/pages/handbook.pdf

LAHDC-Leh DCO (Deputy Commissioner Office). (2011). Disaster Management, Online, Available at http://www.leh.nic.in/Disester%20Management%20Leh.pdf

Lillesand, T. M., & Kiefer, R. W. (2000). *Remote sensing and image interpretation* (p. 95p). John Wiley and Sons. Inc.

Mandal, R. B. (1990). *Land utilization: Theory and practice.* Concept Publishing Company.

Miehe, G., Miehe, S., Kaiser, K., Reudenbach, C., Behrendes, L., La Duo, & Schlütz, F. (2009). How old is pastoralism in Tibet? An ecological approach to the making of a Tibetan landscape, Palaeogeography, Palaeoclimatology, Palaeoecology, *276*(1), 130-147

Mora, O., Le Mouël, C., de Lattre-Gasquet, M., Donnars, C., Dumas, P., Réchauchère, O., Brunelle, T., Manceron, S., Marajo-Petizon, E., Moreau, C., Barzman, M., Forslund, A., & Marty, P. (2020). Exploring the future of land use and food security: A new set of global scenarios. *PLoS ONE, 15*(7), Article e0235597. https://doi.org/10.1371/journal.pone.0235597

Nüsser, M., Dame, J., Kraus, B., Baghel, R., & Schmidt, S. (2018). Socio-hydrology of "artificial glaciers" in Ladakh, India: Assessing adaptive strategies in a changing cryosphere. *Regional Environmental Change, 19*, 1327–1337. https://doi.org/10.1007/s10113-018-1372-0

Nüsser, M., & Gerwin, M. (2008). Diversity, complexity and dynamics: Land use patterns in the central Himalayas of Kumaon, Northern India. In J. Löffler, & J. Stadelbauer (Eds.), *Diversity in Mountain Systems. Sankt Augustin* (= *Colloquium Geographicum* 31), 107–119.

Osmaston, H. (1994). The farming system. In J. Crook & H. Osmaston (Eds.), *Himalayan Buddhist Villages.* Environment, Resources, Society and Religious Life in Zangskar, Ladakh (pp. 139–198). Bristol: University of Bristol.

Osmaston, H. (1989). Agriculture in Ladakh: Building a coherent policy. In *The Future of Agriculture in Ladakh.* Proceedings from an International Workshop organized by the Ladakh Ecological Development Project 22, 23, Ladakh, pp. 4–5.

Santra, P. (2016). Scope of solar energy in cold arid region of India at Leh Ladakh. *Annals of Arid Zone, 54*(3–4), 109–117.

Saxena, S., Bhardwaj, A. K., Kumar, V., Patel, M. K., Kumar, R., & Chaurasia, O. P. (2018). Sustainable utilisation of medicinal plants of Ladakh and Lahaul-Spiti of trans-Himalaya. *Defence Life Science Journal, 3*(2), 120–125. https://doi.org/10.14429/dlsj.3.12566

Sharma, J. P., & Mir, A. A. (2000). *Dynamics of cold arid agriculture.* Kalyani publisher.

Singh, H. (1992). Environmental constraints on agriculture in a cold desert. In N. Mohammad (Ed.), *New Dimensions in Agricultural Geography: The Ecology of Agricultural Systems*, 2, 79–92.

Squires, V. R., & Gaur, M. K. (2020). Food security and land use change under conditions of climatic variability: A multidimensional perspective, Springer Nature Switzerland AG. ISBN: 978-3-030-36761-9.

Squires, V. R., & Gaur, M. K. (2020). Agricultural productivity and food security. In V. R. Squires, & M. K. Gaur (Eds.), Food Security and Land Use Change under Conditions of climate variability: A multidimensional perspective. Springer, pp 33–48.

Squires, V. R., Gaur, M. K., & Feng, H. (2020). The critical role of smallholders in ensuring food security. In V. R. Squires, & M. K. Gaur (Eds.), *Food Security and Land Use Change under Conditions of climate variability: A multidimensional perspective.* Springer, pp. 79–106.

Stanzin, K. (2019). Studies on Some Medicinal Plants of Suru Valley of Ladakh Used in Sowa Rigpa System of Medicine. *International Journal of Current Microbiology and Applied Sciences, 8*(02), 257–269. Accessed Oct 17 2022.

Tiwari, P. C. (2000). Land use changes in Himalaya and their impact on the plains economy: Need for sustainable land use. *Land Use Policy, 17*, 101–111.

Websites

https://www.ipcc.ch/srccl/chapter/chapter-5/

Part III
Coastal Desert Ecosystem

Chapter 16
Management of Desert Ecosystems in Coastal Tamil Nadu

V. Geethalakshmi

16.1 Introduction

Globally, one-fifth of the landmass is occupied by desert regions. Each continent on the planet contains a desert, and each desert is characterised by its own ecosystem, varied as a hot desert, cold desert, coastal desert, etc. In India, coastal deserts are restricted to the states of Tamil Nadu and Gujarat each having diverse peculiarities in terms of their ecosystems. Although the state of Tamil Nadu gets precipitation from southwest as well as northeast monsoons, some places in the Tuticorin district and adjoining areas receive less precipitation, hence being dry and arid, resulting in a desert. Desert regions in Tamil Nadu are characterised by low and inconsistent precipitation and high evaporation, and the invariable occurrence of drought stress is characteristic of deserts owing to moisture scarcity. These areas are sharply demarcated by vegetation comprising plants, which are either short-lived annuals or xerophytes. The sparse vegetation is a consequence of limited water availability and unfavourable climatic conditions. Plants develop adaptation mechanisms in order to grow in this particular zone (Govt. of Tamil Nadu, 2017). The soils are deficient in nutrients and poor in water holding capacity.

16.2 Desert Ecosystem in the Coastal Areas of Tamil Nadu

Tamil Nadu is geographically located along the southeastern coast of India with a coastal line stretching in the order of 1,076 kms (Fig. 16.1). This coastline accounts for over 13% of the total coastline of India. Tamil Nadu has the second largest coastline when compared to other coastal states situated on the Indian subcontinent.

V. Geethalakshmi (✉)
Tamil Nadu Agricultural University, Coimbatore, Tamil Nadu, India
e-mail: geetha@tnau.ac.in

M. S. Rathore et al. (eds.), *Desert Ecosystems*,
https://doi.org/10.1007/978-3-032-03237-9_16

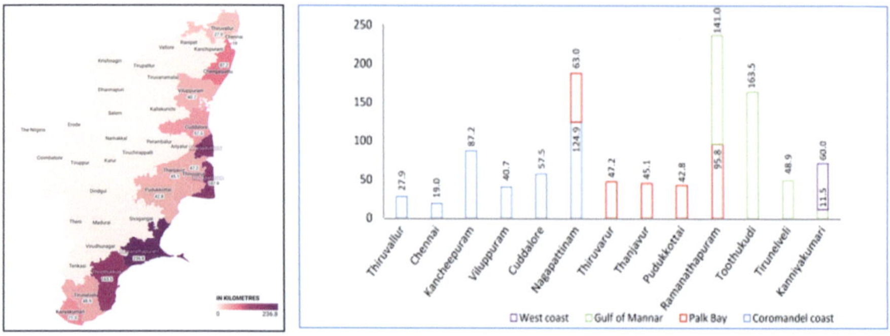

Fig. 16.1 District-wise coast length of Tamil Nadu is also available at www.iomenvis.nic.in (*Source* DoE, 2006)

The coastline of Tamil Nadu spans from the northernmost Thiruvallur district to the southernmost Kanniyakumari district, along the Bay of Bengal and Indian Ocean. Tamil Nadu comprises 13 coastal districts in total, with Chennai having the shortest coastline length of 19 km and Ramanathapuram having the longest coastline length of 237 km. The Tamil Nadu coastline consists of the 365 km long Gulf of Mannar, the Coromandel coast spanning from Chennai to Point Calimere (357 km long), Palk Bay (294 km) and the West coast extending from Kanniyakumari to Neerody (60 km) (State of Environment Report for Tamil Nadu, 2017).

There has been substantial pressure due to ill-planned developmental activities along the east coast of India, particularly in Tamil Nadu, which have already posed challenges to the seashore and its ecology. While ecosystems such as coral reefs, mangroves and seagrass beds are increasingly receiving much recognition as crucial coastal ecosystems and shoreline stabilisers, sandy beaches and sand dunes receive insufficient attention as important coastal ecosystems and their contribution to stabilising the coastline are largely disregarded. Therefore, it is the right time to formulate appropriate strategies for the improvement of coastal regions through an integrated development framework, with the aim of further enhancing the economic and livelihood conditions of Tamil Nadu's desert coastal ecosystem.

16.2.1 Sand Dune in Tamil Nadu

Coastal sand dune ecosystems exist in the Tuticorin District of Tamil Nadu. These ecosystems resemble many characteristics of desert dunes, such as shifting sand, extreme exposure to sun and air, and low moisture retention that impede plant growth. A number of physical and chemical parameters that are characteristic of sand dune ecosystems restrict the distribution of flora and fauna. Such parameters include soil salinity, poor nutrient content, salt spray, lower organic matter and changes in pH. Plants have established mechanisms and traits to cope with harsh environments and enable them to survive successfully. Based on the characteristics of this ecosystem, it is possible to introduce suitable interventions for further development of Sand Dune Ecosystem in Tamil Nadu.

16.2.2 Desert Sandy Soil in Coastal Areas in Tamil Nadu

Desert soil is predominantly sandy soil (90–95%) and is found in regions with low rainfall. The low nitrogen content and organic matter, along with the very high level of calcium carbonate and phosphate, make desert soil infertile. The calcium content in the lower layer is ten times higher than that of the topsoil. Applying fertiliser, proper irrigation and nitrogen availability in the form of nitrates, in addition to the phosphates already present, enable the cultivation of cereals including wheat, pulses, maize, barley, rapeseed and millets.

Coast Desert Sandy Soils present in Ramanathapuram coastal areas are of salinity. This soil is prone to wind erosion and supports only a low-density of population. In India, desert soil covers around 4% of the total land area. This desert soil is found in the regions of Rajasthan, the Rann of Kutch in Gujarat, adjacent regions of Punjab and Haryana situated between the Indus and the Aravallis, and coastal areas of Orissa, Kerala and Tamil Nadu (State Environment Report, 2017).

16.2.3 Red Desert of Tamil Nadu (Theri Kaadu)

In Tamil Nadu, the coastal red sandy mounds locally called "Theri Kaadu" are found in the Tuticorin and Tirunelveli districts spanning an area of 20,171 hectares (Jawahar, 1996) with a substantial red desert area of 16,978 ha in Tuticorin district (Fig. 16.2). The red sand dunes have been formed owing to the deposition of red sand brought from the Nanguneri region in Tirunelveli district by southwest monsoon winds over many years. The sandy texture of the soils and undesired topography are the main constraints in these regions. It was quantified that 77.3% of total *therilands* lack pedogenic development (Jawahar, 1996).

- *In Tamil*, the red dunes are referred to as *theri.*
- Red dunes comprise sediments dated back to the Quaternary era and consist of marine deposits.
- They have poor nutrient retention and water-holding capacity. The sand dunes are vulnerable to aerodynamic lift.
- The petrographic study and X-ray diffraction analysis of the red sand dunes manifest the presence of heavy and light minerals.
- The iron-based heavy minerals such as magnetite, ilmenite, hypersthene, garnet and rutile that exist in the soil were leached by surface water and subsequently oxidised due to the favourable semi-arid climatic conditions.

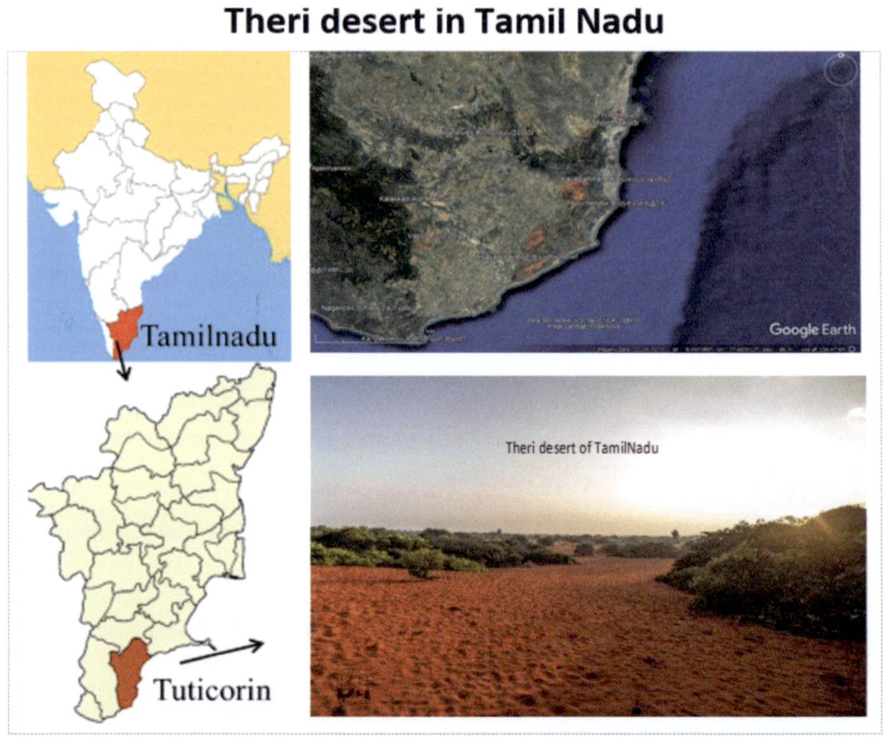

Fig. 16.2 Map of Tamil Nadu showing the red dunes distribution (theri zone) (Udayanapillai Alagaiah Venu and Perumal Velmayil, 2021; Google Earth)

16.2.4 Factors that Led to Sand Dune Formation in Theri Desert

- There appears to be gentle, uneven terrain, and the lithology of the area indicates that it might have an ancient coast in the olden days.
- The present-day study revealed that it might have originated from the localised beach sand confinement subsequent to sea regression.
- An alternative perspective is that these are geological creations that emerged within a few hundred years.
- The red sand is transported from the surface of a wide belt of red loam situated on the plains of the Nanguneri region (about 57 kilometres from this area in the Tirunelveli district) through southwest monsoon winds prevailing from May to September.
- Deforestation and barren lands (no vegetation) in the Aralvaimozhi gap and the Nanguneri plains might be the foremost reasons for wind erosion (Fig. 16.3).
- The processes of erosion, sediment transport and sediment deposit that occur due to wind activity at or near the Earth's surface are termed as aeolian processes.

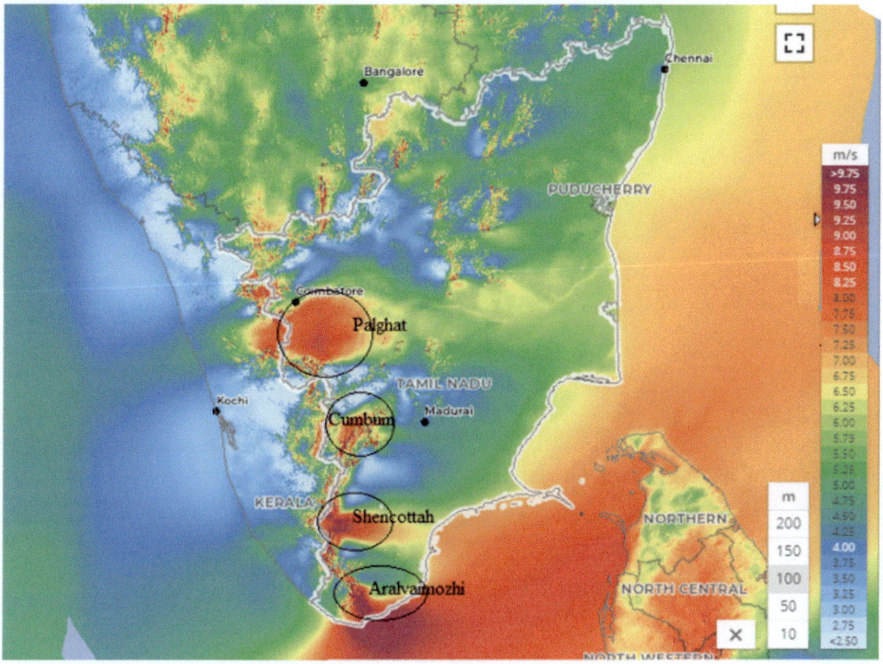

Fig. 16.3 Potential locations of wind passes in Tamil Nadu (Global Wind Atlas-January 2021)

- They result in the continual redistribution of sand (Foote, 1883).
- When the high-velocity wind propels sand above the surface, the sand encounters an obstruction such as a fence post, shrub, bush and any other vegetation. As a result, the sand is accumulated on the leeward side and over time, a dune is formed (Soil Survey Staff, 1989).
- The main land form features include the Thamirabarani, Vaipar, Nambiyar and Karumeniyar river basins, *theri* sand dunes, coasts and plains. River Thamirabarani is perennial. Thamirabarani and Karumeniyar rivers have wide flood plains. The rivers flow west to the east confluence at the Bay of Bengal. The rivers change their paths in the geological past, very often. These changes superimposed by seasonal floods lead to the deposition of sand pockets and lead to the formation of present *therilands* (Jawahar, 1996).

16.3 Characterisation of Climate

The coastal *theries distributed across the* Thiruchendur, Sathankulam and Vilathikulam taluks of Tuticorin district experience an average temperature of 26, 30.7, 29.7 and 27.5 °C during winter, summer, southwest monsoon and northeast monsoon respectively (Table 16.1). They are exposed to a short and mild winter

Table 16.1 Seasonal distribution of temperature and rainfall of the coastal *theries*

Season	Thiruchendur Taluk		Sathankulam Taluk		Vilathikulam Taluk	
	Temperature °C	Rainfall mm	Temperature °C	Rainfall mm	Temperature °C	Rainfall mm
Winter	25.30	75.53	26.65	68.00	26.05	30.14
Summer	30.54	55.34	30.95	90.80	30.68	133.63
Southwest monsoon	28.10	75.31	31.58	89.00	29.55	78.66
Northeast monsoon	26.73	544.51	28.98	454.50	27.00	367.27

which occurs during January and February and summer from May to September. They receive less than 90 mm of rainfall in the southwest monsoon and more than 70% of rainfall during the northeast monsoon season between October and December.

16.4 Soil Formation and Characteristics

The *theri* soils of Tamil Nadu were defined as soils formed in *isomegathermic* and *ustic* regimes from geogenic deposits under a climate with a mean annual temperature of 28 to 32 °C and 700 mm of rainfall (Jawahar et al., 1999b). The soil colour is 2.5 YR with a variation from 5YR to 10 R. Texture ranges from single grain to sandy loam. The structure is fine weak sub-angular blocky to very coarse strong angular blocky. Soils are moderately acidic and salinity is always below 0.1 dS m^{-1}. Organic carbon content is always less than 0.11 g kg^{-1}. Hematite and goethite are dominant minerals with a little hillite, imparting red colour to the soils.

Theri soils are extremely deep, non-calcareous, moderately acidic, excessively drained and non-saline. The composition of mechanical components and a textural class of the most common *theri* soil occurring in 62.5% of *theri* belt is presented in Table 16.2. The soil texture on the surface is sandy and at the subsurface it ranges from loamy sand to clay loam through sandy clay loam depending on the degree of soil development. Clay is a minor component of their environment.

The ion exchange properties of the most common *theri* soils are presented in Table 16.2 (Jawahar, 1996). Their soils are moderately acidic, and non-saline with low ESP values.

Table 16.3 presents the nutrient status of *theri* soils. The carbon and clay contents are very low in all the soils, leading to poor aggregation or single-grain structure. However, in the subsoil of some soils, the presence of a weak structure could be noticed which is attributed to the increased clay and not to organic carbon content as the aggregates have loose and friable constituency.

The tertiary rocks found in fragmented patches in Sathankulam and Thiruchendur taluks of the Tuticorin district are marine, sandstone and limestone (Soil Survey Staff,

Table 16.2 Characterisation of *theri* soils (Jawahar, 1996)

Mechanical components of the soils

Horizon	Depth (m)	Mechanical composition (%)				Textural class
		Clay	Silt	Coarse sand	Fine sand	
A1	0–1.13	12.6	4.5	1.5	80.2	ls
A2	1.13–1.48	12.5	3.5	2.8	80.2	ls
A3	1.48–2.00 +	1.3	4.4	4.8	76.5	ls

Exchange properties of soils

Horizon	pH	EC (dS m^{-1})	CEC	Exchangeable bases cmol (p^+) kg^{-1}				*BSP*	ESP
				Ca^{2+}	Mg^{2+}	Na^+	K^+		
A1	5.00	0.03	2.4	1.4	0.1	0.1	0.1	68.0	3.4
A2	5.03	0.03	2.4	1.2	0.1	0.1	0.1	57.3	2.4
A3	5.14	0.03	2.5	1.5	0.1	0.1	0.1	70.1	3.5

Table 16.3 Nutrient status of the *theri* soils (Jawahar, 1996)

Horizon	Depth (m)	Organic carbon (g kg^{-1})	Available nutrients (kg ha^{+1})			C/N ratio
			N	*P*	*K*	
A1	0–1.13	0.73	49.28	3.58	100.8	10.5
A2	1.13–1.48	0.62	40.32	3.80	91.8	9.3
A3	1.48–2.00 +	0.18	42.56	2.58	69.4	2.5

1991). The recent sediments of the sandy coastal belt have calcareous grits, sandstones and shell limestones. Laterite materials are found with grits and conglomerates that occur in *theri* and seashore areas, in varying thicknesses up to 10 m. *Theri-lands* are classified into the sub-groups of Typic Ustisamments, Typic Ustropepts, Grossarenic Paleustalfs and Typic Rhodustalfs as per the USDA soil taxonomical system (Jawahar, 1996).

16.5 Physiography and Natural Vegetation

The aeolian physiography consists of stabilised and unstabilised landforms in the midst of dunes. The various physiographic positions of the landform are windward plain, stoss side, dune top, downward side or slip face, leeward bottom and leeward plains. The Udangudi, Sathankulam, Surangudi and Nangaimoli theries are sloping to the east. Kudiraimoli theri slopes to the northwest, Ovari theri to the southwest and Pudukkottai theri to the southeast in Tamil Nadu (Jawahar, 1996).

The natural flora of the coastal area includes *Borassus flabellifer, Acacia planifrons, Acacia Arabica, Azardirachta indica, Dodonea viscosa, Prosopis juliflora, Tamarindus indica, Anacardium occidentale, Acacia nilotica, Acacia leucopholea, Eucalyptus spp.*, and herbs including *Borreria hispida, Tephrosia purpurea, Scoparia dulce, Indigofera aspalathoides, Polycarpaea corymbosa, Mullungo disticha and Stylosanthes mucronata* as reported by Jawahar (1996) and Janakiraman et al. (1997). The soil has a subsurface runner *Cypernus arenarius* in the unstabilised sand dunes, trying to bind the soil particles against wind erosion.

16.6 Resources of the Coastal Desert of Tamil Nadu

16.6.1 Mineral-Rich

The analysis of the sand grain mineralogy indicates that the light minerals dominate over the heavy minerals. This is due to the heavy weathering of all the minerals, leaving behind a higher quartz content in the light mineral fraction. Industrial separation of limonite, zircon, rutile, monazite, sillimanite and garnet are commonly seen in the *theriland* areas.

16.6.2 Less Employment Opportunities

The *theri*-dwellers at present have very less employment opportunities, as there is no industry in the area except retting of coir, and mineral separation units from sand, and cashew- and palmyrah-related works.

16.6.3 Land Price

The cost of land is relatively cheaper attracting many investors from other states.

16.6.4 Source of Rare Herbs

The desert location has many wild species of rare traditional Indian herbs like *Borreria hispida, Tephrosia purpuria, Scoparia dulce, Indigofera aspalathoidis* and*Polycarpaea corymbosa.*

16.6.5 Lime Deposits

Lime deposits are common occurrences in *therilands*. Pannamparai near Sathankulam has surface deposits of high-quality lime. Lime is kilned on a large scale.

16.7 Leisure and Tourism

The *theri* locations are desert landscapes with dunes, inter-dune valleys and red sandy desert environments. Karkuvel Iyyanar temple in Therikudiyiruppu is centuries old. It is 9 kms from Thiruchendur. The temple is surrounded by active sand dunes on all three sides, except the east entrance. Hematite and goethite are minerals responsible for the red colour of the soils. The soil colour is bright during summer, due to the roasting of iron minerals at a higher temperature. The nearby Arunjunai Katha Iyyanar temple in Ammanpuram is also a noteworthy picnic spot. The temple is located at a water pond at the Kudiraimoli *theri* location. Thiruchendur, Ovari and Kulasekarapatnam are religious places endowed with popular temples on the coast of Thoothukudi district. Along the coast, Rameswaram temple is also located.

16.8 Coastal Desert Agriculture

Local farmers cultivate groundnut, coconut, cashew and palmyrah by employing concrete-, clay- and polythene-based cultivation methods. The irrigation is done from bore wells through surface temporary cement mortar built channels, by building higher temporary systems to deliver water to elevated areas or by clay-mixed irrigation. Shelter belts were constructed across the wind direction, with *Dodonea viscosa* (first row), cashew, *Acacia planifrons* (second row), causarina (third row), tamarind and jumbulana (fourth row) and again *Dodonea visoca* (fifth row) (Janakiraman et al., 1997).

16.8.1 Scope for Agricultural Enterprise

As palmyrah is the common vegetation of *therilands*, neera and palm sugar are local products. The palm sugar is extracted through a traditional process assuring better returns in organic markets. Improved palm sugar preparation technologies and the introduction of preservation technology for the neera would bring an appreciable response from consumers in the international arena. Cashew has tremendous export potential. Farmers prefer cashews for its suitability for soil. *Moringa* sp., the most profitable crop in *therilands,* is cultivated in areas with water potential. The drumstick vegetable is supplied even to distant markets from *theries*, due to the availability of transport infrastructure.

16.9 Coastal Desert Ecosystem—Challenges

The sustainability and adaptability of desert agriculture are terribly endangered by multiple factors that combined create a storm situation:

- Carbon content is very low in soils leading to poor soil aggregation.
- The physical properties of *theri* soil including bulk density, particle density, water holding property, pore space and volume expansion on wetting are very poor.
- The nutrient-holding property of *theri* soil is low due to low clay and low CEC of the soil. Soil pH is below 6.0 due to the absence of soil bases.
- Exploitation of groundwater and poor water management practices.
- Lack of understanding and insufficient recognition of the ecological importance of the *therilands* and coastal ecosystem.
- Frequent occurrence of droughts and hot summers are common occurrences of *therilands.*
- Wind erosion.
- High evaporation rates.
- Isolation from markets and energy sources.

All these factors merge to cause a bleak outlook and a gloomy future for the farmers who reside there.

16.10 Seawater Intrusion in Coastal Aquifers

Seawater intrusion in coastal aquifers is common in the coastal belt where irrigated agriculture is practiced with groundwater. The intrusion of seawater into groundwater reservoirs can greatly impair the water supply systems of local areas adjacent to coastlines. The exploitation of groundwater for irrigation at Radhapuram taluk resulted in a lowering of the water table and consequent drought and moisture stress on crops forced the farmers to deepen the bore wells. This lowers the water level below the sea level resulting in seawater intrusion into the aquifer and leading to the deterioration of the quality of groundwater (Ammal et al., 2000). Table 16.4 shows the cation and anion sequence of well water from four different zones of Radhapuram Taluk in Tirunelveli district coastal belt, during the study conducted at Agricultural College and Research Institute, Killikulam (Ammal et al., 2000).

The study revealed that among the four zones, the wells occurring <1 km and 1–2 km from the sea coast were affected by seawater intrusion. The decrease in the concentration of HCO_3^- together with an increase in the concentration of Cl^- and

Table 16.4 Cation and anion sequence of well water in different zones of Tirunelveli district coastal belt

Zone	Village	Distance from sea (km)	Cation sequence	Anion sequence	Remarks
Zone I (<1 km)	Kuttam	0.95	$Na^+ > Mg^{2+} > Ca^{2+}$	$Cl^- > SO_4^{2-} > HCO_3^-$	Seawater intruded
	Uvari	0.60			
	Chettikulam	0.80			
Zone II (1–2 km)	Vijayapathi	1.50	$Na^+ > Mg^{2+} > Ca^{2+}$	$Cl^- > SO_4^{2-} > HCO_3^-$	Seawater intruded
	Erukkanthurai	2.00			
	Kudankulam	2.50			
Zone III (2–3 km)	Thiruvambalapuram	2.90	$Na^+ > Ca^{2+} > Mg^{2+}$	$Cl^- > HCO_3^- SO_4^{2-}$	Very close towards seawater intrusion
	K. Puthur	2.20			
Zone IV (>3 km)	Radhapuram	11.1	$Na^+ > Ca^{2+} > Mg^{2+}$	$Cl^- > HCO_3^- SO_4^{2-}$	Very close towards seawater intrusion
	Parameshwarapuram	8.4			

SO_4^{2-} among the anions and the reduction in the concentration of Ca^{2+} together with an increase in the concentration of Na^+ and Mg^{2+} among the cations in the above zones is conclusive proof of seawater intrusion. The wells occurring at 2–3 and >3 km distances from the sea coast were very close to the seawater intrusion, as analysed with the sequence described by Krishna (1988). Seawater intrusion can be controlled by reducing the exploitation of groundwater pumping, and the dependency on groundwater for irrigation in coastal agriculture, through rainwater collection and recharge of aquifers (Ammal et al., 2000).

16.11 Strategies for Development of Coastal Desert Ecosystem

Research and development on process change in coastal desert ecosystems and documentation of various plant and animal systems, their ecology and threats are very limited. It is necessary to conduct effective research in order to protect this ecologically delicate and socially significant ecosystem.

a. **Saline Farming**

Saline farming is one of the most efficacious interventions to improve the desert ecosystems affected by salinity. Field experiments carried out at the Coastal Saline Research Centre, Tamil Nadu Agricultural University in Ramanathapuram district,

Tamil Nadu, demonstrated that the application of organic waste materials like pressmud @ 12.5 t ha^{-1} has a positive influence on soil quality. This is attributed to the considerable reduction in the electrical conductivity of coastal saline soils and the enhanced availability of nutrients and soil organic carbon. Applying the lime and alkaline flyash in an appropriate ratio to the coastal acid sulphate soils can ameliorate the soil to a certain extent.

Amending the coastal sandy soils through the application of 100–120 mg kg^{-1} polyacrylamide can enhance the aggregation of soil, subsequently increase the water-retention capacity of the coastal sandy soil. This plays a crucial role in highly penetrable coastal sandy soils in the summer months.

b. **Integrated Farming System for Saline Coastal Areas**

To optimise land resource utilisation and enhance the standard of living of local inhabitants in coastal areas, an Integrated Farming system (IFS) approach could be practiced, focusing on paddy-cum-fish culture, without affecting soil productivity. This may facilitate the effective utilisation of available resources, minimise risks, accelerate additional income for coastal farmers facing economic hardships and struggling for survival and also uplift their socio-economic status in the coastal regions.

c. **Farming Practices in Coastal Desert Land**

- Rain water harvesting.
- Summer ploughing.
- Growing green manure crops (Fig. 16.4).
- Bio-fertiliser/VAM application.
- Pre-monsoon sowing.
- Micro catchment/percolation ponds/check dams.
- Mixed farming/crop rotation/intercrops.
- Raising heat and drought-tolerant crops/varieties.
- Seed hardening/priming.
- Rainwater harvesting (farm ponds) (Fig. 16.4).
- Reducing evaporation losses from soil—soil mulch or dust mulch, stubble and straw mulch, coir mulching (Fig. 16.4), polythene mulching, etc.
- Land configuration: Ridges and furrows/ploughing across slopes/broad bed furrows/contour farming/contour stone wall/random tied—ridging/basin listing compartmental bunding (Fig. 16.5).
- Drip irrigation/Hydrogel.
- Integrated Nutrient Management (INM)/Integrated Pest Management (IPM).
- Nutrient solution spray (Urea or DAP spray—2% solution) is recommended in the event of revival of rain and release of moisture stress.
- Nanoceria application—Cerium oxide nano-particles (scavenges the reactive oxygen species)—reduces the synthesis of drought stress hormones.
- Reducing transpiration losses—anti-transpirants (Stomatal Closing, Film forming, reflectance type).

Fig. 16.4 Rainwater harvesting, mulching and growing green manure crops in coastal desert land

- Stomatal closing type—Fungicides such as phenyl mercuric acetate (PMA) and herbicides such as atrazine in low concentrations act as anti-transpirants by triggering stomatal closing.
- Film forming type—plastic and waxy materials create a thin layer on the leaf surface that acts as a physical obstruction and prevents water loss, e.g. Mobileaf, hexadeconol and silicone wax.
- Reflectant type—by reflecting radiation, they diminish leaf temperatures and vapour pressure gradient from the leaf to the atmosphere and thereby reduce transpiration—5% kaolin spray is one example.
- In red sand dune (Theri desert), the application of organic amendments such as tank silt, Farm Yard Manure (FYM) and composted coir pith (CCP) can improve the physico-chemical properties of soil (Annadurai, 2011; Sundaram & Annadurai, 2018).

d. **Ecofriendly Interventions in Coastal Saline Areas**

Agro-forestry presents a highly effective remedy to saline lands. It is feasible to establish bio-energy as well as bio-fuel plantations on moderately degraded as well

Fig. 16.5 Land configuration in coastal desert land

as heavily degraded saline soil. This necessitates comprehensive research deploying multidisciplinary teams and addressing core and fundamental researchable issues (Mandal et al., 2021). Rice husk biochar could be utilised as an alternative to liming materials to enhance the quality of acid sulphate soils. It has been extensively documented that the application of rice husk biochar increases the pH of acid sulphate soil.

e. **Sand Dune Stabilisation Techniques**

Dust and sand storms have a significant impact on both natural and human systems. Sand and dust cyclones significantly affect both natural and human systems. The problem can be addressed by application of sand dune stabilisation techniques; it will help in the reduction of sand and dust storms. Sand dune stabilisation aims to prevent and mitigate the frequency of dust and sand storms by employing various techniques (Mainguet & Dumay, 2010). Mechanical methods such as building palisades hinder the sand movement and reducing the deposition of sand on infrastructure.

f. **Palmyrah and Coconut Plantation** can be promoted in the Sand dunes and Therilands of coastal desert agro-ecosystem for protecting the ecosystem and improving the livelihood status of the people living in and around desert agro-ecosystem.

g. **Ley Farming** is another method that involves the rotation of cereals with legume or non-legume forage. This system enhances soil quality besides providing fodder. Pulses can also be cultivated in the coastal desert ecosystem after applying soil amelioration management techniques for improving soil health. Promotion of forage crops will facilitate the development of animal enterprises in the desert ecosystem.

h. **Shelter Belt/Windbreak Plantation**

Across the wind plantation of a 13 m-wide tree belt, interspersed with a 60-m-wide grass belt, it provides beneficial results in the coastal agro-ecosystem (Venkateswarlu, 1993).

i. **Innovative Technologies** could be used in desert environments such as hydroponics, protected cultivation with greenhouse systems, vertical farming (Fig. 16.6), drip irrigation systems and net houses to enhance the yield of crops besides reducing water and energy consumption.

- By incorporating these innovative techniques with indigenous wisdom, and involving stakeholders from the public and private sector, value chain and producers themselves, the impact is higher than any single intervention.

Fig. 16.6 Innovative farming technologies in coastal desert land

- The advances in renewable energy (especially solar), desalination technology, water conservation technology and advancements in alternative feed industries facilitate food production in the barren soil in severe environmental conditions.
- The expanding desert farming, which includes irrigated forages, rangeland rehabilitation, horticulture production, date palms, protected agriculture, livestock production and fish production, is impelled by the keenness of several governments states and countries to establish underpin policies that motivate desert farming in order to lessen their dependency on food imports and to promote value-added products for export like organic produce.
- When there is a lack of rain, desert agriculture and livestock husbandry can only be possible if water is extracted from aquifers for irrigation, making it an indispensable input of desert agriculture and certainly it poses a significant challenge.
- In certain regions, the sole accessible water source is either non-renewable or excessively salty to be used directly on crops.
- The well-structured and sustainable groundwater management should be included in the development of desert agriculture.

j. **Desert Farming: Three Pillars**

- Public sector (and funding) for the development of research components.
- Private sector (public-private partnerships) for designing appropriate technologies and profit-driven scaling up.
- Critical knowledge of desert communities to ensure that modern packages are designed for specific settings.

k. **Integrated Desert Family Farming System (IDFFS)**

IDFFS is crucial when these pillars are enabled to tackle accelerated and intensified climate change.

l. **Enhancing the Livelihood and Economic Resilience of Fragile Communities and Family Farmers Through**

- Increase overall productivity, de-risks vulnerable systems and strengthens value chains.
- Recognises the unexploited potential of youth and women farmers by creating job opportunities.

m. **Creates Flexible, Adaptable and Sustainable Land and Water Resource Management Systems That Include Growing Salinity Concerns**

Owing to the current advancements in research and technology, saline water can be utilised nowadays for aquaculture and desalinated water for crops, specifically cacti, hardy rangeland species and trees such as date palms.

n. **Promotes Intercropping**

It advocates for intercropping as a means to establish windbreaks and shade, thereby improving microclimate conditions conducive to crop growth.

o. **Increasing the heat and drought tolerance of crops (climate-smart crops)—integrates water-conserving strategies with heat-resistant crops, including barley and sorghum, to substantially bolster the resilience and effectiveness of these integrated systems.**
p. **Integrated Farming System (IFS)**

- Conserve and enhance the soil with organic matter and microorganisms
- Increase water retention capacity
- Bio-symbiotic enhancement technologies.

q. **Substitutes solar energy for diesel as an energy source for the abstraction of ground-water and desalination processes.**
r. **Increased accessibility to agriculture and improving market accessibility through initiatives such as value-added livestock products or by developing effective technologies for alternative feed resources from food, fruit and vegetables to generate income.**
s. **Conducting capacity-building workshops.**

16.12 Turn Deserts into Resilient Farmland

- To ensure its resilience and sustainability and optimise socio-economic impacts, desert farming needs to be crafted and maintained as a long-term and multifaceted, complex integrated agricultural system.
- Numerous elements interact with each other in this system, and facilitation mechanisms among these components build resilience and efficacy.
- These can be straightforward, like afforestation to reduce wind-induced soil erosion, and crop evapo-transpiration or they can be more intricate like instituting the concept of circular economies to facilitate the recycling of nutrients among fish, crops, livestock, and soil.
- Integrate diverse technologies and engage in farm and community-level initiatives to create additional synergies such as energy and water conservation.
- One potential approach is to integrate the goals of creation of job, improved nutrition and resilient livelihoods with better social policies and effective value chains that encourage, promote and facilitate the inclusion of displaced communities, women and youth.
- **Promotion of FPOs in the Desert Ecosystem:** Formation of Farmers Producers Organisation (FPO) can be one of the potential interventions to promote coastal desert areas in Tamil Nadu. The intervention of the FPO will encourage agripreneurship, guarantee a steady and regular supply of inputs, and improve marketing. FPOs are contemporary institutional developments that aim to offer a

single window solution for multiple challenges faced by farmers and ensure better price realisation by creating direct market linkages.

- **Village-Level Weather Station:** Consistent and seamless long-term weather and climate monitoring system is vital for an in-depth comprehension of the problem and status of desertification in the coastal desert agro ecosystem. In addition to that, coastal areas are also frequently affected by natural disasters due to unpredictable heavy winds, rains, etc. To provide precise weather advisory services to the local people to protect from natural disasters or mitigate the effects of disasters, establishment of a mini automatic weather station at village level especially in and around coastal desert ecosystem is needed.
- **Promotion of Agro Tourism in Coastal Desert Ecosystem:** Agrotourism supports sustainable development, local cultures and traditional livelihoods while preserving the desert biome. Desert regions offer numerous opportunities to enhance ecotourism, such as distinctive geological features, archaeological sites exhibiting past cultures, a wider range of ethnic groups and costumes, and biodiversity conserved by reserves and national parks.
- Capacity-building on dry land technologies.

16.13 Conclusion

Therilands occurring in the south coastal districts of Tamil Nadu, Thoothukudi and Tirunelveli have active and stabilised sand dunes. The unfavourable relief in the top unstabilised dunes has resulted in little pedogenesis in 77.3% of *theri* areas. The soils have physical constraints such as low water retention, a higher presence of sand, less aggregation, low organic carbon content, the presence of unstabilised sand dunes, low plant nutrient reserves and associated problems due to poor physical properties. Wind erosion, sandy nature and excessive drainage, and unfavourable topography are serious constraints on the soil as indicated through various interpretative groupings for use in agriculture. The productivity of these soils is extremely poor, and the coefficient of improvement is very high, indicating the soils have tremendous scope for improvement in productivity (Janakiraman, 1993).

The exploitation of groundwater has led to the seepage of brackish seawater into the coastal area, deteriorating the quality of the groundwater aquifer along the coast of Tirunelveli district, indicating the necessity of groundwater policies in the coastal ecosystem.

The soils can be transformed into considerable productive soils through proper soil management techniques and the selection of suitable plants or silvi-mix. Improvement of soil organic carbon, irrigation water management to minimise irrigation water efficiency, use of microbial cultures and inputs to create favourable biological conditions in soil, and improving the soil moisture holding capacity could be the objectives of soil management strategies. The selection of crop and tree species can also help in the profitable management of *theri* soils.

References

Ammal, U. B., Mahendran, P. P., & Arunachalam, G. (2000). Influence of amendments on soil properties and yield of rice irrigated by sea water. *Madras Agricultural Journal, 87*(7/9), 534–535.

Annadurai, B. (2011). Theri-soil reclamation using tank silt in Tuticorin district. *Indian Journal of Environmental Protection, 31*, 679–683.

Foote, R. B. (1883). On the geology of the Madura and Tinnevelly Districts. Chapter 10. *Memoirs of the Geological Survey of India, 20*, 87–90.

Janakiraman, M. (1993). Pedogenesis and soil resource inventory of theriland (Sirappur). M.Sc.(Ag) Thesis. Tamil Nadu Agricultural University, Coimbatore, Tamil Nadu.

Janakiraman, M., Arunachalam, G., & Jawahar, D. (1997). Soil survey interpretation for land use planning in the Thzeri-soils of Tamil Nadu. *Journal of the Indian Society of Soil Science, 45*(2), 329–332.

Jawahar, D. (1996). Studies on the sand dunes (theries) in the coastal belt of V.O. Chidambaranar and Nellai Kattabomman districts. Ph.D. Thesis. Tamil Nadu Agricultural University, Coimbatore, Tamil Nadu.

Jawahar, D., Arunachalam, G., & Janakiraman, M. (1999). Theries—red coastal sand dunes of Tamil Nadu and their characteristics. *Journal of the Indian Society of Soil Science, 47*(1), 125–128.

Krishna, G. R. (1988). Identification of sea water intrusion in coastal aquifers based on chemical parameters. *Geophysical Research Bulletin, 26*(3), 103–107.

Mainguet, M., & Dumay, F. (2010). Fighting wind erosion. One aspect of the combat against desertification. Les dossiers thématiques du. CSFD/Agropolis International, Montpellier, France. 44 pp.

Mandal, U. K., Mandal, S., Lama, T. D., & Burman, D. (2021). Innovative Approaches for Agriculture in Coastal Fragile Ecosystem in India. In *Innovations in Agriculture for a Self-Reliant India* (pp. 421–464). CRC Press.

Soil Survey Staff. (1989). Soil of Tirunelveli Kattabomman District, Tamil Nadu, District report No. 2. Published by Soil Survey and Land Use Organisation, Publn. Palayankottai 627007, India

Soil Survey Staff. (1991). Soil of V.O.Chidambaranar District, Tamil Nadu, District report No. 3. Published by Soil Survey and Land Use Organisation, Publn. Palayankottai 627007, India

Sundaram, S., & Annadurai, B. (2018). Comparative study of soil reclamation using vermicomposting and tank silt amended theri soil in Tuticorin District. *World Wide Journal of Multidisciplinary Research and Development (WWJMRD), 4*(1), 117–121.

Udayanapillai, A. V., & Velmayil, P. (2021). Texture, mineralogy and geochemistry of Teri sediments from the Kuthiraimozhi deposit, Southern Tamilnadu, India: Implications on provenance, weathering and palaeoclimate. *Arabian Journal of Geosciences, 14*, 364.

Venkateswarlu, J. (1993). Problems and prospects in desertification control: Role of Central Arid Zone Research Institute. In A. K. Sen & A. Kar (Eds.), *Desertification and its control in the Thar, Sahara and Sahel regions* (pp. 249–267). Scientific Publishers.

Chapter 17
Climate Dynamics and Agriculture System in Deserts

Nabansu Chattopadhyay, K. Malathi, and Laxman Singh Rathore

17.1 Introduction

Deserts are created by aridity plus exceptionally high or low temperatures characterised by more evaporation than precipitation. It can be either hot or cold depending on the atmospheric temperature. As a desert is defined by low precipitation, many experts concur desert as area of land with annual precipitation of less than 25 cm (10 inches). The unique plants and animals in desert ecosystems have adapted to harsh environments in order to survive (Chouhan & Sharma, 2009; Gupta, 2015).

Deserts are created by the perpetual existence of dry air advection and/or subsidence over a region. Reduced rainfall leads to decreased soil moisture, limited vegetation growth, and reduced plant productivity. The extreme temperatures and very high thermal variability coupled with poor water availability culminate in very harsh conditions for the survival of biota. Some deserts are hot all year-round (e.g., Sahara Desert), whereas the others become quite cold in winter like Thar and Gobi Deserts. The deserts generally experience very high diurnal as well as seasonal variations in temperature with extreme values. Even in polar deserts, temperatures rarely exceed 10 °C. The water vapour present in the air is less resulting in low humidity, particularly during the daytime. Rainfall in deserts is typically sparse, erratic, unreliable, and localised both in terms of space and duration. Drought conditions, or effective aridity, define desert environments, with varying durations from temporary to

N. Chattopadhyay (✉)
South Asian Forum for Agricultural Meteorology, Hoogly, West Bengal, India
e-mail: nabansu.nc@gmail.com

K. Malathi
Agricultural Meteorology Division, India Meteorological Department Pune, Pune, Maharashtra, India

L. S. Rathore
Formerly Director General of Meteorology, India Meteorological Department, Ministry of Earth Sciences, New Delhi, India

M. S. Rathore et al. (eds.), *Desert Ecosystems*,
https://doi.org/10.1007/978-3-032-03237-9_17

permanent across different desert zones. High wind velocities are common in deserts. The cloud cover is thin resulting in intense solar radiation resulting in near-maximum sunshine hours. These conditions lead to very high potential evapotranspiration (PET) rates in deserts.

The soil in deserts is typically sandy and contains very low organic matter. Desert landscapes exhibit diverse topographies, including shimmering sand dunes, hard, flat "pavements," twisting chasms, stark mountains, and scattered boulders. The annual season for plant growth in deserts is extremely short due to the scarcity of water and harsh environmental conditions. Hot deserts face water scarcity, while cold deserts have water present but in frozen form (ice), making it inaccessible to plants and animals. Animals in deserts often face food shortages due to limited vegetation and water resources.

17.2 Genesis on the Formation of Desert Climate Across the World

Many studies have been conducted on the climate and feedback processes in desert locations (see, e.g., Mahmood, 2014 for a general overview). Self-stabilisation effects in subtropical deserts were first described by Charney (1975) and were linked to feedback processes between precipitation and surface albedo. The feedback mechanisms between vegetation and the hydrological cycle were also taken into account by Prentice et al. (1992). Two stable regimes have been identified in the Sahara: a desert equilibrium with little precipitation, and another through conceptual models (Brovkin et al., 1998) and the linking of a climate model to a dynamical vegetation model (Claussen, 1994).

The Earth's deserts, which cover approximately one-fifth of the land area, have formed based on the global circulation pattern, as illustrated in Fig. 17.1. The world's deserts can be categorised into five types depending on their aridity: subtropical, coastal, rain shadow, interior, and polar.

Subtropical deserts are usually found between 15° and 30° north of the Equator or along the Tropic of Capricorn between 15° and 30° south of the Equator. These deserts form mainly because of the sinking of air resulting from the global circulation patterns of air masses along the Tropic of Cancer, leading to arid conditions. In contrast, deserts in middle latitudes are influenced by the flow of the jet stream from the Arctic and subtropical regions. These jet streams influence high-pressure and low-pressure systems, causing fluctuations in weather patterns at any given location over a period of days. In the tropics and subtropics, stable high-pressure and low-pressure ridges dominate. The stable high-pressure ridges in the subtropics contribute to the formation of deserts.

Interior deserts form because they are shielded from moisture-laden winds, which predominantly originate from oceanic areas. This shielding effect often occurs due to the presence of mountain ranges that create a rain shadow effect leading to arid

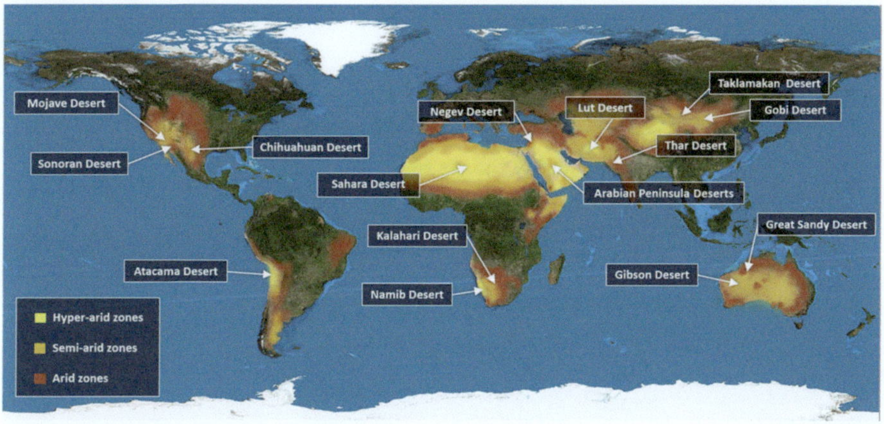

Fig. 17.1 Deserts across the world. *Source* Wiam et al., (2020)

conditions and the formation of deserts in the interior regions. Deserts located around 30 degrees north and south of the equator are typically hot. This is due to the descending air from the Hadley circulation, creating semi-permanent high-pressure areas. Deserts in the Arctic region are formed mostly due to lack of evaporation and stable high-pressure systems (anticyclones) in the troposphere.

17.3 Aridity and Desert

Aridity is a crucial criterion for defining a desert, often measured using the Aridity Index. Thornthwaite (1948) introduced this concept using the ratio between mean annual precipitation (P) and mean annual potential evapotranspiration (PET). Hyper-arid and arid regions can be identified and stated that aridity index (P/PET) lower than 0.20 means these areas receive less than 20% of the water required for optimal plant growth (UNEP, 1997).

PET is calculated using Thornthwaite's equations, which mainly depend on temperatures and daylight hours in different temporal scales. In contrast, rainfall is directly measured from weather stations. The most arid regions, based on this index, are the Saharan and Chilean-Peruvian deserts, Arabian, East African, Gobi, Australian, and South African deserts.

An Aridity Anomaly Index (AAI) provides information about the moisture stress experienced by plants throughout their life cycle. By visualising the anomalies in aridity, farmers and agricultural planners can identify areas where plants are experiencing significant water stress, allowing for timely interventions to improve plant health and yields, plan irrigation schedules more effectively, ensuring that water is provided at critical times when plants need it the most, allocate water resources more efficiently, focusing efforts on regions with the greatest need, and make informed

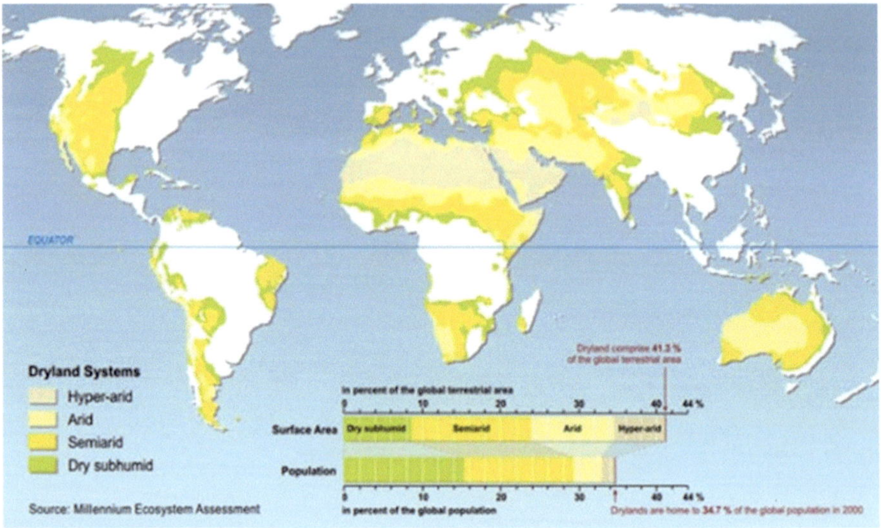

Fig. 17.2 World aridity map. *Source* Millennium Ecosystem Assessment (https://www.millenniumassessment.org/en/index.html)

predictions about crop yields based on the extent of moisture stress observed in different regions. India Meteorological Department has generated AAI for operational use to monitor the incidence, spread, intensification, and cessation of drought (near-real-time basis) on different temporal scales over the country. Several indices are also used widely across the world in defining the aridity of a region (Alley, 1984; Cammalleri et al., 2016; Martínez-Fernández et al., 2015). Figure 17.2 indicates the world aridity map.

17.4 Drought Monitoring in Desert

The impact of drought is far-reaching and recurrent, affecting numerous aspects of society on a global scale. Tracking drought presents challenges due to its intricate spatial and temporal nature, as well as its severity, which is frequently exacerbated by rising temperatures leading to higher rates of evapotranspiration, even as rainfall patterns may remain relatively unchanged.

Operational drought monitoring, especially in desert regions, is a crucial activity for various National Meteorological and Hydrological Services (NMHS) across the globe. These services display drought information on their websites and other social media and update it periodically. Effective drought monitoring in deserts is essential for making tactical and strategic decisions, particularly in agriculture, to make optimal use of the limited rainfall in these regions.

Integrated Drought Management Programme (WMO, 2006) of World Meteorological Organisation has identified three main methods for monitoring drought for early warning assessment:

- Using a single indicator or index
- Using multiple indicators or indices
- Using composite or hybrid indicators.

17.5 Drought Monitoring Indices

a. Thornthwaite's concept of Aridity describes (Thornthwaite & Mather, 1955) water deficiency experienced by plants. The formula for computing AI is

$$AI = \frac{PE - AE}{PE} \times 100$$

PE is potential evapotranspiration, AE is actual evapotranspiration, and (PE−AE) denotes the water deficit. PE is computed by Penman's equation. AE is calculated from the water balance procedure.

Plants use rainfall for evapotranspiration initially. If the potential evapotranspiration (PET) fulfils the plants' evapotranspirative demands completely, any extra rainfall starts to percolate and replenish the soil. This cycle continues until the soil reaches its field capacity, which is the maximum moisture it can hold. Any additional rainfall results in water surplus, as surface or deep drainage runoff.

In contrast, when evapotranspirative demands of plants are high due to deficient rainfall, it extracts moisture from the soil. This process continues until the soil is depleted of its moisture, affecting plant growth and potentially leading to drought conditions. Soil loses moisture as per the empirical law:

$$S = fc \cdot \exp(Acc(P - PET)/fc)$$

where S = moisture remaining in the soil as storage, fc is field capacity, P is precipitation, and PET potential evapotranspiration.

The Aridity Index is computed weekly or biweekly to indicate the water stress that plants endure because of insufficient moisture from rainfall and soil. Deviation from the normal value indicates water shortage. Table 17.1 indicated the aridity anomaly index and the corresponding drought intensity.

Table 17.1 Classification of aridity index

Anomaly of aridity index	Agricultural drought intensity
1–25	Mild
26–50	Moderate
>50	Severe

Table 17.2 Classification of SPI

SPI value range	Drought class
>2	More extreme wet
1.5–1.99	Severe wet
1–1.49	Moderately wet
0–0.99	Mildly wet
−0.99–0	Mildly dry
−1.49 - −1.00	Moderately dry
−1.99 - −1.5	Severely dry
>−2.00	Less extremely dry

17.5.1 Standardised Precipitation Index (SPI)

The Standardised Precipitation Index (SPI) is a widely used indicator for assessing and monitoring drought conditions (Sergio et al., 2010; Sepulcre-Canto, 2012). It represents rainfall occurrence over a specific period in a region and is calculated using a single parameter: rainfall and its long-term average. The SPI is suitable for monitoring and evaluating climatic conditions on a monthly or longer time frame. McKee et al. introduced the SPI in 1993 and utilised it to evaluate climate and drought variations through the application of an incomplete gamma probability density function to the precipitation series' frequency distribution. The index is particularly valued for its simplicity and effectiveness in characterising meteorological drought. Many National Meteorological Services use SPI as a major indicator to declare a drought and monitor it on different time scales. The real-time drought monitoring and bulletins are issued worldwide and can be seen on the website https://www.droughtmanagement.info/drought-monitors-and-products/. Classification of SPI is given in Table 17.2.

17.5.2 Combined Drought Index (CDI)

The use of combined indices is a valuable tool for addressing the constraints of basic drought indices. The development of a single combined drought indicator (CDI) integrating meteorological data, soil moisture information, and remote sensing-derived data enhances the ability to monitor agricultural drought's onset and its development over time and space more efficiently. By combining various data sources, the CDI provides a comprehensive view of drought conditions, capturing different aspects of drought such as meteorological deficits, soil moisture shortages, and vegetation stress. Intensity of drought condition with respect to the CD classes is shown in Table 17.3. Many researchers across the world have used different single indices together and generated a combined drought index which is gaining popularity in NMSs. Some of the few CDI developed are as follows:

Table 17.3 Classification of CDI

Classes	Classification
1–3	No drought
4–5	Mild drought
6–7	Moderate drought
8	Severe drought
9	Extreme drought

- **Spain**: Developed Combined Drought Indicator (CDI) that integrates rainfall, soil moisture, and vegetation dynamics (Jiménez Donaire et al., 2020).
- **Philippines**: Developed a CDI to monitor the transition from meteorological to agricultural drought, combining monthly land surface temperature (LST), normalised difference vegetation index (NDVI), and the three-month rainfall represented by the Standardised Precipitation Index.
- **Sri Lanka**: Developed an agricultural drought index (agCDI) using remote sensing and model-based agroclimatic input parameters—NDVI, LST, three-month precipitation z-score (stdPCP), and evaporative demand drought index (EDDI) (Yared et al., 2022).
- **Jordan**: Explores and evaluates a CDI based on precipitation (PDI), temperature (TDI), and vegetation (VDI) drought indices to characterise drought variability and trends (Al Adaileh, 2019).
- **India**: Developed a comprehensive CDI (CCDI) using both satellite observations (RET, NDVI, VCI, VHI, TCI, and ESI) and land-based observations of rainfall and soil moisture, providing a better indicator compared to indices that use only two or three parameters. Figure 17.3 depicts the flowchart of the Combined Drought Indicator of India (Chattopadhyay et al., 2020).

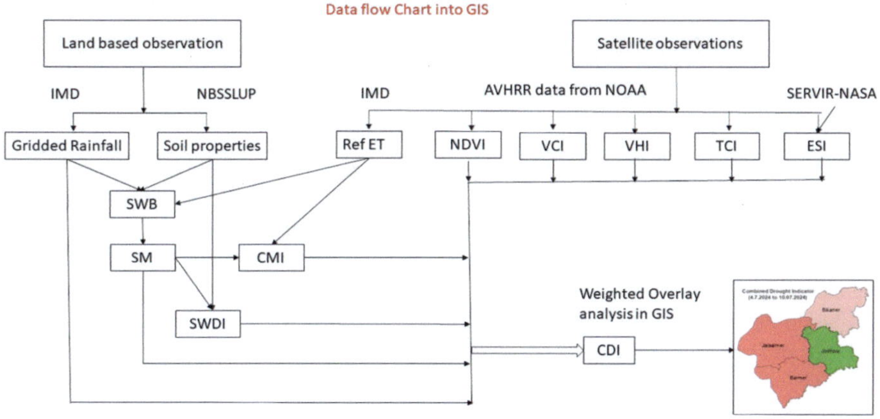

Fig. 17.3 Flowchart of combined drought indicator of India

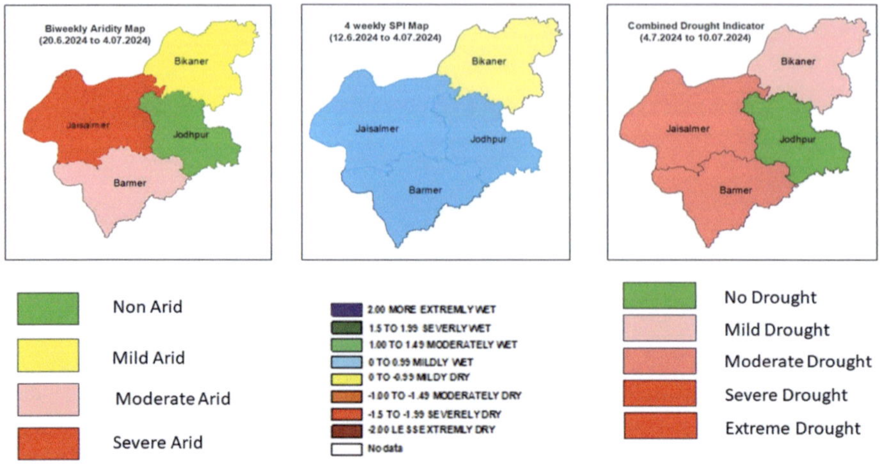

Fig. 17.4 Aridity, SPI and CDI indices

The Standardised Precipitation Evapotranspiration Index (SPEI) is an emerging drought indicator that incorporates both precipitation and temperature components, making it particularly suitable for arid and desert regions where temperature plays a critical role in drought conditions. The SPEI measures the balance between precipitation and potential evapotranspiration, providing a comprehensive assessment of drought severity (Fig. 17.4). Its ability to account for the effects of temperature on water balance makes it a valuable tool for monitoring and analysing drought in diverse climatic regions, especially those prone to high temperatures and variable rainfall.

Traditional drought indices often lack the spatial resolution needed to capture local-scale variations in drought severity, offering a broad, generalised view that may not accurately reflect specific landscape conditions. This limitation can reduce their effectiveness for localised early warning systems and proactive drought planning.

Indices-Based Remote Sensing Data: Remote sensing information is extremely useful for monitoring drought over large areas as it can offer detailed and frequent observations (Anderson et al., 2011; Kogan, 1995; Peng et al., 2019; Quiring & Ganesh, 2010). Several satellite-derived vegetation indices, such as the Normalised Difference Vegetation Index (NDVI), Vegetation Condition Index (VCI), and Temperature Condition Index (TCI), have been developed to monitor drought from local to global scales. These indices help assess drought-related vegetation stress and are often validated against ground observations. Figure 17.5 indicates the satellite-based indices for monitoring drought in deserts.

Hybrid Drought Indices: Hybrid drought indices have become more prevalent in the last few years, combining climate data, satellite observations, and environmental parameters. This integration results in a more thorough and precise evaluation of drought conditions.

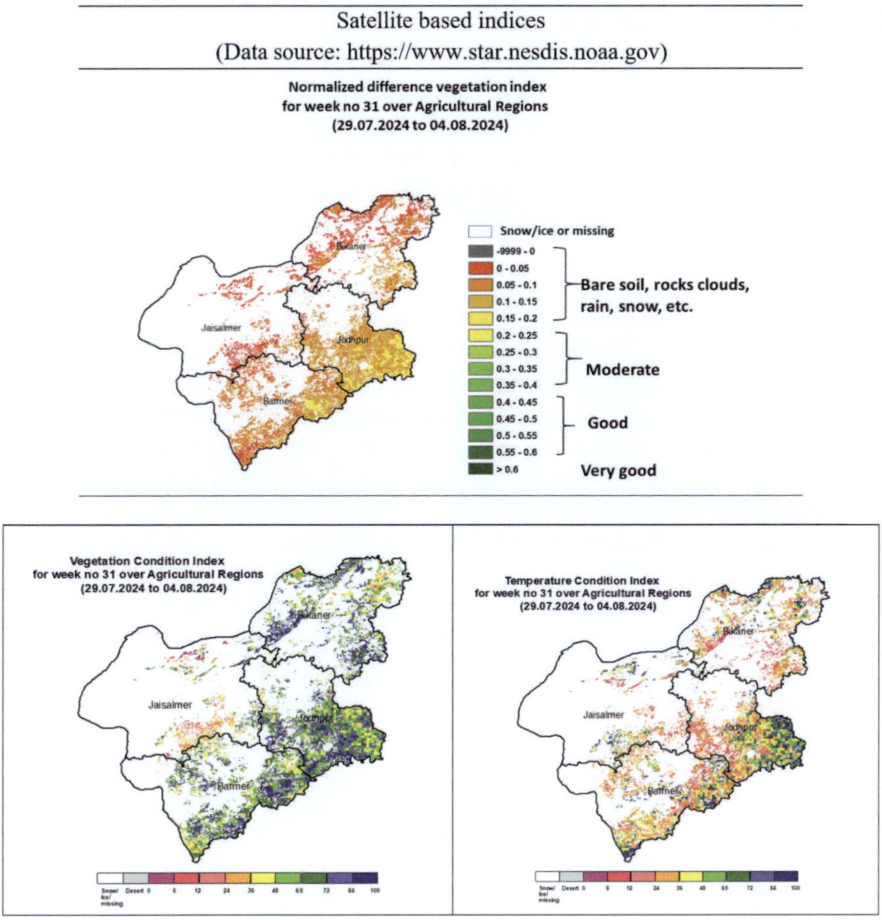

Fig. 17.5 Satellite-based indices for monitoring drought

Satellite-Based Instruments: Advances in satellite technology have enabled the estimation of key drought-related variables, including:

- Land Surface Temperature (LST)
- Evapotranspiration (ET)
- Soil Moisture
- Precipitation.

Satellite data is increasingly utilised for weather monitoring, leading to more frequent use of microwave and radar technology to gauge soil moisture and rainfall. This improved approach provides crucially accurate information for effective drought monitoring.

17.6 Climate of Indian Desert

High evaporation, scant, and very unpredictable precipitation are characteristics of Indian desert regions. Drought is a common occurrence in deserts because of the scarcity of moisture. These areas are easily defined by their plant life, which mainly comprises xerophytes and short-lived annuals. The sparse vegetation is a result of limited water resources and challenging weather conditions.

i. **Hot Desert**

The Thar Desert of Rajasthan, also known as the Great Indian Thar Desert, is a large arid region situated in the northwestern part of India (Fig. 17.6). It is a natural boundary running along the border between India and Pakistan, covering an area of more than 100,000 square kilometres in the Indian state of Rajasthan covering the districts of Jaisalmer, Barmer, Bikaner, and Jodhpur, and extends into some regions of Punjab and Haryana. The north boundary is the large thorny steppe to the north is ill-defined and bordered by the Aravalli ranges in the East, towards the south salt marsh known as the Great Rann of Kutch, which is included in the Thar, and river Indus in Pakistan marks the western boundary. The Thar Desert is characterised by its arid climate, unique flora and fauna, and significant cultural and historical heritage, making it an important geographical and ecological region in India.

The districts under the Thar Desert face significant challenges due to its arid climate and dependency on rainfall for agriculture and daily water needs. Southwest Rajasthan receives the lowest annual normal rainfall in the country. The region's economy, primarily based on agriculture and animal husbandry, heavily relies on this limited rainfall. The monsoon season in this region is very short, lasting only about two months and it exhibits large variability both in terms of amount and spatial distribution. This variability impacts agricultural productivity, as kharif crops (monsoon crops) are entirely rain-fed. Understanding the characteristics of rainfall in this region is essential for effective water management, agricultural planning, and industrial development. It helps in determining irrigation strategies, crop selection, and resource allocation. The arid conditions pose challenges for sustainable

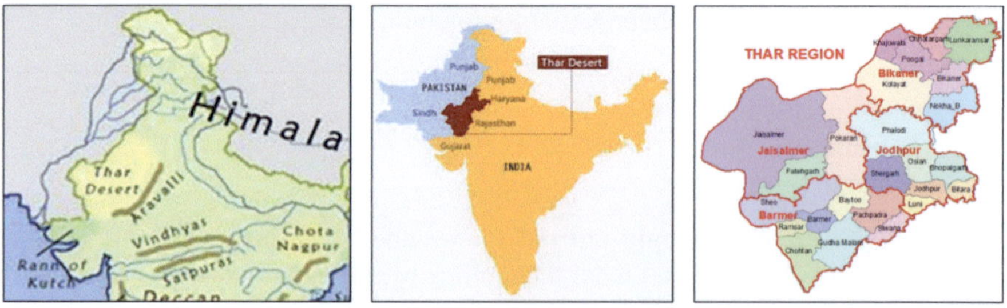

Fig. 17.6 The position of Thar desert. *Source* (1). https://www.pinterest.com/pin/658299670520101779/ (2). https://www.beautifulworld.com/asia/india/thar-desert

development, but the region's inhabitants have developed resilience strategies over generations to cope with the climatic constraints.

Weather in Thar Desert

The Thar Desert of Rajasthan experiences a dry and hot climate characterised by three main seasons: summer, rainy, and winter.

Temperatures: The Summer season begins in mid-March and continues up to the end of June with May and June being the hottest months. During summers, typically maximum temperatures range from 45 deg. Celsius to 49 deg. Celsius on many days causing heat waves and hot day conditions. In contrast, nights are quite pleasant because of its soil characteristics. The unusually high temperatures lead to the occurrence of strong hot winds that raise dust during the afternoon, known as *andhis*. Figure 17.7 gives a picture of monthly normal maximum, minimum temperature, and mean temperature over the region.

The Thar region's temperature data over the past decade shows interesting patterns. It has been observed that long-period average temperature variation ranges from 0 °C to 0.5 °C with an exception in 2020. Some pockets in this region experienced temperatures exceeding 42 °C during certain periods (Table 17.4). Extreme temperatures occurred in May 2020 and 2021 wherein in many areas, temperatures exceeded 42 °C. During the crop growing season (July to September/October), general trend of maximum temperatures did not exceed 40 °C with exceptions in July 2018 and 2019 which experienced temperatures exceeding 42 °C. This data suggests

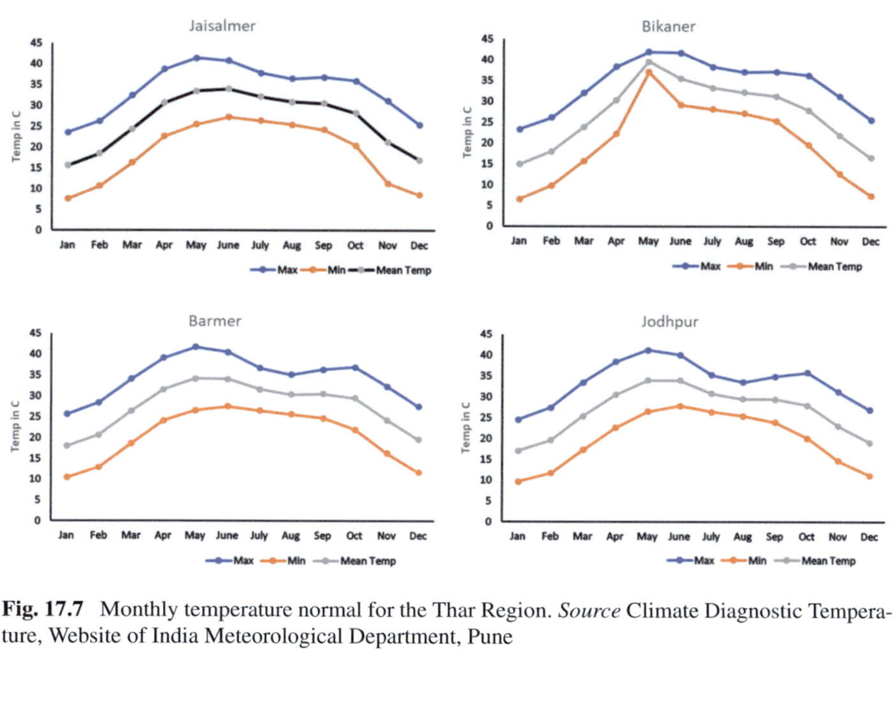

Fig. 17.7 Monthly temperature normal for the Thar Region. *Source* Climate Diagnostic Temperature, Website of India Meteorological Department, Pune

a relatively stable temperature range during April over the past decade, with occasional spikes in certain years and areas. However, there has been a noticeable increase in extreme temperatures during May in recent years, and occasional extreme heat during the crop-growing season, particularly in July 2018 and 2019.

Monsoon in this region advances during first week of July and withdrawal occurs by the first week of September. The monsoon season is a critical period for the Thar region, with most rainfall influenced by complex interactions between low pressure systems and geographical oscillations. Both the Bay of Bengal and the Arabian Sea play crucial roles in contributing to the region's rainfall, especially during interactions with western disturbances. Understanding these patterns helps in anticipating rainfall distribution for agriculture and water management in the region.

Rainfall: Despite annual and monsoon rainfall showing an increasing trend, variability persists across different districts (Singh et al., 2014). Jodhpur is an exception in receiving sufficient rainfall during the 2014 monsoon when other districts faced deficits. Barmer district exhibits a unique pattern of alternating between deficit and adequate monsoon rainfall every other year. This variability highlights the complexity of rainfall distribution in the Thar region, influenced by geographical, atmospheric, and possibly local factors. Understanding these patterns is crucial for effective water resource management and agricultural planning. Figure 17.8 indicates the annual and district-wise variation of rainfall in the Thar region for the past decade.

The Thar Desert, primarily known for its arid and harsh climate, has been experiencing noticeable changes in its monsoon patterns over the past decade. The Southwest Monsoon, which is the principal rainy season for the region, has shown a trend of setting in earlier than usual. Normally, the monsoon onset in the Thar Desert occurs around July 1. However, recent observations indicate that the monsoon is arriving 4–5 days earlier than this traditional date.

Moreover, the withdrawal of the monsoon has also been delayed, extending the rainy season by an additional 9–10 days beyond the typical withdrawal date of

Table 17.4 Actual maximum temperature range

Year	Maximum temperature in °C					
	April	May	June	July	August	September
2015	38–40	40–44	38–42	36–40	38–40	36–38
2016	38–40	40–44	40–44	36–40	36–40	36–38
2017	38–40	40–44	40–42	38–42	36–40	>38
2018	38–42	40–44	40–44	38–42	36–38	36–40
2019	38–42	40–42	40–42	38–40	34–38	36–38
2020	38–40	>42	40–42	36–40	36–40	38–40
2021	38–40	40–42	38–42	38–40	36–40	34–38
2022	NA					
2023	36–39	36–39	40–44	38–40	32–36	34–38

Source IMD

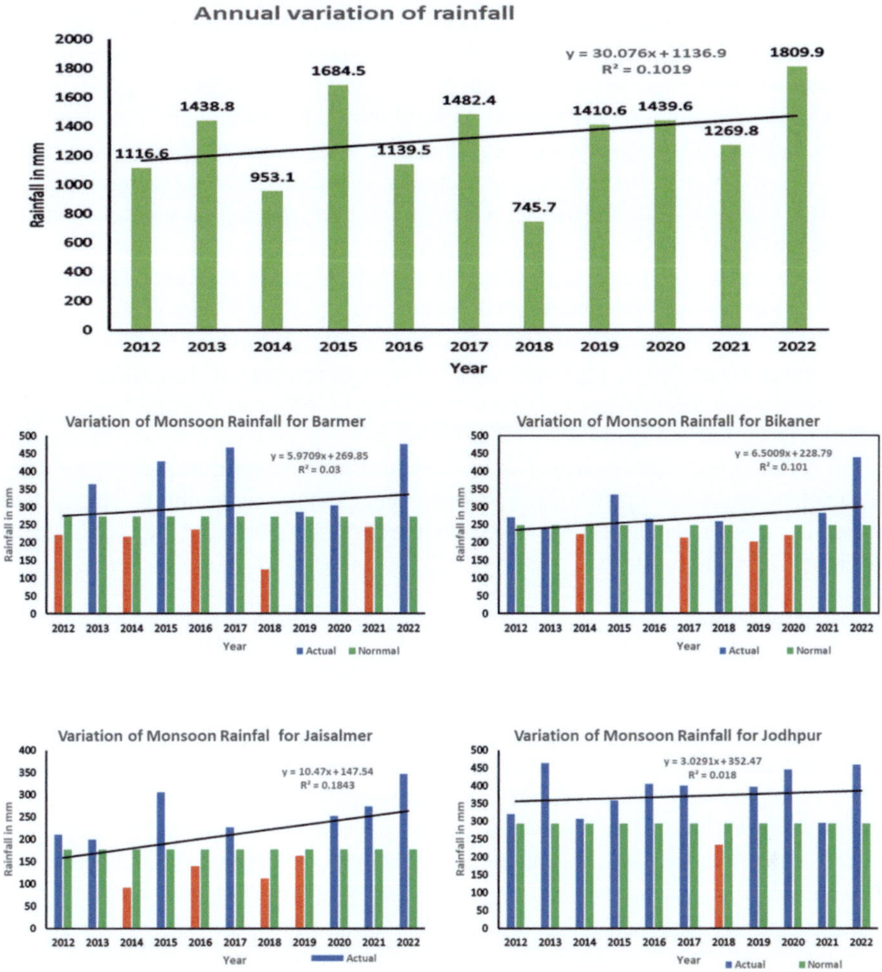

Fig. 17.8 Variation of monsoon rainfall. *Source* Rainfall Statistics of India, 2012–2022, India Meteorological Department

September 20. This extension has contributed to an increase in the duration of the rainfall period in the Thar Desert during monsoon season.

Climatologically, there has been an overall increase in monsoon rainfall activity, with the region experiencing approximately a 7% rise in rainfall during the monsoon season.

b. **Cold Desert in India**

In the northernmost part of India, between latitudes 32°–36°N and longitudes 75°–80°E (ASSOCHAM, 2021), cold desert of India is situated called Ladakh which is bordered by Tibet to the east, Kargil to the west, China to the north, and Himachal

Pradesh to the south (Fig. 17.9). Ladakh is situated 3000 to 8000 m above mean sea level and has extreme climatic conditions, including very low temperatures, minimal precipitation, and intense sunlight. The geography includes high mountains and elevated plateaus, and the climate is characterised by significant diurnal temperature variations. The region's low atmospheric pressure and oxygen levels present challenges for human habitation, necessitating physiological adaptations. Water is primarily sourced from snowmelt, which is vital for the region's ecosystem and human activities.

Ladakh faces significant agricultural and hydrological challenges. The narrow crop growing period is restricted to valleys and coincides with scant rainfall, leading to severe limitations in water availability. The region's precipitation is less than 15% of the mean annual PET, highlighting a large moisture deficit. Evapotranspiration is minimal when temperatures are below freezing, resulting in negligible vegetation cover as indicated in Fig. 17.10. PET is higher in Kargil than in Leh and Drass. The major water supply during winter is snow, with little change in rainfall from April to October, exacerbating the gap between water demand and supply. The general increasing trend of monsoon rainfall is observed during the past 10 years indicated in Fig. 17.11.

c. **Coastal Desert**

In the southern peninsula of the country, the Thoothukudi district of Tamil Nadu which is notable for its distinctive red colour sand dunes has unique formation history, originating from the marine sediments of the Quaternary Period. These dunes have very low water and nutrient retention capacities, making them inhospitable for most types of vegetation. The dunes are highly susceptible to aerodynamic lift, which allows sand particles to be easily moved and reshaped by the wind. This dynamic nature of the dunes highlights the ongoing influence of wind on their structure and distribution.

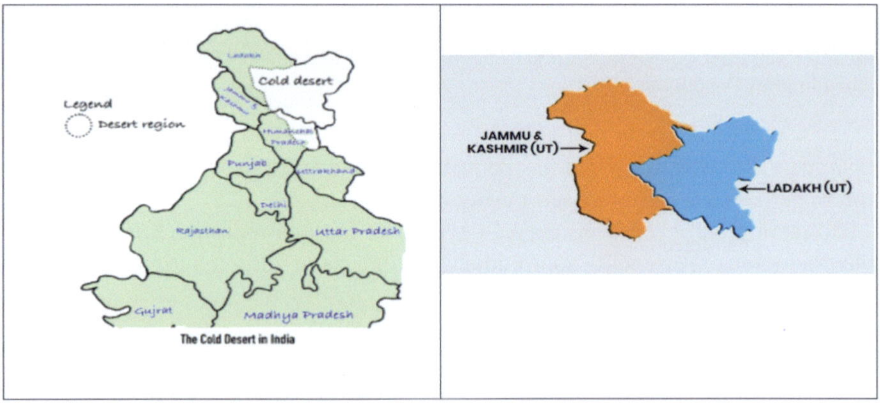

Fig. 17.9 Cold desert. *Source* (1). https://www.studyiq.com/articles/deserts-of-india/

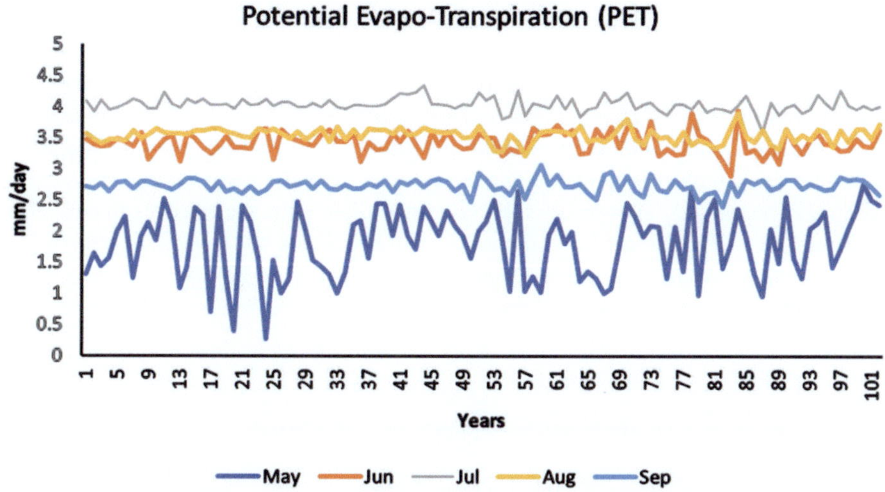

Fig. 17.10 PET Trends over past 100 years in Ladakh. *Source* Consolidate studies on the weather pattern of Leh, Ladakh from 1901 to 2000. India Meteorological Department

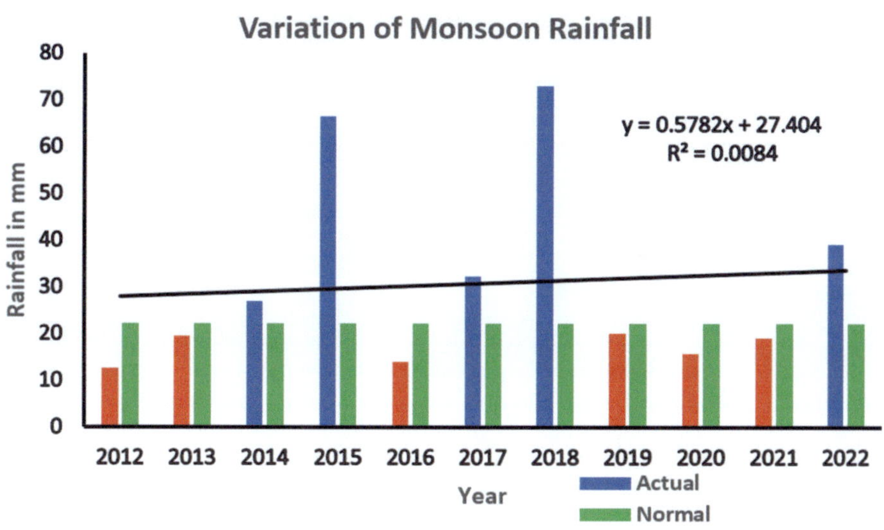

Fig. 17.11 Variation of monsoon rainfall over Ladakh region. *Source* Rainfall Statistics of India, 2012–2022, India Meteorological Department

The distinctive red-coloured dunes near Thiruchendur owe their unique hue to iron-rich minerals such as ilmenite, magnetite, garnet, hypersthene, and rutile, which undergo leaching and oxidation in the semi-arid climate. The dunes are located in the Kuthiraimozhi theri and Sathankulam reserves forest areas along the Bay of Bengal. The red loam is carried to Thiruchendur during the southwest monsoon season (May to September) by dry winds that have lost moisture over the Mahendragiri peak and the Aralvaimozhi gap of the Western Ghats, which are roughly 75 km away. The lack of vegetation in the Aralvaimozhi gap and the Nanguneri plains exacerbates wind erosion, facilitating the transportation of red loam.

As the monsoon winds blow with high velocity, they lift and carry the red loam in large columns until they encounter sea breezes near the coast, where the sand is eventually deposited. This process, known as Aeolian activity, involves the erosion, transport, and deposition of sediments by wind. It leads to continuous redistribution of sand, forming characteristic sand dunes when wind-blown sand encounters obstructions such as vegetation or man-made structures.

17.7 Agriculture in Different Deserts in India

Desert environments are typically characterised by sparse vegetation due to the lack of water and the harsh climatic conditions. Trees are generally absent, and shrubs or herbaceous plants provide only limited ground cover. In some deserts, flora is almost non-existent due to extreme aridity. This aridity can be exacerbated by human activities, such as excessive cattle grazing, which put additional stress on an already fragile environment.

Human disturbance, particularly overgrazing, can lead to further degradation of the land, reducing the already limited vegetation cover. This, in turn, can increase erosion and reduce soil quality, making it even harder for plants to survive and grow. The combination of natural aridity and human impact creates a challenging environment for flora to thrive, resulting in the characteristic sparse vegetation of desert landscapes.

a. **Agriculture in Thar Desert**

The Thar Desert, located in Rajasthan, India, is home to about 40% of the state's population and is the most densely populated desert in the world with a density of more than 80 people per square kilo meter compared to the average of 7 in other deserts. The primary occupations here are agriculture, tourism, and animal husbandry. However, due to increasing human and animal populations, there has been significant degradation of vegetation resources and soil fertility. As agricultural activities in this region are rain-fed crops, they are sown after the onset of monsoon. Major crops grown are bajra, pulses (such as guar), jowar, maize, sesame, and groundnuts. However, due to rainfall variability, about 33% of these crops fail annually. With the development of canals and tube wells, the region has also started producing rabi crops like wheat, mustard, cumin, and other cash crops. Opium is significantly produced in the Thar region.

The major crops grown along with area (in 1000 ha) sown are indicated in Table 17.5, and Fig. 17.12 indicates the total area, net area, and double area sown in the Thar desert.

Table 17.5 Major crops grown along with area sown in Thar region

District	Wheat	Kharif sorghum	Sorghum	Pearl millet	Barley	Chickpea	Minor pulses	Groundnut
Barmer	17.61	6.61	6.61	795.06	0.02	2.01	329.38	4.06
Bikaner	113.18	0.04	0.04	117.99	4.52	226.59	424.75	174.95
Jaisalmer	9.99	3.13	3.13	98.69	0	152.93	58.14	19.79
Jodhpur	77.94	44.47	44.47	417.35	0.38	19.07	387.5	115.45
	Sesamum	Rapeseed and mustard	Castor	Cotton	Fruits and veg	Onion	Fodder	
Barmer	6.28	14.43	22.83	0	0.52	0.12	419.46	
Bikaner	9.4	71.79	0.1	0.66	1.83	1.08	829.01	
Jaisalmer	6.39	43.82	4.25	0.04	0.49	0.03	599.16	
Jodhpur	28.55	143.64	26.19	40.55	24.5	18.77	228.71	

Source ICRISAT District level crop data

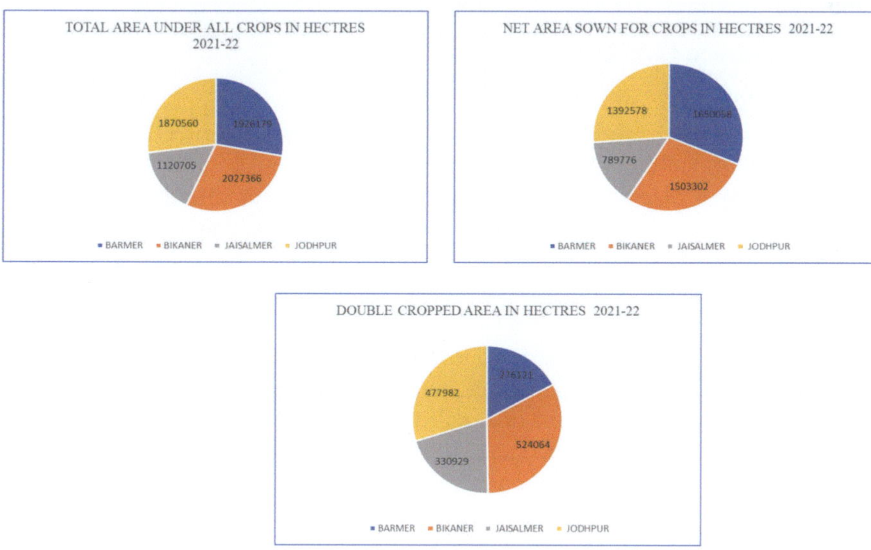

Fig. 17.12 Total area of the crops sown with net area and double cropped area in hectares over Thar region. Source Directorate of Economics and Statistics Rajasthan, 2021–22

Challenges in Agriculture in the Region: The region is prone to frequent droughts, high animal populations lead to overgrazing, which, combined with wind and water erosion and mining activities, results in severe land degradation and due to inconsistent rainfall, agriculture is not always dependable. Given the challenges in crop production, animal husbandry, along with agroforestry practices like intercropping trees, grasses, vegetables, and fruit trees, is considered a more viable livelihood model for the region. Understanding these features and challenges is crucial for effective planning in water management, agricultural operations, and industrial development in the Thar Desert (Kar, 2019; Kaur & Singh, 2023).

In recent years, there has been an increase in both human and animal populations in the Thar Desert of Rajasthan challenging farming conditions and making animal husbandry a popular and crucial livelihood for many residents. Compared to the other farming areas in the country, farmers in the region have ten times more animals per person contributing to issues like overgrazing and environmental degradation. The primary livestock includes cows, buffaloes, sheep, goats, camels, and oxen. Barmer district has the highest cattle population, with sheep and goats being the most common. Thar region is the largest wool-producing area in India, contributing 40–50% of the country's total wool production. *Proposis cineraria* species, known locally as Khejri, is vital for the livelihood of Thar Desert communities. It provides fodder, fuel, and other resources essential for daily life. The dependency on animal husbandry and the significant wool production highlight the importance of sustainable practices to mitigate environmental degradation and support the local economy.

ii. **Agriculture in Cold Desert**

Ladakh, a cold arid desert region, faces unique challenges and opportunities in agriculture due to its harsh climate. Roughly 70% of the people are directly or indirectly dependent on agriculture. The cropping season in this region is only four to five months every year (April to October) depending on different altitudes. Single cropping is the norm in the region as agriculture is entirely dependent on irrigation with a cultivable area of 10,223 ha. The major crops cultivated included Grim (Naked Barley), wheat, peas, and alfa-alfa for fodder. The workforce is entirely focused on agriculture and social structures and customs are all based on an agrarian way of life. Mostly agricultural lands are restricted to the comparatively lower regions that align with the river valleys. This region's economy has always been centred on cattle raising and subsistence farming. In this area, cultivation primarily depends on snowmelt, which provides irrigation water. The residents in the area conduct self-sustainable agriculture in spite of severe environmental restrictions. In this area, polyhouses are used primarily by farmers to cultivate vegetables. Utilising polyhouses lowers the area's irrigation costs.

Ladakh's cold desert environment presents unique agricultural challenges, but with the adoption of innovative techniques, the region can significantly improve its agricultural productivity and sustainability. Here are some techniques that can change the face of agriculture in Ladakh:

- Water management and solar energy utilisation
- Protected cultivation and precision farming
- Emphasise research and training programmes to promote natural farming practices.
- About soil health and nutrient management.

These suggestions collectively form a comprehensive strategy to address the unique challenges of agriculture in Ladakh, contributing to sustainable development and the well-being of resource-poor farmers in this cold desert region.

iii. **Agriculture in Coastal Desert**

The coastal desert is situated in Kuthiraimozi Theri Reserved Forest. Soil in this area is lost due to its close proximity with the Bay of Bengal causing wind erosion. As a protective measure to reduce wind erosion, it is necessary to maintain vegetative cover over the land by constructing wind break or shelter belts.

Based on the soil type, climate, rainfall, and availability of soil moisture, vegetation cover should be selected in this region. Rainstorms in these areas temporarily form lakes and ponds. As it is a reserved forest land, government is encouraging the farmers to grow drumstick plants, palm trees, and cashew nuts in this region.

17.8 Summary and Conclusion

Deserts are characterised by unique climatic, topographic, edaphic, and biotic features, with aridity being a common feature across all deserts for most or all of the year with significant fluctuations in day and night temperatures, and between seasons. Temperatures across the polar desert also do not go beyond 10 °C. Low humidity during the day and relatively higher humidity at night are typical. Precipitation deficiency is a defining feature, characterised not only by the low amount of rainfall but also by its unpredictability in timing and quantity. Rain in the desert is often seasonal, unpredictable, inconsistent, and restricted to specific areas and time periods.

Deserts located around 30°N and 30°S are typically hot due to the global circulation patterns, particularly the descending air from the Hadley circulation. This descending air creates semi-permanent high-pressure areas at these latitudes, leading to abundant sunshine and high surface temperatures. Subtropical deserts form due to these air circulation patterns. Deserts in rain shadows develop on the side of mountains that don't face the prevailing wind, causing arid conditions due to blocked wet weather systems. Deserts in the interior of continents form because they are not reached by moisture-carrying winds. Coastal deserts, affected by cold ocean currents, may receive very little rain but can be moist with fog as the chilly ocean air warms up over the land. Hyper-arid and arid regions are defined by an aridity index (P/PET) lower than 0.20, indicating less than 20% of the water requirement is met by rainfall for optimal plant growth. Key meteorological indicators for monitoring agricultural drought include the Standardised Precipitation Index, aridity anomaly index, and combined drought indicator.

Desert areas in India, such as the Thar Desert in Rajasthan, have low and unpredictable rainfall, along with high rates of evaporation, resulting in frequent occurrences of droughts. The Thar Desert experiences very high daytime temperatures, ranging from 45 °C to 49 °C. Despite increasing trends in annual and monsoon rainfall, variability persists across districts. Ladakh, a cold desert, features high altitudes, extreme climatic conditions, very low temperatures, minimal precipitation, and intense sunlight. It includes high mountains and elevated plateaus.

The Thar Desert has seen significant changes in recent years, with the increase in both human and animal populations. With the onset of monsoon (in June and July), *Kharif* crops like bajra, pulses (guar), jowar, maize, sesame, and groundnuts are sown and harvested in September and October. Animal husbandry is crucial due to challenging farming conditions, leading to issues like overgrazing and environmental degradation. The region has a high livestock-to-person ratio, with cows, buffaloes, sheep, goats, camels, and oxen being the primary livestock. Sustainable practices are essential to mitigating environmental degradation and supporting the local economy.

In the coastal desert of Tamil Nadu's Thoothukudi district, the soil has low water and nutrient retention capacities, making it inhospitable for most vegetation. Agricultural areas are typically confined to river valleys, especially in central and western Ladakh.

To protect the land from wind erosion, maintaining vegetative cover and constructing wind breaks or shelter belts is crucial. Tree species for wind breaks should be selected based on soil type, climate, rainfall, and moisture availability.

Ladakh's cold desert environment presents unique agricultural challenges. However, innovative techniques can significantly improve agricultural productivity and sustainability in the region. These include:

- Water Management and Solar Energy Utilisation: Efficient use of water resources and harnessing solar energy for agricultural activities.
- Protected Cultivation and Precision Farming: Using greenhouses and precision farming techniques to optimise resource use and increase crop yields.
- Research and Training Programmes: Promoting natural farming practices and educating farmers about soil health and nutrient management.

These strategies aim to address the unique challenges of agriculture in Ladakh, contributing to sustainable development and improving the well-being of resource-poor farmers in this cold desert region.

Global climate change is having a pronounced impact on desert regions, affecting them more significantly than many other areas. Deserts are experiencing higher temperature increases compared to other regions. This intensifies evapotranspiration rates, further drying out already arid areas. While deserts typically receive minimal rainfall, climate change is causing more erratic and extreme precipitation events leading to flash flooding and increased erosion, disrupting the fragile desert ecosystem. The frequency and severity of droughts in desert regions are increasing. Increased temperatures and prolonged droughts put additional stress on desert vegetation, which is already adapted to harsh conditions. Higher temperatures and reduced vegetation cover can lead to soil degradation. Communities living in or near desert regions are facing greater challenges due to water scarcity, reduced agricultural productivity, and increased heat stress. This can lead to migration and socio-economic instability. Efforts to monitor and mitigate these impacts include advanced remote sensing technologies, incorporating new data, methodologies, and metrics to improve the capability to monitor regional-scale agricultural drought conditions accurately. The use of geospatial technology allows for precise mapping and tracking of drought conditions across different regions, facilitating targeted interventions.

Websites

https://www.ipcc.ch/srccl/chapter/chapter-5/.

References

Alley, W. M. (1984). The Palmer drought severity index: Limitations and assumptions. *Journal of Applied Meteorology, 23*, 1100–1109. https://doi.org/10.1175/1520-0450(1984)0232.0.CO;2

Alsharif, W., Saad, M.M., & Hirt, H. (2020). Desert microbes for boosting sustainable agriculture in extreme environments. *Frontiers in Microbiology, 11*, 1666. https://doi.org/10.3389/fmicb.2020.01666

Al Adaileh, H., Al Qinna, M., Barta, K., Al-Karablieh, E., Al Bakri, J., & Rakonczai, J. (2019). Applicability of a combined drought index to monitoring drought in Jordan. *Journal of Engineering Research and Application, 9*(7), 2248–9622.

Anderson, M. C., Hain, C. R., Wardlow, B., Mecikalski, J. R., & Kustas, W. P. (2011). Evaluation of a drought index based on thermal remote sensing of evapotranspiration over the continental U.S. *Journal of Climate, 24*, 2025–2044. https://doi.org/10.1175/2010JCLI3812.1

ASSOCHAM, September Ladakh Unleashing Potential 2021.

Bayissa, Y., Srinivasan, R., Joseph, G., Bahuguna, A., Shrestha, A., & Ayling, S. (2022). Developing a combined drought index to monitor agricultural drought in Sri Lanka. *Water, 14*(20), 3317. https://doi.org/10.3390/w14203317

Bhawan, Y., Scheme, C., & Marg, T. Directorate of Economics and Statistics Rajasthan: Agricultural Statistics of Rajasthan. Jaipur, Rajasthan. www.gov.instatistics.

Brovkin, V., Claussen, M., Petoukhov, V., & Ganopolski, A. (1998). On the stability of the atmosphere–vegetation system in the Sahara/Sahel region. *Journal of Geophysical Research, 103*, 31613–31624. https://doi.org/10.1029/1998JD200006

Cammalleri, C., Micale, F., & Vogt, J. (2016). A novel soil moisture-based drought severity index (DSI) combining water deficit magnitude and frequency. *Hydrological Processes, 30*, 289–301. https://doi.org/10.1002/hyp.10578

Chattopadhyay, N., Malathi, K., Tidke, N, Attri, S. D., & Ray, K. (2020). Monitoring agricultural drought using combined drought index in India. *Journal of Earth System Sciences, 129*, 155. Indian Academy of Sciences.

Charney, J. G. (1975). Dynamics of deserts and drought in the Sahel. *Quarterly Journal of the Royal Meteorological Society, 101*, 193–202. https://doi.org/10.1002/qj.49710142802

Claussen, M. (1994). On coupling global biome models with climate models. *Climate Research, 4*, 203–221. https://doi.org/10.3354/cr004203

Chouhan, T. S., & Sharma, U. K. (2009). Desert ecosystem in India. Tropical Biology and Conservation Management (Vol. IX).

Gupta, S. R. (2015). Desert ecosystem. In *Ecosystem Structure and Functions*. EPGP Book,

Jiménez-Donaire, M. P., Tarquis, A., & Giráldez, J. V. (2020). Evaluation of a combined drought indicator and its potential for agricultural drought prediction in Southern Spain. *Natural Hazards and Earth System Sciences, 20*, 21–33. https://doi.org/10.5194/nhess-20-21-2020

Kar, A. (2019). Adaptation of the agri-based society to environmental changes in Thar desert. In *Climate Change, Disaster and Adaptations* (pp. 151–171). Central Arid Zone Research Institute (CAZRI), Jodhpur.

Kaur, S., & Singh, J. (2023). Changing cropping pattern of oilseed crops and its diversification: The case of Thar Desert, Rajasthan (1985–1986 to 2015–2016). *Published by EDP Sciences*. https://doi.org/10.1051/ocl/2023017

Kogan, F. N. (1995). Application of vegetation index and brightness temperature for drought detection. *Advances in Space Research, 15*(11), 91–100.

Mahmood, R. (2014). Land cover changes and their biogeophysical effects on climate. *International Journal of Climatology, 34*, 929–953. https://doi.org/10.1002/joc.3736

Martínez-Fernández, J., González-Zamora, A., Sánchez, N., & Gumuzzio, A. (2015). A soil water-based index as a suitable agricultural drought indicator. *Journal of Hydrology, 522*, 265–273. https://doi.org/10.1016/j.jhydrol.2014.12.051

McKee, T. B., Doesken, N. J., Kleist, J. (1993). The relationship of drought frequency and duration of time scales. In *Eighth conference on applied climatology. American Meteorological Society* (pp. 179–186). Anaheim, CA.

Mishra, A. K., & Singh, V. P. (2011). Drought modelling—A review. *Journal of Hydrology, 403*(1–2), 157–175.

Peng, J., Muller, J.-P., Blessing, S., Giering, R., Danne, O., Gobron, N., Kharbouche, S., Ludwig, R., Müller, B., Leng, G., You, Q., Duan, Z., & Dadson, S. (2019). Can we use satellite-based FAPAR to detect drought? *Sensors, 19*, 3663. https://doi.org/10.3390/s19173662

Prentice, I. C., et al. (1992). A global biome model based on plant physiology and dominance, soil properties, and climate. *Journal of Biogeography, 19*, 117–134. https://doi.org/10.2307/2845499

Quiring, S. M., & Ganesh, S. (2010). Evaluating the utility of the vegetation condition index (VCI) for monitoring meteorological drought in Texas. *Agricultural and Forest Meteorology, 150*, 330–339.

Rossi, S., & Niemeyer, S. (2012). Drought monitoring with estimates of the fraction of absorbed photosynthetically-active radiation (fAPAR) derived from MERIS. In *Remote Sensing for Drought: Innovative Monitoring Approaches*, CRC Press (pp. 95–120).

Sepulcre-Canto, G., Horion, S., Singleton, A., Carrao, H., & Vogt, J. (2012). Development of a combined drought indicator to detect agricultural drought in Europe. *Natural Hazards and Earth System Sciences, 12*, 3519–3531. https://doi.org/10.5194/nhess-12-3519-2012

Singh, S.S., Chauhan, M. and R.N. Kumawat (2014). Rainfall Structure of Thar "The Great Indian Desert". Meteorological Centre, Jaipur, India Meteorological Department, New Delhi.

Thornthwaite, C. W. (1948). An approach toward a rational classification of climate. *Geographical Review, 38*, 55–94.

Thornthwaite, C. W., & Mather, J. R. (1955). The water balance. *Climatology, 8*(1). C.W. Thornthwaite & Associates, Centerton, New Jersey.

World Meteorological Organization—WMO (2006). Drought monitoring and early warning: Concepts, progress and future challenges. WMO No. 1006.

UNEP. (1997). United nations list of protected areas, 20 January 1997. https://www.unep.org/resources/report/1997-united-nations-list-protected-areas.

Vicente-Serrano, S. M., Beguería, S., & López-Moreno, J. I. (2010). Multiscalar drought index sensitive to global warming: The standardized precipitation evapotranspiration index. *Journal of Climate, 23*(7), 1696–1718. https://doi.org/10.1175/2009JCLI2909.1

Index

M. S. Rathore et al. (eds.), *Desert Ecosystems*,
https://doi.org/10.1007/978-3-032-03237-9

D

E

M

MIX
Papier aus verantwortungsvollen Quellen
Paper from responsible sources
FSC® C105338

If you have any concerns about our products,
you can contact us on
ProductSafety@springernature.com

In case Publisher is established outside the EU,
the EU authorized representative is:
Springer Nature Customer Service Center GmbH
Europaplatz 3, 69115 Heidelberg, Germany

Printed by Libri Plureos GmbH
in Hamburg, Germany